T0134503

Lecture Notes in Networks and Systems

Volume 64

The series "Lecture Notes in Networks and Systems" publishes the latest developments in Networks and Systems—quickly, informally and with high quality. Original research reported in proceedings and post-proceedings represents the core of LNNS.

Volumes published in LNNS embrace all aspects and subfields of, as well as new challenges in, Networks and Systems.

The series contains proceedings and edited volumes in systems and networks, spanning the areas of Cyber-Physical Systems, Autonomous Systems, Sensor Networks, Control Systems, Energy Systems, Automotive Systems, Biological Systems, Vehicular Networking and Connected Vehicles, Aerospace Systems, Automation, Manufacturing, Smart Grids, Nonlinear Systems, Power Systems, Robotics, Social Systems, Economic Systems and other. Of particular value to both the contributors and the readership are the short publication timeframe and the world-wide distribution and exposure which enable both a wide and rapid dissemination of research output.

The series covers the theory, applications, and perspectives on the state of the art and future developments relevant to systems and networks, decision making, control, complex processes and related areas, as embedded in the fields of interdisciplinary and applied sciences, engineering, computer science, physics, economics, social, and life sciences, as well as the paradigms and methodologies behind them.

Advisory Board

More information about this series at http://www.springer.com/series/15179

Salim Chikhi · Abdelmalek Amine
Allaoua Chaoui · Djamel Eddine Saidouni
Editors

Modelling and Implementation of Complex Systems

Proceedings of the 5th International Symposium, MISC 2018, December 16–18, 2018, Laghouat, Algeria

Editors
Salim Chikhi
Faculty of New Information
and Communication Technologies
University of Constantine 2
Constantine, Algeria

Allaoua Chaoui
Faculty of New Information
and Communication Technologies
University of Constantine 2
Constantine, Algeria

Abdelmalek Amine
Faculty of Technology, GeCoDe Laboratory
University of Saida
Saida, Algeria

Djamel Eddine Saidouni
Faculty of New Information
and Communication Technologies
University of Constantine 2
Constantine, Algeria

ISSN 2367-3370 ISSN 2367-3389 (electronic)
Lecture Notes in Networks and Systems
ISBN 978-3-030-05480-9 ISBN 978-3-030-05481-6 (eBook)
https://doi.org/10.1007/978-3-030-05481-6

Library of Congress Control Number: 2018963068

This Springer imprint is published by the registered company Springer Nature Switzerland AG
The registered company address is: Gewerbestrasse 11, 6330 Cham, Switzerland

Preface

This volume contains research papers accepted and presented at the 5th international symposium on Modelling and Implementation of Complex Systems (MISC 2018), held on December 16–18, 2018, in Laghouat, Algeria. As the previous editions (MISC 2010, MISC 2012, MISC 2014 and MISC 2016), this symposium is intended as a tradition offering open forum and meeting space for researchers working in the field of complex systems science.

This year, the MISC symposium received 109 submissions from seven countries: Algeria, Canada, France, Saudi Arabia, Singapore, United Arab Emirates, and Yemen. In a rigorous reviewing process, the Program Committee selected 27 papers, which represents an acceptance rate of 25%. However, we have not received the camera-ready copies of two papers. So, the volume includes 25 papers. The PC included 98 researchers and 24 additional reviewers from ten countries.

The accepted papers were organized into sessions as follows: Cloud Computing and IoT, Metaheuristics and Optimization, Computational Intelligence and Software Engineering and Formal Methods.

We would like to thank the co-chairs of the program committee and all its members for their effort in the review process and the selection of the papers.

We are grateful to the Organizing Committee members from the University of Laghouat and the University of Constantine 2 for their contribution to the success of the symposium.

Our thanks also go to the authors who submitted papers to the symposium for their interest to our symposium.

Enough thanks cannot be expressed to Dr Nabil Belala for managing *EasyChair system* for MISC'2018 from submissions to proceedings elaboration.

October 2018

Salim Chikhi
Abdelmalek Amine

Organization

The 5th international symposium on Modelling and Implementation of Complex Systems (MISC 2018) was co-organized by Abdelhamid Mehri, University of Constantine 2, and Amar Telidji, University of Laghouat, and took place in Laghouat, Algeria (December 16–18, 2018).

Honorary Chairs

Mohamed Benbertal	University of Laghouat, Algeria
Mohamed Elhadi Latreche	University of Constantine 2, Algeria

General Chairs

Salim Chikhi	University of Constantine 2, Algeria
Abdelmalek Amine	University of Saïda, Algeria
Mohamed Bachir Yagoubi	University of Laghouat, Algeria

Steering Committee

Allaoua Chaoui	University of Constantine 2, Algeria
Salim Chikhi	University of Constantine 2, Algeria
Mohamed-Khireddine Kholladi	University of El Oued, Algeria
Djamel Eddine Saïdouni	University of Constantine 2, Algeria

Invited Speakers

Ahmed Al-Dubai	Edinburgh Napier University, Scotland
Mario Rosario Guarracino	Italian National Research Council, Italy
Saad Harous	United Arab Emirates University, UAE

Organizing Committee Chairs

Tahar Bendouma	University of Laghouat, Algeria
Mourad Bouzenada	University of Constantine 2, Algeria

Organizing Committee

Mohamed Djoudi	University of Laghouat, Algeria
Mohamed Lahcen Bensaad	University of Laghouat, Algeria
Younes Guellouma	University of Laghouat, Algeria
Noureddine Chaib	University of Laghouat, Algeria
Lakhdar Oulad Djedid	University of Laghouat, Algeria
Mohamed El Habib Maicha	University of Laghouat, Algeria
Mustapha Bouakkaz	University of Laghouat, Algeria
Sarah Benkouider	University of Laghouat, Algeria
Laradj Chellama	University of Laghouat, Algeria
Raida Elmansouri	University of Constantine 2, Algeria
Ilham Kitouni	University of Constantine 2, Algeria
Said Labed	University of Constantine 2, Algeria
Ishak Benmohammed	University of Constantine 2, Algeria

Publication Chairs

Nabil Belala	University of Constantine 2, Algeria
Ahmed Chawki Chaouche	University of Constantine 2, Algeria
Chaker Mezioud	University of Constantine 2, Algeria

Publicity and Sponsor Chairs

Hacène Belhadef	University of Constantine 2, Algeria
Abdelkrim Bouramoul	University of Constantine 2, Algeria
Mohamed Gharzouli	University of Constantine 2, Algeria

Program Committee Chairs

Allaoua Chaoui	University of Constantine 2, Algeria
Djamel Eddine Saïdouni	University of Constantine 2, Algeria

Program Committee

Takoua Abdellatif	Ecole Polytechnique Tunis, Tunisia
Abdelkrim Abdelli	LSI Laboratory, USTHB, Algeria
Wahabou Abdou	University of Bourgogne, Dijon, France
Nabil Absi	Ecole des Mines, Saint-Etienne, France
Abdelkader Adla	University of Oran 1, Algeria
Otmane Ait Mohamed	Concordia University, Canada
Ali Al-Dahoud	Al-Zaytoonah University, Jordan
Abdelmalek Amine	University of Säida, Algeria
Abdelkrim Amirat	University of Souk Ahras, Algeria
Baghdad Atmani	University of Oran 1, Algeria
Mohamed Chaouki Babahenini	University of Biskra, Algeria
Abdelmalik Bachir	University of Biskra, Algeria
Amar Balla	ESI, Algiers, Algeria
Kamel Barkaoui	CNAM, Paris, France
Faiza Belala	University of Constantine 2, Algeria
Nabil Belala	University of Constantine 2, Algeria
Ghalem Belalem	University of Oran 1, Algeria
Hacéne Belhadef	University of Constantine 2, Algeria
Mohamed Benmohammed	University of Constantine 2, Algeria
Hammadi Bennoui	University of Biskra, Algeria
Azeddine Bilami	University of Batna, Algeria
Salim Bitam	University of Biskra, Algeria
Karim Bouamrane	University of Oran 1, Algeria
Rachid Boudour	University of Annaba, Algeria
Mahmoud Boufaida	University of Constantine 2, Algeria
Zizette Boufaida	University of Constantine 2, Algeria
Kamel Boukhalfa	USTHB, Algiers, Algeria
Abdelkrim Bouramoul	University of Constantine 2, Algeria
Zine Eddine Bouras	EPST, Annaba, Algeria
El-Bay Bourennane	University of Bourgogne, Dijon, France
Mourad Bouzenada	University of Constantine 2, Algeria
Ahmed Chawki Chaouche	University of Constantine 2, Algeria
Allaoua Chaoui	University of Constantine 2, Algeria
Salim Chikhi	University of Constantine 2, Algeria
Abdallah Chouarfia	University USTO, Oran, Algeria
Lakhdar Derdouri	University of Oum El Bouaghi, Algeria

Lynda Dib	University of Annaba, Algeria
Karim Djemame	University of Leeds, UK
Mahieddine Djoudi	University of Poitiers, France
Amer Draa	University of Constantine 2, Algeria
Cédric Eichler	INSA Bourges, France
Zakaria Elberrichi	University of Sidi Bel Abbes, Algeria
Cherif Foudil	University of Biskra, Algeria
Khedoudja Ghanem	University of Constantine 2, Algeria
Salim Ghanemi	University of Annaba, Algeria
Mohamed Gharzouli	University of Constantine 2, Algeria
Nacira Ghoualmi-Zine	University of Annaba, Algeria
Said Ghoul	Philadelphia University, Jordan
Zahia Guessoum	University of Paris 6, France
Djamila Hamdadou	University of Oran, Algeria
Redha Mohamed Hamou	University of Saïda, Algeria
Saad Harous	UAE University, UAE
Jean-Michel Ilié	University of Paris 6, France
Okba Kazar	University of Biskra, Algeria
Mohamed-Khireddine Kholladi	University of El Oued, Algeria
Kamel Khoualdi	University of King Abdulaziz, Djeddah, KSA
Mohamed Tahar Kimour	University of Annaba, Algeria
Ilham Kitouni	University of Constantine 2, Algeria
Said Labed	University of Constantine 2, Algeria
Yacine Lafifi	University of Guelma, Algeria
Abdesslem Layeb	University of Constantine 2, Algeria
Kahloul Laïd	University of Biskra, Algeria
Ramdane Maamri	University of Constantine 2, Algeria
Derdour Makhlouf	University of Tebessa, Algeria
Mimoun Malki	University of Sidi Bel Abbes, Algeria
Smaine Mazouzi	University of Skikda, Algeria
Kamal Eddine Melkemi	University of Biskra, Algeria
Salah Merniz	University of Constantine 2, Algeria
Hayet-Farida Merouani	University of Annaba, Algeria
Souham Meshoul	University of Constantine 2, Algeria
Djamel Meslati	University of Annaba, Algeria
Chaker Mezioud	University of Constantine 2, Algeria
Sihem Mostefai	University of Constantine 2, Algeria
Abdelouahab Moussaoui	University UFAS Sétif, Algeria
Youcef Ouinten	University of Laghouat, Algeria
Mathieu Roche	CIRAD, Montpellier, France
Djamel Eddine Saidouni	University of Constantine 2, Algeria
Abdelhak-Djamel Seriai	University of Montpellier 2, France
Hamid Seridi	University of Guelma, Algeria
Noria Taghezout	University of Oran 1, Algeria

Hichem Talbi	University of Constantine 2, Algeria
Sadek Labib Terrissa	University of Biskra, Algeria
Chouki Tibermacine	University of Montpellier 2, France
Salah Toumi	University of Annaba, Algeria
Belabbas Yagoubi	University of Oran, Algeria
Mohamed Bachir Yagoubi	University of Laghouat, Algeria
Nacereddine Zarour	University of Constantine 2, Algeria
Nadia Zeghib	University of Constantine 2, Algeria
Mounira Zerari	University of Constantine 2, Algeria

Co-editors

Abdelmalek Amine	University of Saïda, Algeria
Allaoua Chaoui	University of Constantine 2, Algeria
Salim Chikhi	University of Constantine 2, Algeria
Djamel Eddine Saïdouni	University of Constantine 2, Algeria

Additional Reviewers

Nabiha Azizi
Ezedin Barka
Hichem Ben Abdallah
Imen Bensetira
Meriem Bensouyad
Mohamed Berkane
Mustapha Bouakkaz
Akila Djebbar
Raida El Mansouri
Brahim Farou
Nousseiba Guidoum
Meziane Hassina
Elhillali Kerkouche
Ilhem Kitouni
Bouchera Maati
Mohammed Amine Mami
Mohammad Mehedy Masud
Ahlam Melouah
Farid Mokhati
Brahim Nini
Merad Boudia Omar Rafik
Ziouel Tahar
Hichem Talbi
Aymen Yahyaoui

Contents

Cloud Computing and IoT

Enhanced RECCo Controller with Integrated Removing Clouds Mechanism

Oualid Lamraoui(✉) and Hacene Habbi(✉)

Applied Automation Laboratory, FHC, M'hamed Bougara University of Boumerdès, Av. de l'indépendance, 35000 Boumerdès, Algeria
lamraoui.oualid@gmail.com, habbi_hacene@hotmail.com

Abstract. The original RECCo controller algorithm evolves with data streams by adding new clouds and tuning the controller parameters in the consequent part autonomously. While performing the control of a given plant, useless information might be involved in the process of evolving the controller structure, which is a problematic issue with regard to control protocol implementation and big data processing. To deal with, in this work, a RECCo controller with removing clouds mechanism is designed. The enhanced RECCo controller is checked for performance from structural viewpoint and compared to the original RECCo controller by considering the problem of temperature control in a parallel heat exchanger.

Keywords: Robust evolving cloud-based controller · Removing clouds · Self evolving controller · Heat-exchanger

1 Introduction

Intelligent evolving systems are actually the subject of an increasing number of research works covering various fields of interest. The evolving concept relies on the data streams processing mechanisms based on which structural variations are introduced simultaneously into the evolving model for satisfactory or improved performance. The Evolving Cloud-based Controller (RECCo) belongs in fact to the class of intelligent evolving systems whose design procedure is built upon the concept of "data cloud" which allows handling non parametric data distributions. Basically, the controller is of Angelov-Yager (ANYA) type [1,2], which has non-parametric antecedent part. Application studies of this data-based fuzzy system showed good performance on evolving with data streams of different shapes. The RECCo controller employs a fuzzy rule-based evolving model which has the ability to start with zero fuzzy rules, and as so it does not require any offline training. It evolves with each input/output data while performing the control of the plant by adding new rules when detecting shifts in the distribution of data, otherwise it updates the existing structure. This way of evolving with data

S. Chikhi et al. (Eds.): MISC 2018, LNNS 64, pp. 3–15, 2019.
https://doi.org/10.1007/978-3-030-05481-6_1

streams can be involved into the controller structure with sensitive impact on the control implementation phase.

Obviously, the fact of adding a cloud removing mechanism to the original control algorithm, as detailed in [3], might be of major interest. This issue is precisely the main subject of the present contribution. The proposal of including a cloud removing mechanism to the RECCo controller is introduced and then checked for performance from operational as well as implementation aspects on a heat exchanger temperature control problem [4–7]. The suggested mechanism is designed to discard the "less active" and "less informative" clouds from the controller operational structure by avoiding new rules creation or removing non-significant existing ones.

The paper is organized as follows. Section 2 presents the self-evolving RECCo algorithm. The modified control algorithm with the involved cloud removing mechanism is described in Sect. 3. Section 4 provides an analysis study of the performance of the proposed evolving control algorithm on the operation of a thermal heat exchanger process. Finally, concluding remarks are made in Sect. 5.

2 RECCo Controller Design

As depicted in Fig. 1, the RECCo control algorithm consists of three main blocks: the reference model, the evolving law, and the adaptation law [8].

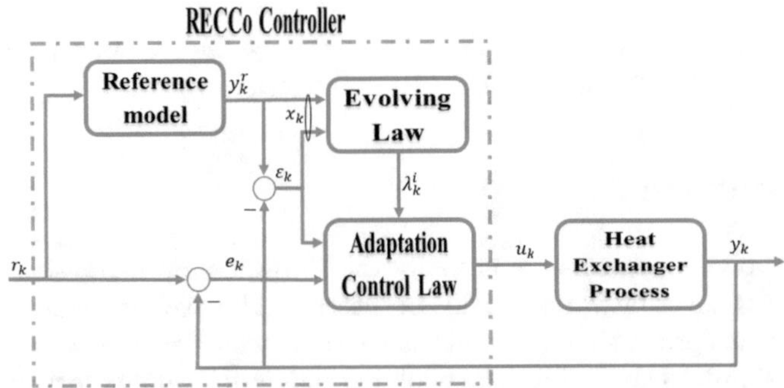

Fig. 1. The three main parts of the RECCo algorithm.

One of the most important advantage of using the RECCo control concept relies in the fact that no prior knowledge about the controlled process is needed. Thus the control algorithm can be initialized directly from the first received sample data. However, of course, any existing information can be used to properly initialize the controller parameters. Starting from the second received data sample and according to pre-defined conditions, the controller evolves its structure by updating an existing cloud (fuzzy rule) or adding a new cloud, followed by an adaption of the conclusion parameters. The concept of cloud is used to describe a fuzzy set or fuzzy rule, where local density calculation is used to link the recorded data with the nearest cloud. We can define the local density as a distance between a data and all the data associated to a cloud. As described in [9,10], the Euclidean norm is used to measure the distance between the data samples. The structure of the RECCo fuzzy rule-based system is of the following form [9]:

$$Rule^i : \textbf{If} x \sim X^i \ \textbf{Then} \ u^i, \quad i = 1, 2, \ldots, N_R \tag{1}$$

where the symbol $\sim$ refers to the fuzzy membership expressed linguistically as *is associated with*. X^i represents the ith cloud in the data space where $x = [x_1, x_2, \ldots, x_n]^T$ is the sample of data which represents the input of the controller and u^i designates its output in the ith fuzzy domain. The contribution of a controller output in a certain fuzzy domain to the current control is given by the normalized relative density:

$$\lambda_k^i = \frac{\gamma_k^i}{\sum_{j=1}^{N_R} \gamma_k^j}, \quad i = 1, 2, \ldots, N_R \tag{2}$$

where γ_k^i is the local density of the ith cloud for the current sample x_k. The algorithm measures the distance between the new arrived sample data to all the existing clouds by computing the N_R distinct densities γ_k^i. Depending on the maximum density, it decides to which cloud the current data has to be associated. In [1,10] the local density can be written recursively as follows:

$$\gamma_k^i = \frac{1}{1 + \left\|x_k - \mu_k^i\right\|^2 + \sigma_k^i - \left\|\mu_k^i\right\|}, \quad i = 1, 2, \ldots, N_R \tag{3}$$

with μ_k^i is the mean value of the clouds data points and σ_k^i is the mean-square length of the data vectors in the ith cloud. Both of those quantities can be calculated recursively using the following equations for mean value and mean-square length, respectively:

$$\mu_k^i = \frac{M_i - 1}{M_i} \mu_{k-1}^i + \frac{1}{M_i} x_k, \quad i = 1, 2, \ldots, N_R \tag{4}$$

$$\sigma_k^i = \frac{M_i - 1}{M_i} \sigma_{k-1}^i + \frac{1}{M_i} \left\|x_k\right\|, \quad i = 1, 2, \ldots, N_R \tag{5}$$

The initial value ($M_i = 1$) of the mean value is $\mu_1^i = x_1$ and the mean-square length is $\sigma_1^i = \|x_1\|$. It is recommended to use a first order linear reference model with a time constant approximately similar to the time constant of the controlled process, which gives better results, particularly in terms of robustness and minimal error [10]. In our case study, the reference model is defined as follows:

$$y_{k+1}^r = a_r y_k^r + (1 - a_r) r_k, \quad 0 < a_r < 1. \tag{6}$$

where r_k is the reference signal set by the operator. The parameter a_r is the pole of the first order discrete reference model and defines the transient dynamics. It can be approximated by $a_r = 1 - \frac{T_s}{\tau}$, where T_s is the sampling period of the process and τ is the time constant of the reference model which is slightly smaller than the estimated system time constant.

Different rule consequents with different control laws were used in the RECCo algorithm [1,11]. However, regarding real-world applications, the PID-based rule consequents are mostly preferred. The PID-type control consequent has the following form:

$$u_k^i = P_k^i \epsilon_k + I_k^i \Sigma_k^\epsilon + D_k^i \Delta_k^\epsilon + R_k^i \tag{7}$$

where $\epsilon_k = y_k^r - y_{k-1}$ is the tracking error (The difference between the output of the reference model and the output of the plant). P_k^i, I_k^i, and D_k^i are the controller gains and R_k^i is the compensation of the operating point for each cloud $i = 1, \ldots, N_R$. The discrete integral Σ_k^ϵ and the discrete derivative Δ_k^ϵ of the tracking error ϵ_k are computed recursively as follows:

$$\Sigma_k^\epsilon = \sum_{\kappa=0}^{k-1} \epsilon_\kappa = \Sigma_{k-1}^\epsilon + \epsilon_{k-1} \tag{8}$$

$$\Delta_k^\epsilon = \epsilon_k - \epsilon_{k-1} \tag{9}$$

For ANYA-type fuzzy rule-based systems, the weighted average is used for the defuzzification, the output of the controller becomes:

$$u_k = \sum_{i=1}^{N_R} \lambda_k^i u^i = \frac{\sum_{i=1}^{N_R} \gamma_k^i u^i}{\sum_{j=1}^{N_R} \gamma_k^j} \tag{10}$$

where u^i represents the consequence of the ith rule which is evaluated together with the normalized relative density (2) for global control output calculation. The input of the RECCo controller is chosen as follows:

$$x_k = \left[\frac{\epsilon_k}{\Delta\epsilon}, \frac{y_k^r - r_{min}}{\Delta r} \right]^T \tag{11}$$

note that $\Delta r = r_{max} - r_{min}$, which depends on the operating range of the system and $\Delta\epsilon = \frac{\Delta r}{2}$. In this case, we are mostly interested in the region where we expect the majority of the data samples. For the consequent part of the RECCo controller, the PID controller is used [2,8,10] and each cloud (fuzzy rule) has its own parameters of the consequent part. The parameter vector is denoted $\theta_k^i = [P_k^i, I_k^i, D_k^i, R_k^i]^T$ and the parameters of the first cloud are initialized with zeros $\theta_0^1 = [0,0,0,0]^T$, while all subsequently added clouds are initialized with the mean value of the parameters of all previous clouds as follows:

$$\theta_0^{N_R} = \frac{1}{N_R - 1} \sum_{j=1}^{N_R - 1} \theta_k^j \tag{12}$$

where N_R is the index of the added cloud. After associating the sample to one of the clouds, only the parameters of the consequent part of this cloud are adapted while the parameters of the other clouds are kept constant:

$$\theta_k^i = \theta_k^i + \Delta\theta_k^i \tag{13}$$

A better adaptation of the RECCo controller is obtained when the absolute value of the error is used. A new proposed adaptation method is used only in the start-up phase of the evolving system RECCo and then the adaptation continues as originally proposed in [10,12]. In the start-up phase, the control gains are adapted as follows:

$$\Delta P_k^i = \alpha_P G_{sign} \lambda_k^i \frac{|e_k \epsilon_k|}{1 + {r_k}^2} \tag{14}$$

$$\Delta I_k^i = \alpha_I G_{sign} \lambda_k^i \frac{|e_k \Sigma_k^{\epsilon}|}{1 + {r_k}^2} \tag{15}$$

$$\Delta D_k^i = \alpha_D G_{sign} \lambda_k^i \frac{|e_k \Delta_k^{\epsilon}|}{1 + {r_k}^2} \tag{16}$$

$$\Delta R_k^i = \alpha_R G_{sign} \lambda_k^i \frac{\epsilon_k}{1 + {r_k}^2} \tag{17}$$

where r_k is the reference signal, $e_k = r_k - y_{k-1}$ is the system error (The difference between the reference signal and the process output), and λ_k^i is the normalized relative density. The constants α_P, α_I, α_D, and α_R are the adaptation gains, and G_{sign} represents the sign of system monotony.

The adaptation mechanism can be the source of some potential instability problems caused by the parameter drift [8]. To make the RECCo controller more robust against this problem, several techniques were already applied in [1,10]. These techniques are used in our control design, and they are summarized as follows;

- Dead zone in the adaptive law (d_{dead}):

$$\Delta \overline{\theta}_k^i = \begin{cases} \Delta \theta_k^i & |\epsilon| \geq d_{dead} \\ 0 & |\epsilon| < d_{dead} \end{cases} \tag{18}$$

- Parameter projection ($[\underline{\theta}, \overline{\theta}]$):

$$\theta_k^i = \begin{cases} \theta_{k-1}^i + \Delta \theta_k^i & \underline{\theta} \leq \theta_{k-1}^i + \Delta \theta_k^i \leq \overline{\theta} \\ \underline{\theta} & \theta_{k-1}^i + \Delta \theta_k^i < \underline{\theta} \\ \overline{\theta} & \theta_{k-1}^i + \Delta \theta_k^i > \overline{\theta} \end{cases} \tag{19}$$

- Leakage in the adaptive law (σ_L):

$$\theta_k^i = (1 - \sigma_L)\theta_{k-1}^i + \Delta \theta_k^i \tag{20}$$

- Interruption of adaptation ($[umin, umax]$):

$$\Delta \overline{\theta}_k^i = \begin{cases} \Delta \theta_k^i & u_{min} \leq u_k \leq u_{max} \\ 0 & \text{otherwise} \end{cases} \tag{21}$$

where, $i = 1, \ldots, N_R$.

As presented in [2,10], the evolving law adopts a cloud adding mechanism. Once a new sample of data is received, the control protocol computes N_R different local densities and associates the data sample to the cloud with the maximum local density ($max_i \gamma_k^i$). Also, the parameters of that cloud are updated using equations (4) and (5). If the sample has the same density at two or more clouds, the current data sample is associated with the oldest cloud (The one that was added before the others). However, if the maximum local density ($\max_i \gamma_k^i$) is less than a threshold γ_{max} (The data sample is far away from all existing clouds), a new cloud is added. Some other criteria must be met before adding a new cloud (Like the time elapsed since the last change n_{add}).

3 Clouds Removing Mechanism

Using only the adding clouds condition, a useless information can alter the RECCo structure starting from the first arrived sample. In [3,13], the useless information may belong to one of the two concepts: a) The "less active" that present a small relative number of data samples associated with the cloud, counted from the moment of the cloud's, b) The "non-informative", which is a new property of the cloud and expresses the number of data samples that are most far away from particular cloud. Based on those two concepts, the whole mechanism for removing cloud is a logical combination of the two conditions C_{rem}^1 and C_{rem}^2;

$$C_{rem}^1 = \left(\frac{M_i}{k - k_i} < \zeta \frac{1}{N_R} \right), \qquad i = 1, \ldots, N_R \tag{22}$$

$$C_{rem}^2 = \left(1 - \frac{\tilde{M}_i}{k - k_i} < \frac{\zeta}{2} \right), \qquad i = 1, \ldots, N_R \tag{23}$$

where $\zeta \in [0, 1]$ is a constant parameter. If $\zeta = 0$ then the removing mechanism is disabled, while if $\zeta = 1$ then with each new added cloud the previous one is removed. Choosing $\zeta = 1$ is not a reasonable solution, and therefore in practice the constant ζ should be within $[0, 1)$. The two quantities $\frac{M_i}{k-k_i}$ and $\frac{\tilde{M}_i}{k-k_i}$ are called the "activity" and the "non-informative", respectively. For each data point x_k we calculate all the local densities and find the one with the minimum value ($min_i \gamma_k^i$). That point is associated as "anti-data" to the cloud with minimal density [3]. The number of such points for each cloud is denoted as $\tilde{M}_i$. We can summarize the whole self-evolving procedure presented above in the pseudo algorithm represented below [2,10];

Algorithm 1. Pseudo code of the RECCo PID control algorithm with the integrated clouds removing mechanism.

1: Initialize process parameters: τ, T_s, u_{min}, u_{max}, r_{min}, r_{max}.
2: Initialize evolving parameters: N_R^{max}, N_R, n_{add}, γ_{max}, ζ.
3: Initialize adaptation parameters: α_P, α_I, α_D, α_R, α_L, d_{dead}, $\underline{\theta}$, $\overline{\theta}$.

Repeat
  Measurement: y_k.
  Compute: y_k^r.
  Compute: e_k, ϵ_k, Σ_k^ϵ, Δ_k^ϵ.
  Compute: x_k.
  If $N_R = 0$ **Then**
    Increment: N_R.
    Store: K_{add}.
    Initialize: μ_0^1, θ_0^1, σ_0^1.
  Else
    Calculate: γ_k^i, λ_k^i, $i = 1, \ldots, N_R$.
    Associate the sample x_k with the cloud $min_i \gamma_k^i$.
    If $max_i \gamma_k^i < \gamma_{max}$ **and** $k > k_{add} + n_{add}$ **Then**
      Increment: N_R.
      Store: k_{add}, k.
      Initialize: $\mu_0^{N_R}$, $\theta_0^{N_R}$, $\sigma_0^{N_R}$.
    Else If $\frac{M_i}{k-k_i} < \zeta \frac{1}{N_R}$ **and** $1 - \frac{\tilde{M_i}}{k-k_i} < \frac{\zeta}{2}$ **Then**
      Decrement: N_R.
      Clear: μ_k^i, θ_k^i, σ_k^i.
    Else
      Associate the sample x_k with the cloud $max_i \gamma_k^i$.
      Update μ_k^i, θ_k^i, σ_k^i for the cloud $max_i \gamma_k^i$.
    End If
  End If
  Adaptation of the PID controller gains.
  Computation of the control law.
Until: End of data stream.

4 Application to Heat Exchanger Temperature Control

In this application study, the controlled process is a co-current (parallel flow) heat exchanger, which consists of three subsystems: the heater, the air circuit, and the water circuit [14]. As depicted in Figure 2 the air enters to the heater with temperature T_{13} and is heated up to temperature T_{14}. The electric heater is controlled via a power electronic device which can provide a variable heating power P from 0 to $10kW$. The hot air is used to increase the water temperature from temperature T_{33} to temperature T_{34}. Two motor-driven valves V_r and V_e are used to control the recycling of air over the range [0% 100%]. A single leak flow from 0 to $80l/h$ can be simulated on the real plant by using a bypass valve installed in the water circulation pipe. A PC-based data acquisition system is installed to collect measurements from the real plant.

Firstly, we aim to find an appropriate and transparent model structure using real recorded data for the exchanger. Based on prior knowledge and previous works [14,15], we can conclude that the behavior of the heat exchanger can be divided into two subsystems with a final output T_{34}. It is known that this output

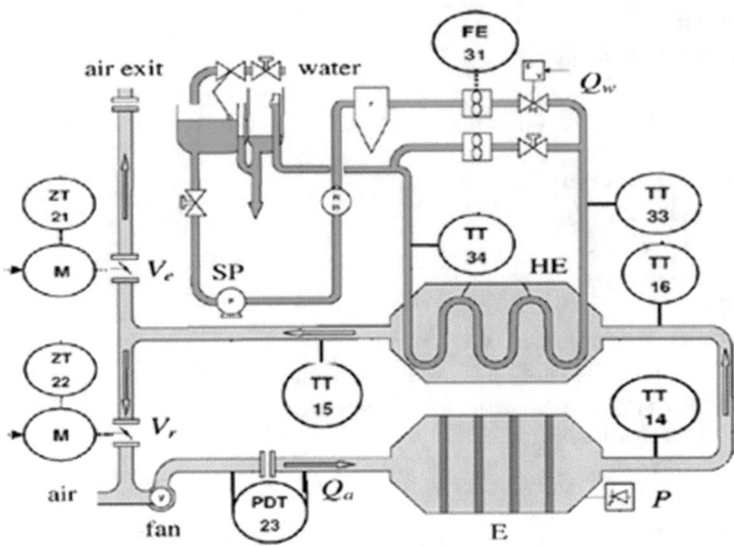

Fig. 2. Schematic representation of the pilot heat exchanger.

depends mainly on the air temperature after the heater T_{16}, which depends on the heating power P. Here, it must be noted that the air recycling valve position V_r and the air evacuation valve V_e are kept constant during measurement experiments. Analysis of the recorded data leads to choose a linear first-order model for the heater subsystem. However, a nonlinear first-order local model [5,16] is adopted for the parallel flow exchanger process. The constructed models are described by using the following structures:

$$\begin{cases} x_{16}(k+1) = A_{16}x_{16}(k) + B_{16}P(k) \\ T_{16}(k+1) = C_{16}x_{16}(k) + D_{16}P(k) \end{cases} \tag{24}$$

and

$$T_{34}(k+1) = f_{34}(T_{34}(k), T_{16}(k)) \tag{25}$$

The numerical values of the matrices A, B, C, and D are given as follows;

$$A = \begin{bmatrix} 0.9229 & -0.0473 \\ -0.2008 & 0.8462 \end{bmatrix}, B = \begin{bmatrix} 0.002003 \\ 0.005146 \end{bmatrix}$$
$$C = \begin{bmatrix} 57.23 & -1.109 \end{bmatrix}, D = 0 \tag{26}$$

where x_{16} is the state-space of the heater model and it is calculated by using an efficient numerical algorithm for subspace state-space identification (N4SID) [17], and f_{34} is a dynamic TS fuzzy model [4] constructed with the subtractive algorithm, and it can be described by a set of If-Then fuzzy rules of the form:

$$Rule^i : \textbf{If}\, T_{34}\, is\, A_1^i \text{ and } T_{16}\, is\, A_2^i \text{ and}$$
$$\textbf{Then}\, T_{34}^i(k+1) = b_1^i + a_1^i\, T_{34}(k) + a_2^i\, T_{16}(k) \tag{27}$$

where A_j^i is the fuzzy set associated with the jth variable for the ith rule and the constants a_j^i and b_j^i are the rule-consequents parameters.

To check the performance of the RECCo controller with the integrated removing mechanism, we consider the control of the heat exchanger model described by (24) and (27). The original RECCo controller is also considered for operational performance checking. In the simulation study, the reference signal is chosen to cover a wide range of the operating space. Both controllers start with an empty structure and then evolve and adapt their parameters during the control of the process plant. The control parameters are set as follows:

- Sampling time: $T_s = 0.1\text{s}$,
- The reference model pole: $a_r = 0.98$,
- Local density threshold: $\gamma_{max} = 0.96$,
- Maximum number of clouds: $N_R^{max} = 100$,
- Minimum samples between two clouds: $n_{add} = 20$,
- Constant of the removing condition: $\zeta = 0.15$,
- Minimal reference signal: $r_{min} = 0°\text{C}$,
- Maximal reference signal: $r_{max} = 100°\text{C}$,
- Adaptive gains: $\alpha_P, \alpha_I, \alpha_D, and\ \alpha_R = 0.5$,
- Dead zone: $d_{dead} = 0$,
- Parameter projection: $]-\infty, \infty[$,
- Leakage: $\sigma_L = 0$,
- Minimal control signal: $u_{min} = 0\text{kW}$,
- Maximal control signal: $u_{max} = 10\text{kW}$.

In this simulation, a staircase reference signal with a step value equal to 10°C is used. Figures 3, 4, and 5 depict a good tracking performance for both controllers. As can be seen, the RECCo controller shows similar behavior as the enhanced RECCo controller. This result is obviously expected since the enhancement is targeted at the structural level of the controller. In fact, the removing condition deleted the useless information without altering the operational performance of the controller. This can be concluded from Fig. 5, where it can be seen that the removing mechanism has suppressed the initial cloud, which becomes less informative after some time instants. To further explain the cloud removing operated by the integrated removing mechanism, we show in Fig. 6 the activity and the non-informative functions generated under the enhanced RECCo controller protocol. From that figure, it can be clearly noticed the decreasing in the weight of the first cloud, which continues until reaching the thresholds for both functions. As a consequence, the first cloud is marked as less informative and then found deleted by the removing condition. The main impact of the removing condition can be perceived at the implementation phase. As described in Sect. 2, the original RECCo algorithm may evolve with more complex structure than the enhanced RECCo controller does, which has an important impact on practical implementation.

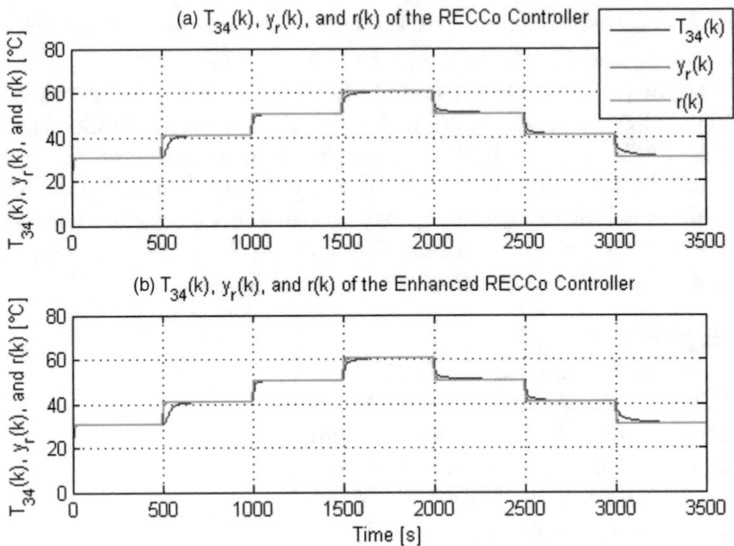

Fig. 3. The reference $r(k)$, model reference $y_r(k)$, and the output signal tracking $y(k)$ for the heat exchanger.

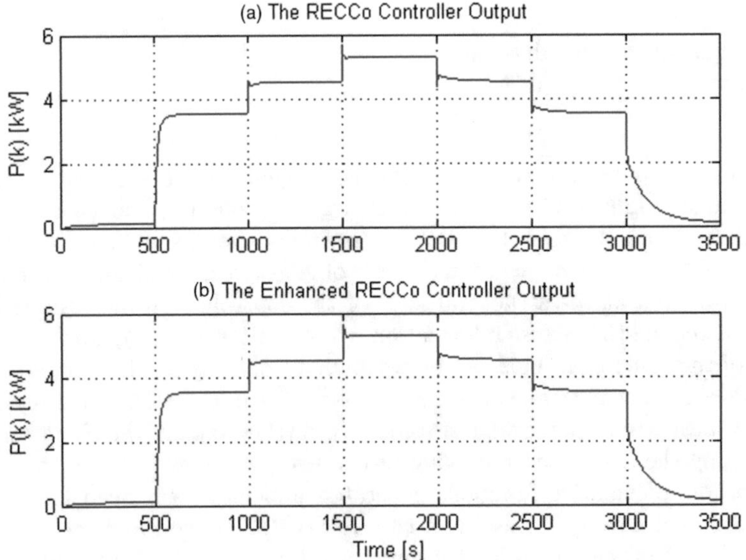

Fig. 4. The control signal $u(k)$ for the heat exchanger.

5 Conclusion

In this paper, a RECCo control algorithm with integrated clouds removing mechanism is proposed. In order to check the closed loop performance of the enhanced

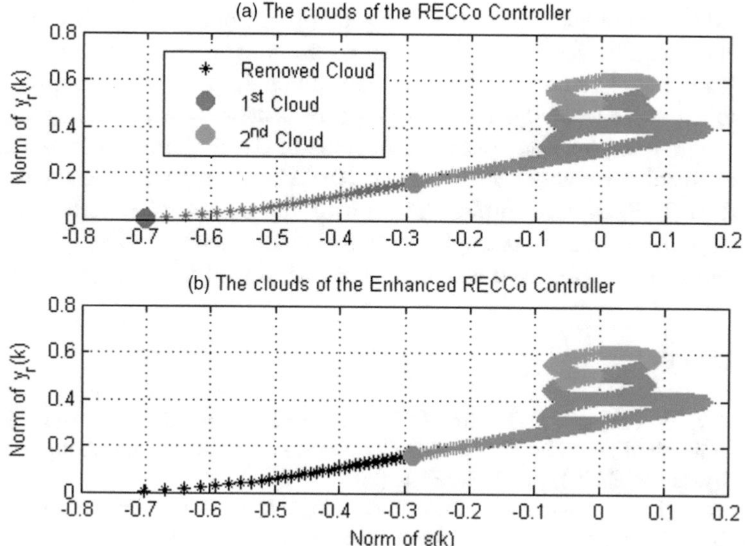

Fig. 5. The data clouds created in the input space.

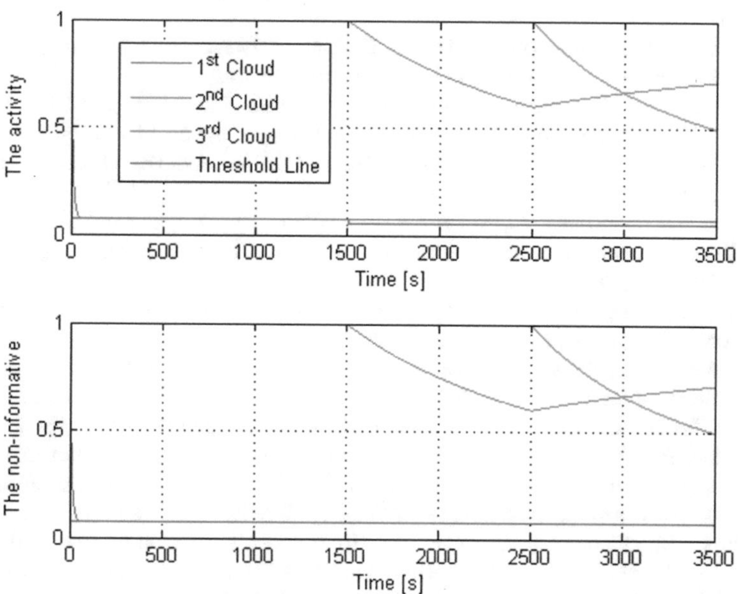

Fig. 6. The activity and non-informative functions for each data cloud.

RECCo control, an application study to the problem of temperature control in a heat exchanger process is considered. Firstly, we started by modeling the plant, where a nonlinear identification based on the subtractive clustering is conducted.

The closed loop simulation results show good operational performance under reference changes. The structure of the enhanced controller has also been altered with the process of removing clouds, which might impact considerably well the implementation issue of the RECCo protocol. Future works might extend the study to more complex control configurations of the exchanger process in order to check the robustness of the enhanced RECCo controller against parameters shift and drift. Real-time implementation of the enhanced control protocol is also an important issue that might have to be conducted, as the integration of the removing clouds proposal is specifically introduced for that purpose.

References

1. Angelov, P., Škrjanc, I., Blažič, S.: Robust evolving cloud-based controller for a hydraulic plant. In: The 2013 IEEE Conference on Evolving and Adaptive Intelligent Systems (EAIS), pp. 1–8. IEEE (2013)
2. Andonovski, G., Angelov, P., Blažič, S., Škrjanc, I.: A practical implementation of robust evolving cloud-based controller with normalized data space for heat-exchanger plant. Appl. Soft Comput. **48**, 29–38 (2016)
3. Andonovski, G., Mušič, G., Blažič, S., Škrjanc, I.: On-line evolving cloud-based model identification for production control. IFAC-PapersOnLine **49**(5), 79–84 (2016)
4. Lamraoui, O., Boudouaoui, Y., Habbi, H.: Data-driven approaches for fuzzy prediction of temperature variations in heat exchanger process. In: The 2017 International Conference on Control, Automation and Diagnosis (ICCAD'17). IEEE (2017)
5. Habbi, H., Boudouaoui, Y.: Hybrid artificial bee colony and least squares method for rule-based systems learning. Waset Int. J. Comput. Control Quantum Inf. Eng. **8**(12), 1968–1971 (2014)
6. Oravec, J., Bakošová, M., Mészáros, A., Míková, N.: Experimental investigation of alternative robust model predictive control of a heat exchanger. Appl. Therm. Eng. **105**, 774–782 (2016)
7. Jamal, A., Syahputra, R.: Heat exchanger control based on artificial intelligence approach. Int. J. Appl. Eng. Res. (2016)
8. Andonovski, G., Bayas, A., Sáez, D., Blažič, S., Škrjanc, I.: Robust evolving cloud-based control for the distributed solar collector field. In: The 2016 IEEE International Conference on Fuzzy Systems (FUZZ-IEEE), pp. 1570–1577. IEEE (2016)
9. Angelov, P., Yager, R.: Simplified fuzzy rule-based systems using non-parametric antecedents and relative data density. In: The 2011 IEEE Workshop on Evolving and Adaptive Intelligent Systems (EAIS), pp. 62–69. IEEE (2011)
10. Škrjanc, I., Blažič, S., Angelov, P.: Robust evolving cloud-based pid control adjusted by gradient learning method. In: The 2014 IEEE Conference on Evolving and Adaptive Intelligent Systems (EAIS), pp. 1–8. IEEE (2014)
11. Costa, B., Skrjanc, I., Blazic, S., Angelov, P.: A practical implementation of self-evolving cloud-based control of a pilot plant. In: The 2013 IEEE International Conference on Cybernetics (CYBCONF), pp. 7–12. IEEE (2013)
12. Andonovski, G., Blažič, S., Angelov, P., Škrjanc, I.: Analysis of adaptation law of the robust evolving cloud-based controller. In: The 2015 IEEE International Conference on Evolving and Adaptive Intelligent Systems (EAIS), pp. 1–7. IEEE (2015)

13. Dovžan, D., Logar, V., Škrjanc, I.: Implementation of an evolving fuzzy model (eFuMo) in a monitoring system for a waste-water treatment process. IEEE Trans. Fuzzy Syst. **23**(5), 1761–1776 (2015)
14. Habbi, H., Kidouche, M., Kinnaert, M., Zelmat, M.: Fuzzy model-based fault detection and diagnosis for a pilot heat exchanger. Int. J. Syst. Sci. **42**(4), 587–599 (2011)
15. Habbi, H., Kinnaert, M., Zelmat, M.: A complete procedure for leak detection and diagnosis in a complex heat exchanger using data-driven fuzzy models. ISA Trans. **48**(3), 354–361 (2009)
16. Habbi, H., Boudouaoui, Y., Karaboga, D., Ozturk, C.: Self-generated fuzzy systems design using artificial bee colony optimization. Inf. Sci. **295**, 145–159 (2015)
17. Van Overschee, P., De Moor, B.: N4sid: Subspace algorithms for the identification of combined deterministic-stochastic systems. Automatica **30**(1), 75–93 (1994)

Virtual Machines Allocation and Migration Mechanism in Green Cloud Computing

Nassima Bouchareb([✉]) and Nacer Eddine Zarour

LIRE Laboratory, Faculty of New Information and Communication Technologies, Department of Software Technologies and Information Systems, University of Constantine 2 - Abdelhamid Mehri, Constantine, Algeria
{nassima.bouchareb,nasro.zarour}@univ-constantine2.dz

Abstract. The resource allocation in dynamic environments presents numerous challenges since the requests of consumers are important. Unfortunately, the maximization of accepted Cloud user's requests, and at the same time, the reduction of energy consumption is a conflicting problem. In this paper, we propose a mechanism for the VMs allocation and reallocation in the Cloud data centers, using the VM migration aspect. Our mechanism is based on Min/Max thresholds, to avoid the emerging of underloaded/overloaded hosts and to keep servers relatively stable after VM consolidation. Finally, this paper presents some tests and simulation results, using the CloudSim simulator.

Keywords: Cloud computing · Virtual machine allocation/reallocation
Virtual machine migration · Energy efficient · Green computing

1 Introduction

Cloud Computing (CC) is a recent paradigm which provides ubiquitous access to shared computing resources for organizations and end users. So, CC providers should increase their performances to overcome difficulties that they may encounter, especially resource limitation. The resource management in CC encounters difficulties, particularly when there are a big number of requests. The selection of suitable resources and provider cost are among these problems. It has been argued that energy costs are among the most important factors impacting on provider total cost [1], which result also in the rising of Cloud users' costs. Therefore, the Cloud providers should minimize their energy consumption as much as possible while satisfying the consumers' requests. The consequences of improper resource management may result into underutilized resources and wastage of them. It is a big challenge for Cloud datacenters to reduce operating costs and improve resource utilization while meeting the service-level agreement (SLA) of applications.

As a solution, Cloud providers use the concept of virtualization, which is becoming popular in CC environments due to the advantage of server consolidation, resource isolation and live migration. Virtualization helps in partitioning of one Physical Machine (PM) into numerous Virtual Machines (VMs) that run concurrently. So, VM refer to one instance of an operating system along with one or more applications running in an isolated partition within the computer. Allocation of VMs is divided

S. Chikhi et al. (Eds.): MISC 2018, LNNS 64, pp. 16–33, 2019.
https://doi.org/10.1007/978-3-030-05481-6_2

in two parts: the first one is admission of new requests for VM provisioning and placement VMs on hosts, whereas the second part is optimization of current allocation of VMs.

Sometimes, it may be required to dynamically transfer certain amount of load of a machine to another machine with minimal interruption to the users. This process is termed as *VM migration*. It is used to *reduce power consumption* in Cloud data centers, and for *load balancing*, which reduces the inequality of resource usage levels across all the PMs in the cluster. VM migration can also be used in case of sudden failure "*physical machine fault tolerant*"; it allows the VM to continue its job even any part of system fails [2, 3]. VM migration methods are divided into two types: (1) Hot (live) migration- VM keeps running while migrating and does not lose its status (without shutting down.). User doesn't feel any interruption in service in hot migration. (2) Cold (non-live) migration- The status of the VM loses and user can notice the service interruption [3, 4].

An idle machine unnecessarily uses 70% power of data centers [5]. So, the power is wasted because of the underutilization of resources at data centers, on one side. On the other side, if the total requested resource is larger than the PM's capacity and the host may increase response time and reduce throughput, the Cloud users may not get their expected QoS, which is defined in terms of the service-level agreement (SLA), and the service provider must get a penalty. So, live Migration for load balancing is done for two types of VMs: *Underloaded VMs* are VMs which their CPU capacities are underutilized. All the VMs of such node are migrated to nodes which residual capacity is big enough to hold them. So, the underloaded node is switched off to save power. *Overloaded VMs* are VMs which have already crossed their utilization capacities. In this case, migration is done to underloaded VM. In our approach, we have considered these factors to save energy.

The process of live migration uses some computing resources (it produces large CPU overhead and network traffic) and consumes energy and time. So, a large number of VM migrations can consume a large amount of system resources and may cause SLA violations. Therefore, before initiating a migration, the reallocation controller had to ensure that the cost of migration does not exceed the benefit, and it is necessary to design proper policies to optimize the VMs migration process. The optimization includes reducing the number of VM migration, as well as migrating small VM instead of big VM.

In this paper; we study the allocation and reallocation of VMs in a Cloud, in order to save energy, which can be a small step towards Green technology. Our mechanism is based on Min/Max thresholds, to avoid the emerging of underloaded/overloaded hosts. We explain how do Clouds allocate VMs, and especially how do they select VMs to migrate and which ones receive them. We particularly treat the VM migration case of overloaded resources.

The remainder of this paper is organized as follows. First, we discuss similar works in Sect. 2. In Sect. 3, we present the principle of our mechanism. Then, the rules for the VMs allocation and migration are presented in Sects. 4 and 5. Finally, we evaluate the mechanism and we present the experimental results in Sect. 6, before concluding and giving some future directions in Sect. 7.

2 Related Work

Cloud platform needs to dynamically deploy and relocate VMs onto proper physical hosts in order to meet different needs, such as avoid hotspot, power saving and load balancing. Therefore, how to dynamically and efficiently schedule virtual machines (VMs) among physical hosts (PMs) to meet the needs of different targets becomes a problem. To address this issue, many optimization theory based VM placement approaches are used to solve the resource management problem in IaaS Cloud.

A lot of researches are being conducted in the area of CC to reduce the power consumption in the data centers as surveyed in [6]. Different techniques to overcome the power wastage have been proposed and devised with and without VM migration.

In [5], authors focus on the problem of VM placement appeared in initialization plan stage of datacenters management with the goal of minimizing the number of PMs. They have first formulated the problem of VM placement as a variant of multi-dimensions bin packing problem, and then exploited constraint solver to solve this problem. Finally, they have implemented VMs placement algorithm using constraint programming technique in CloudSim. There is no aspect of migration in this work.

Authors in [7] investigate the VM reallocation inside a Cloud in order to save energy while maintaining the resources required by users, in order to minimize the number of hosts powered on, while limiting concurrent migrations of VMs, and in a reasonable computational time. Their approach has been implemented in OpenNebula and experimentally compared with its default approach. OpenNebula scheduler gives the best QoS to the VMs, as it tries to allocate each VM to the host where there are the most resources. In this paper, migrations are only enforced if and only if the host may be fully unloaded. It is an optimization choice. The aim is to avoid migrations that could irrelevant because another decisions could be better in the next scheduling loop and would only generate overhead.

Authors in [4] and [8] have presented the various VM migration techniques, especially, those of the live migration. In [4], the authors only talked about migration in case of "hotspot". When one physical host gets overloaded, it may be required to dynamically transfer certain amount of its load to another machine with minimal interruption to the users, without even explaining the principle of thresholds. In [8], no migration case was processed.

In [9], authors focused on the optimization of the multiple VMs migration process. The purpose of this optimization is to minimize the overhead of CPU and network cost (i.e., the migration of VMs with smaller size will generate less network traffic) and adjust VM migration steps to avoid conflicts during the migration process. Authors proposed That VMs which have same source or destination host would migrate one by one, because the node migration performance of host would decreased significantly due to simultaneous execution of multiple VM migration. VMs which their migration do not conflict with each other, could migrate at same time. In their model, VMs are divided into clusters (groups of VM that will be migrated together). After, they choose the best location of each cluster by affecting it on all the hosts and they calculate the minimum cost. At the end, they choose the min cost. Then, the algorithm determines the migration order of these virtual machines to be migrated. Firstly, the scheduler gives priority to migrate VM on high-load host to the low-load host. But on what basis do they form clusters "the choice of groups of VMs"? In addition, there is no aspect of threshold. But what is positive in this work, is that they treat the reallocation of VMs.

In [3], the authors also deal with the migration of VMs in CC data centers. They proposed an algorithm that allows to select the VM that will support the received request according to the free memory "the one that has more free memory" to minimize the number of migrations. Then, if the free memory is lower than the requested memory, they seek the process to migrate, it is the one that is nearest to X (the minimum of greater than or equal to x), x is the rate of memory that must be liberated. Finally, they determine the inviting VM; it is the one that also has the Min of memory sufficient to accommodate the migrated process. They did not talk about the case where all the free memory at the level of all the VMs is lower than the required memory, how do they proceed? (Do they activate a new resource, search for a coalition, cancel an old request, or simply refuse the new request).

Authors in [10] have proposed a remaining utilization-aware algorithm for VM placement, and a power-aware algorithm is proposed to find proper hosts to shut down for energy saving. These two algorithms have been combined and applied to Cloud data centers for completing the process of VM consolidation. When there exists more than one host with the same least CPU utilization (underloaded hosts) in a heterogeneous Cloud data center, they select the host with higher power consumption and a lesser number of VMs as the underloaded host to switch it off. They have considered the number of VMs running on the host as another key indicator. The host with more VMs has fewer chances to be selected as the underloaded host. So, choose the host that has a lesser number of VMs and more power consumption among the underutilized hosts as the underloaded host.

In our work, we also want to minimize the number of migrations, but we have not treated in this paper the case of the underloaded hosts. Also, it depends on the execution time, we can have only one VM on a host, but with a long run time, so it is better to migrate it and switch off the machine. In conclusion we must see the number of VMs and their execution time in the underloaded hosts).

In [15], a machine learning technique for VMs live migrations based on adaptive prediction of utilization thresholds is presented. It is a heuristic based predictive model where future SLA violation is to be predicted, then migration decision will be made.

In [11], again a threshold based approach is proposed using Single Threshold (ST) value, which is based on the idea of setting upper utilization threshold for hosts and placing VMs while keeping the total utilization of CPU below this threshold. The aim is to preserve free resources in order to prevent SLA violation due to consolidation in cases when resource requirements by VMs increase. But the node has to remain active even if the load is much less than threshold value. Authors have proposed also, three other policies for choosing VMs that have to be migrated from the host: (1) Minimization of Migrations (MM) – migrating the least number of VMs to minimize migration overhead; (2) Highest Potential Growth (HPG) – migrating VMs that have the lowest usage of CPU relatively to requested in order to minimize total potential increase of the utilization and SLA violation; (3) Random Choice (RC) – migrating the necessary number of VMs by picking them according to a uniformly distributed random variable.

MM policy achieves the best energy saving, that's why we adopt it in our work. In [11], the authors also presented the double thresholds aspect which is more detailed in [12].

In [12], authors have proposed VMware Distributed Power Management operates based on thresholds, with the lower and upper utilization thresholds set to 45% and 81% respectively. However, fixed values of the thresholds are not suitable for systems with dynamic and unpredictable workloads. So, in [13] and [14], authors have proposed an approach to set the CPU utilization thresholds values dynamically, depending on a current set of instantiated VMs and historical data of the resource usage by each VM. They have not only met energy efficiency requirement but also ensured quality of service to the user by minimizing the SLA violations. Dynamicity has also reduced human labor of defining the threshold limit for the host and made the machine capable to calculate threshold value itself. They have also showed the cost structure at both user and data center side. But the authors have not explained how do they allocate VMs, and especially how do they select VMs to migrate and which ones receive them.

From all these results, we have based our work on this last one "dynamic double thresholds", and we give more detail on the allocation and migration of VMs.

3 Mechanism Principle

Let's start with an initial allocation. In our solution, we opted for using two dynamic thresholds Min and Max (cold and hot thresholds), as was shown in [13] and [14]. The Min threshold is used because an underloaded resource will consume a lot of power for a minimum of VMs. The Max threshold is used to avoid the violation of SLAs which requires that the service provider must get a penalty and avoid also increasing response time and reducing throughput. Therefore, some policies should be made to avoid the emerging of overloaded hosts and to keep servers relatively stable after VM consolidation.

Then, if we obtain resources that are overloaded or underloaded, we will proceed to the optimization of the VMs allocation, which is the process of reallocation of VMs. This latter is based on the migration of some VMs in order to have a balanced load. For the strategy of choosing VMs to be migrated, we are always trying to minimize the number of migrations, because migrating VMs also consumes power, and the large number of VM migrations can cause SLA violations. Therefore, it is necessary to design proper policies to decrease the number of migrations when conducting VM consolidation.

So, to solve the complex problem of dynamic server consolidation, we have used the four main events or steps proposed in [16]:

- *Host Overload Detection:* The scheduling technique must set a threshold limit in order to decide when a certain host/server is over-utilized. This limit can be termed as *'Hot Threshold'* and when this limit is crossed, some of this host's VMs need to be migrated to other hosts.

- *Host Under-load Detection:* If a certain server is under-utilized, i.e. it has reached below the *'Cold Threshold',* the aim of server consolidation is to identify that under-utilized server and migrate all of its VMs to other active hosts. Thus the under-utilized server is freed up and it can be switched to sleep/idle mode to save power.
- *VM Selection and Migration:* Appropriate candidates (VMs) are selected either from overloaded or under-loaded host for migration.
- *VM Placement:* The VMs selected in previous step are then placed on some other PMs according to a suitable VM to PM mapping criteria.

The VM allocation problem may appear to be simple at a glance, but a closer examination unmasks its complication. In this paper, we based on the dynamic Thresholds strategy by adding the allocation, migration and placement rules presented in Sects. 4 and 5, and we called it Min /Max thresholds strategy. We just give examples of minimum and maximum thresholds equal to 20% and 80% respectively, to explain the mechanism.

4 VMs Allocation Rules

Summarizing, here are the initial rules of the VMs allocation (see Fig. 1):

- In (t_0), when receiving a request:
  - The load of each resource must $\in$ [Min, Max].
- In (t_1), all resources are at max, and the Cloud receives a new request:
  - Overload a resource (before activating a new one), choosing the one that will be liberated as soon as possible (the resource that will be below max as soon as possible). Liberated resource is a machine that has some free VMs, (partially liberated not necessarily completely released). If we notice that there can be a violation of SLAs because of the overloading, we directly activate a new resource and migrate to it few VMs.
- In (t_2), we have overloaded α resources, and the Cloud receives a new request:
  - Activate a resource (if we have α overloaded resources "not all the resources", because in this case, the energy consumed by a new resource is lower than the energy consumed by the overloading of $\alpha + 1$ resources, in addition, there is no risk of SLA violations, or the other disadvantages « long response time and low throughput).
- In (t_3), after activating all the resources, and the Cloud receives a new request: we overload the other resources.
- In (t_4), we have no resource to activate and all the resources are overloaded:
  - Cancel an old request, to release resources and reuse them to satisfy the new request.
  - Solicit another Cloud provider to use its own resources by forming a coalition.
  - Refuse the new request.

```
(Res)n : activated resources
(Res)m: non activated resources
(Res)i = overloaded resources

Begin
  If new request Then
    ∀ Res: Load (Res)n ∈ [Min, Max]
  End if
  If (∀ Res: Load (Res)n = Max) Λ (new request) Then
    Repeat
      ∃ (Res)n = (Res)i   (Load (Res)i > Max)   / T(Res)i = Min (T)
      If ∃ SLA violation in (Res)i Then
        Activate a new resource
        Migrate few VMs to it
      End if
    Until   nbr (Res)i = α
  End if
  If (nbr (Res)i = α / Load (Res)i > Max) Λ (new request) Then
    Activate a new resource
  End if
  If (∄ Res: Load (Res)m = 0) Λ (new request) Then
    Overload the other resources
  End if
  If (∄ Res: load (Res)m = 0) Λ (∀ Res: Load (Res)n > Max) Λ (new request) Then
    (Cancel an old request)  V  (form a coalition)  V  (refuse the new request)
  End if
End
```

Fig. 1. VMs allocation rules.

5 VMs Migration Rules

After a short time, some requests are completed or can be canceled, which frees some VMs. In addition, as we have some overloaded or underloaded resources, we will migrate their VMs to the freed resources by following the rules presented in this section. We remind that we treat in this paper the case of the overloaded machines:

(a) The selection of VMs to migrate

- We migrate the request that occupies the exact number of VMs (X) required for the migration.
- If it does not exist, then select the one that occupies the minimum number of VMs greater than X (the nearest one > X).
- Otherwise, we migrate several requests (while minimizing the number of migrations), such that these requests require "X" VMs or the minimum number of VMs greater than X.

(b) The selection of the receiving machines « VM placement »
   1. There is at least one resource with free VMs that can perform the request. We choose the one with the minimum number of free VMs, greater than or equal to X (see Fig. 2):

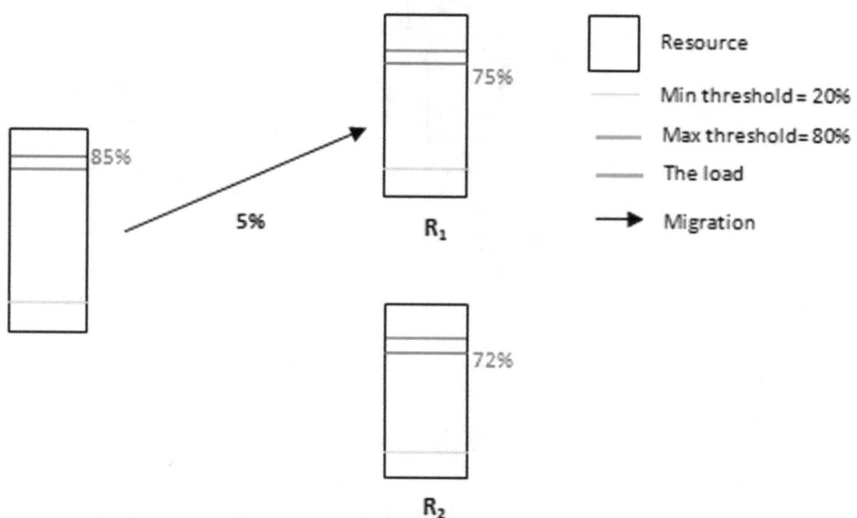

Fig. 2. Case 1.

   2. No resource has free VMs to complete the request. We share the request on resources with free VMs starting with the one with the maximum number (see Fig. 3):

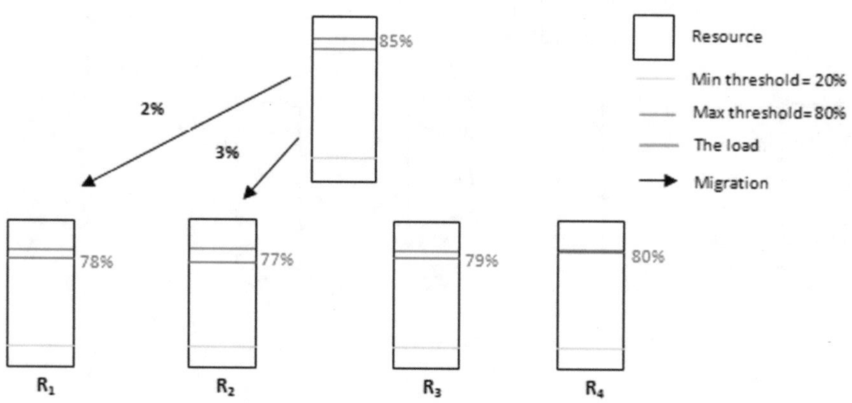

Fig. 3. Case 2.

3. The number of free VMs in the resources is exactly the same. First, we select the resource that will be released as soon as possible to not be activated for a negligible number of VMs (see Fig. 4):

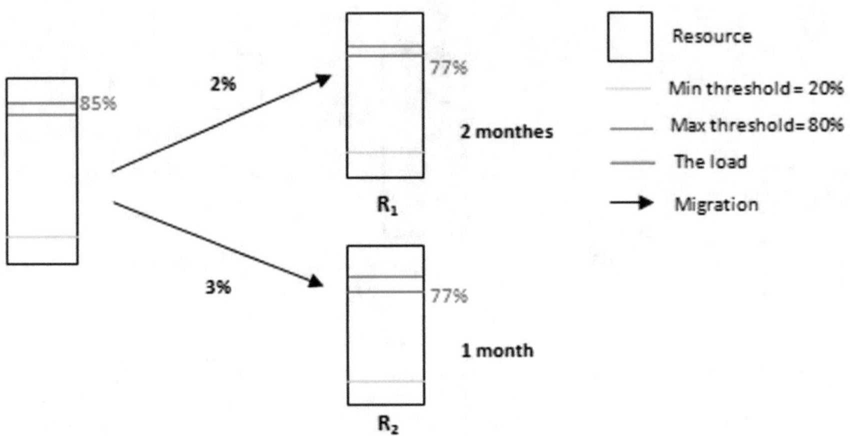

Fig. 4. Case 3.

4. Minimize the number of migrations. Although we just need to migrate 5% to not have an overloaded resource, we prefer to migrate the request which occupies 10%, than migrating two requests 2% and 3% (see Fig. 5):

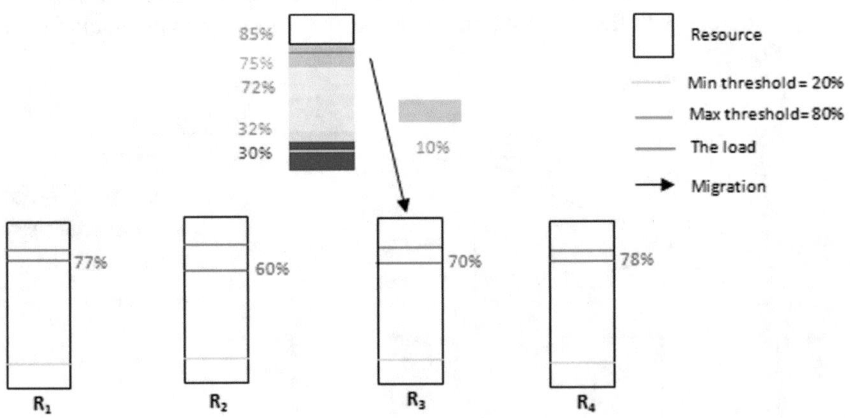

Fig. 5. Case 4.

- Unless there are not enough resources, no 10% free in all resources. Obligation to make several migrations (see Fig. 6):

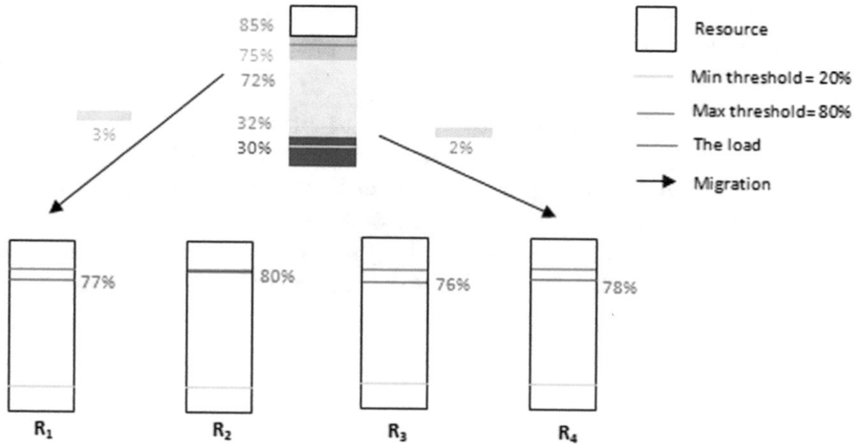

Fig. 6. Case 5.

5. Minimize the sharing of requests to avoid communication that requires more time, energy and consistency between machines. Privilege to make two migrations of two different requests (we have just two migrations), to do a sharing of a single request and to migrate it on two different machines, because by sharing the request in two, we will finally make two migrations for this same request (so there will be two migrations + one share). (Ex. we do not migrate the request that occupies 10%, because it must be shared on both resources by making two migrations) (see Fig. 7):

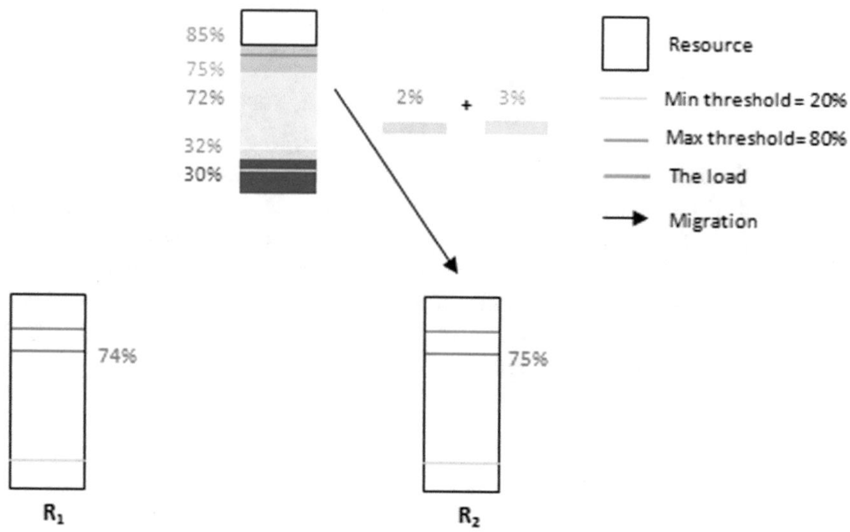

Fig. 7. Case 6.

6. If the number of free VMs in the other machines is less than the number of VMs we want to migrate, we must see in t' (after a moment) the state of all the machines:

   - If the overloaded resource will not be overloaded, we do not migrate anything. We prefer to wait for a moment (because in t' the resource will not be overloaded anymore) than to make several migrations plus a sharing of the request. (Whatever the state of the other resources, even if they will also be free at t', the essential thing is to obtain the number of free VMs free on the initial resource before or at the same time as the other resources) (see Fig. 8):

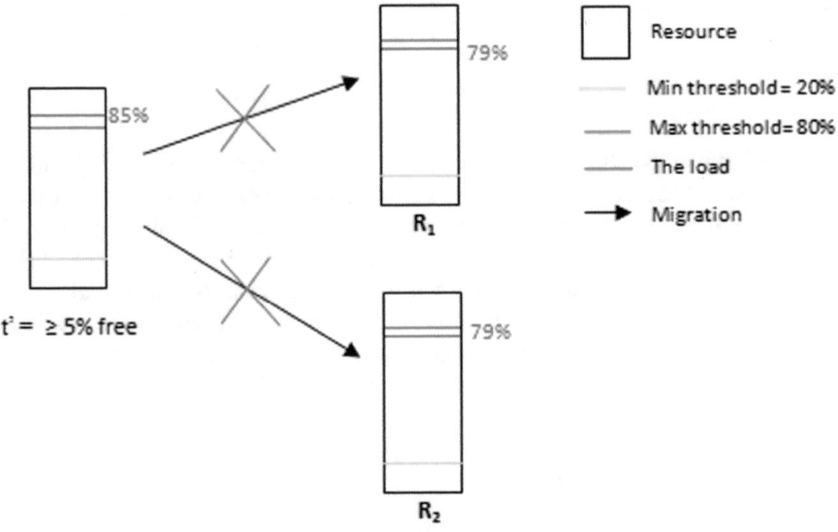

Fig. 8. Case 7.

   - If migrating the VMs from the overloaded machine to the other machines, in t' this machine will not be overloaded, then we have to migrate. i.e. a part of the request will be migrated to the other machines, and a part will run on the initial machine (we migrate the maximum of VMs possible, and in t' the initial resource liberates some VMs and it will not be overloaded, we do not migrate the rest of VMs, even if the other resources will also liberate some VMs) (see Fig. 9):

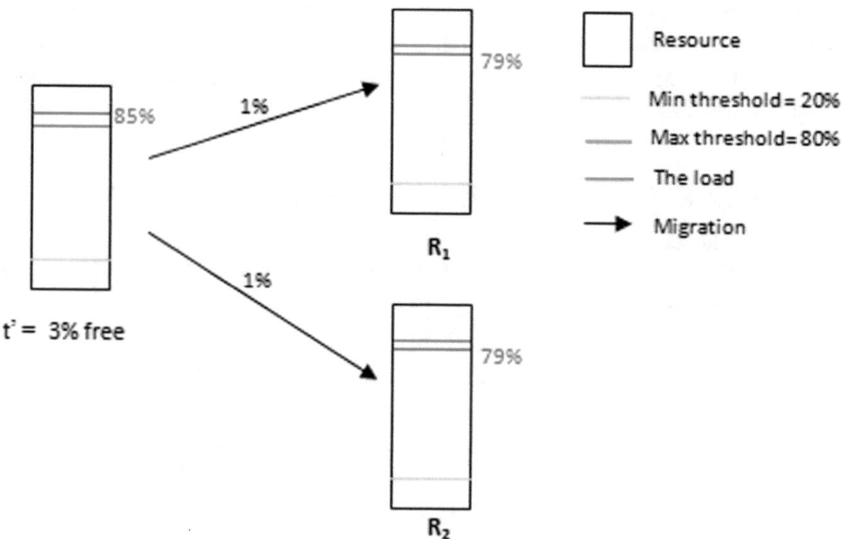

Fig. 9. Case 8.

- If the other machines are free before the overloaded machine, we also do the migration (i.e., we migrate at the beginning to the free VMs, then at t', we migrate the rest) (see Fig. 10):

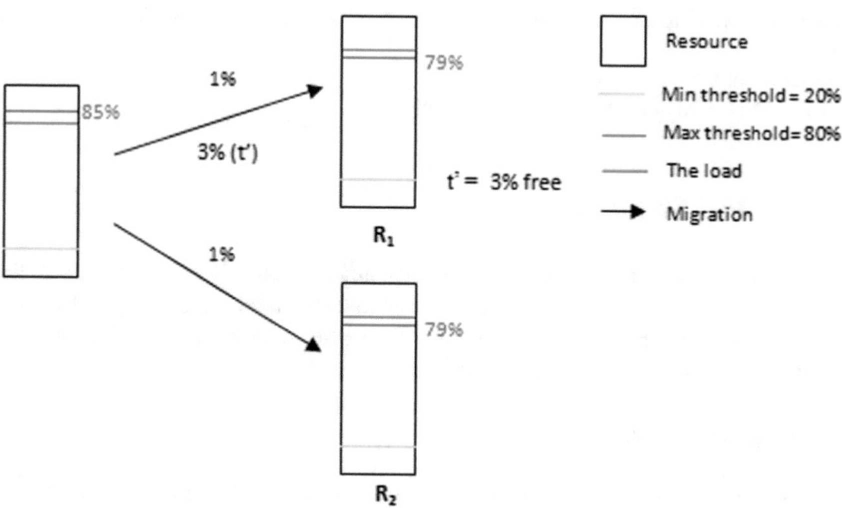

Fig. 10. Case 9.

All the approaches presented in related work are akin to ours, as the main focus is to save energy when accepting the maximum of requests and minimizing SLA violations. However some differences are seen when managing VMs migrations, the main ones are shown in this Table 1.

Table 1. Differences between our mechanism and some related works based on VMs migration.

	MM[a]	Migrate the nearest[b]	Place on the nearest[c]	Minimum number of hosts	Minimize request sharing	Minimize coldspots[d]	Minimize hotpots[e]	Min Th	Max Th
[7]	×			×					
[9]	×						×		
[3]	×	×	×			×	×		
[4]						×	×		
[10]	×			×		×	×	×	
[11]				×			×		×
[12] [13] [14]	×			×		×	×	×	×
Min/ Max Th	×	×	×	×	×	×	×	×	×

[a]Minimization of Migrations
[b]Migrate the request that occupies the min number of VMs greater than or equal to X (the nearest to X).
[c]Place in the PM which has the minimum number of VMs greater than or equal to X (the nearest to X).
[d]Without using a Minimum threshold (Min Th)
[e]Without using a Maximum threshold (Max Th)

6 Tests and Simulation Results

This section evaluates the proposed mechanism and compares it with two other policies using the CloudSim toolkit. In order to evaluate the effectiveness and the performances of our proposal, we have used two metrics: the power consumption of the data center and the number of the performed migrations.

6.1 Simulation Parameters

In this simulation, we have created a data center containing homogeneous hosts; each one has 4 processors with a speed of 2400 MIPS, 40 GB of RAM, 1 TB of storage and consumes 175 W with 0% CPU utilization, to 250 W with 100% CPU utilization. Each VM has 1 processor with a speed of 1000 MIPS, 512 MB of RAM, 1 GB of storage. We have considered the thresholds of the Dynamic Thresholds approach proposed in [13] (0.8 (80%) as upper threshold and 0.2 (20%) as lower threshold), and we have estimated the simulation time equal to 60 min.

6.2 Experience

In order to study the behavior of our proposal and to analyze the results obtained by the simulation, we have compared our results with two approaches. The Non Power Aware (NPA) approach or still called Without Migration policy (WM). This approach uses no energy optimization techniques and all PMs operate at 100% CPU utilization. In CloudSim, the task of VM placement "WM" is mainly completed by *VmAllocationPolicySimple* class which selects the PM with the most number of available process units. The VM placement policy will use up all PMs in the datacenter. The second policy is Single Threshold Policy (ST); this approach is based on the idea of setting only a higher utilization threshold for hosts and placing VMs while keeping the total CPU usage below this threshold. In our experience, the data center has 6 PMs and 30 VMs, distributed equitably over two customers. We would like to point out that the two CPU utilization thresholds for our approach are fixed throughout the simulation. The results are presented in the following graphs.

(a) Consumed Energy

As shown in Fig. 11, the energy consumption in the WM approach that runs at 100% CPU utilization remains stable at 48 KW throughout the simulation. For the ST approach, the consumption of energy starts at 48 KW, then decreases at (time = 8 min) and remains stable at 32 KW during the rest of the simulation (see Fig. 12). For our Min /Max Thresholds policy and after setting the two CPU utilization thresholds (Min = 20% and Max = 80%), the energy consumption is visibly reduced, it remains

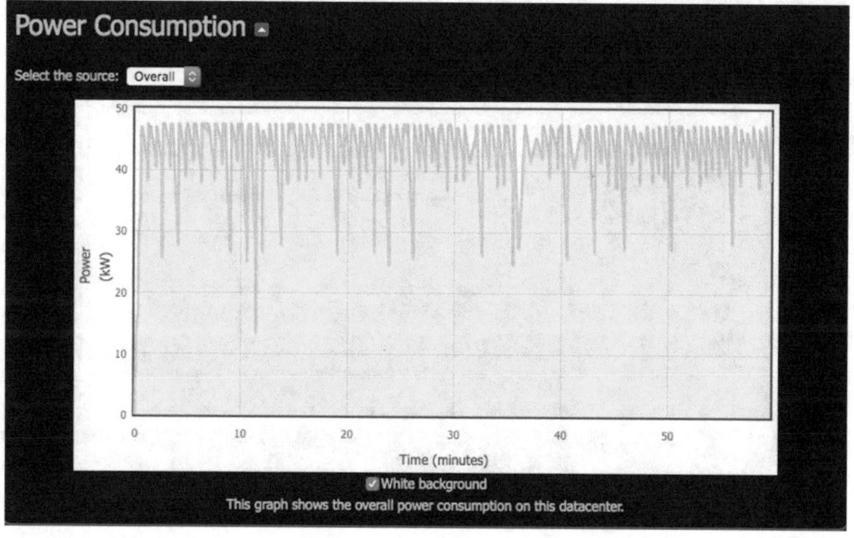

Fig. 11. Energy consumption in WM approach.

stable at (time = 12 min) 12 KW throughout the rest of the simulation (see Fig. 13). Following the simulation of the different approaches, we note that the consumed energy by our Min /Max Thresholds approach is lower than the energy consumed by the WM and ST approaches. Therefore, it is considered as a Green Computing approach because it saves energy.

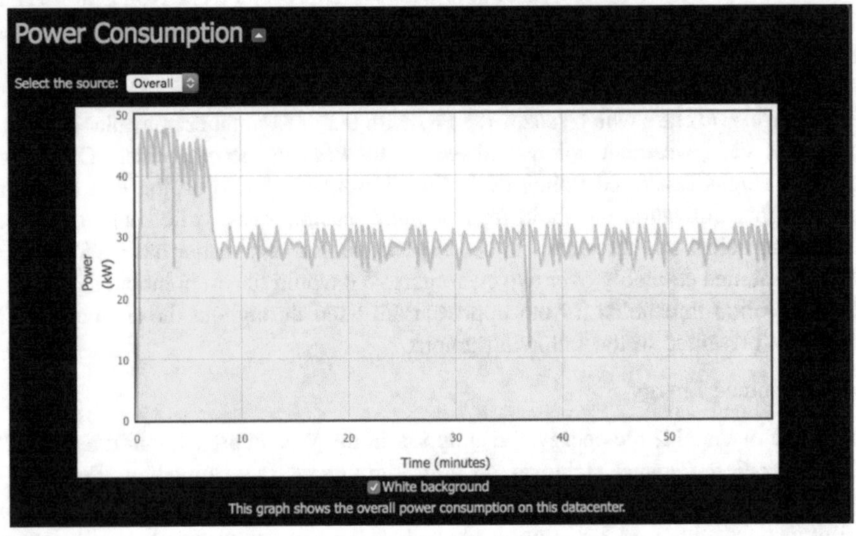

Fig. 12. Energy consumption in ST approach.

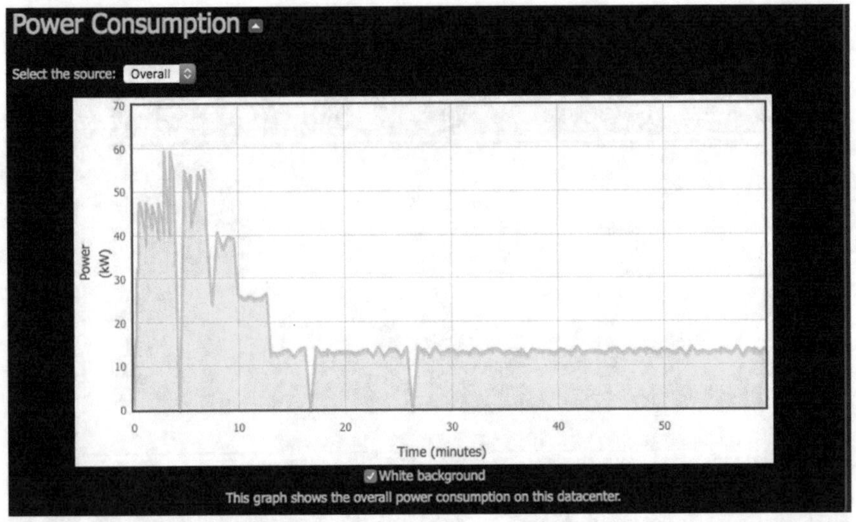

Fig. 13. Energy consumption in our Min/Max thresholds approach.

(b) Number of Migrations

Live (hot) migration has a negative impact on the performance of applications running in a VM. Each migration can lead to some SLA violation, so it is essential to reduce the number of VM migrations. The Min /Max Thresholds policy that we have adapted uses a reduced number of migration compared to the Single Threshold approach (Figs. 14 and 15) since our migration process is based on minimizing the number of migrations.

Fig. 14. Number of migrations in Min/Max thresholds approach.

Fig. 15. Number of migrations in ST approach.

7 Conclusion

The proposed approach, principally aims to achieve energy savings with the help of VM migration in the Cloud data centers, using minimum and maximum thresholds, to avoid the emerging of underloaded/overloaded hosts. We have first proposed the principle of our mechanism, then we have presented the different rules for the allocation of VMs as well as for their reallocation "the migration rules". The optimization of VMs allocation is carried out in two steps: at the first step we select VMs that need to be migrated, at the second step chosen VMs are placed on the appropriate hosts. In this paper, we have treated the VM migration on overloaded resources. Our approach has been implemented in CloudSim and experimentally compared with the *Without Migration* and *Single Threshold* policies. We have seen that our approach gives the best results: minimum number of migrations and minimum energy consumption, which can be a small step towards Green technology.

For our future work, we would like to investigate this technique on real Cloud setup and check what will be its exact reaction on real environment, with more resources and more processors in PMs and also in VMs, and with a big number of costumers and requests. We want also to propose a strategy to fix α which represents the number of machines that can be overloaded before activating a new resource, to minimize the energy consumed. Another important point is the evaluation of the SLAs violations in our mechanism comparing to the others. We also want to deal with the VM migration of the underloaded resources to shut them down. Also, security is an important concept when VMs are migrated from the source machine to the destination machine.

References

1. Guazzone, M., Anglano, C., Canonico, M.: Energy-efficient resource management for cloud computing infrastructures. In: Proceeding of the 3rd IEEE International Conference on Cloud Computing Technology and Science, pp. 424– 431 (2011)
2. Leelipushpam P.G.J., Sharmila, J.: Live VM migration techniques in cloud environment—a survey. In: Proceedings of the IEEE Conference on Information and Communication Technologies (2013)
3. Chandramouli R., Suchithra R.: Virtual machine migration in cloud data centers for resource management. Int. J. Eng. Comput. Sci. **5**(9), 18029–18034 (2016)
4. Kaur, P., Rani, A.: Virtual machine migration in cloud computing. Int. J. Grid Distrib. Comput. **8**(5), 337–342 (2015)
5. Yu, Y., Gao, Y.: Constraint programming-based virtual machines placement algorithm in datacenter. In: Proceedings of the 7th International Conference on Intelligent Information Processing (IIP), Guilin, China. Intelligent Information Processing VI. IFIP Advances in Information and Communication Technology, AICT-385, pp. 295–304. Springer (2012)
6. Usmani, Z., Shailendra Singh, S.: A survey of virtual machine placement techniques in a cloud data center. In: Proceedings of the International Conference on Information Security & Privacy, Nagpur, INDIA (2015). Procedia Comput. Sci., pp. 491– 498 (2016)
7. Borgetto, D., Stolf, P.: An energy efficient approach to virtual machines management in cloud computing. In: Proceedings of the 3rd International Conference on Cloud Networking, Luxembourg, Luxembourg (2014)
8. Patel, P.D., Karamta, M., Bhavsar, M.D., Potdar, M.B.: Live virtual machine migration techniques in cloud computing: a survey. Int. J. Comput. Appl. **86**(16) (2014)
9. Zhang, Z., Xiao, L., Chen, X., Peng, J.: A scheduling method for multiple virtual machines migration in cloud. In: The 10th International Conference on Network and Parallel Computing, Guiyang, China. Lecture Notes in Computer Science, LNCS-8147, pp. 130–142. Springer (2013)
10. Han, G., Que, W., Jia, G., Shu, L.: An efficient virtual machine consolidation scheme for multimedia cloud computing. J. Sens. **16**(2), 246 (2016)
11. Beloglazov, A., Buyya, R.: Energy efficient allocation of virtual machines in cloud data centers. In Proceedings of the 10th IEEE/ACM International Conference on Cluster, Cloud and Grid Computing, pp. 577–578 (2010)
12. VMware Inc: VMware distributed power management concepts and use (2010)
13. Sinha, R., Purohit, N., Diwanji, H.: Energy efficient dynamic integration of thresholds for migration at cloud data centers. Int. J. Comput. Appl. Commun. Netw. **11**, 44–49 (2011)

14. Maheshwari, D., Gandhi, P., Sinha, R.: Energy efficient threshold based approach for migration at cloud data center. Int. J. Eng. Res. Technol. (IJERT) **1**(10) (2012)
15. Hassan, M.K., Babiker, A., Amien, M.B.M., Hamad, M.: SLA management for virtual machine live migration using machine learning with modified kernel and statistical approach. Eng. Technol. Appl. Sci. Res. **8**(1), 2459–2463 (2018)
16. Beloglazov, A., Buyya, R.: Optimal online deterministic algorithms and adaptive heuristics for energy and performance efficient dynamic consolidation of virtual machines in cloud data centers. J. Concurr. Comput. Pract. Exp. **24**(13), 1397–1420 (2012)

Secure Data Transmission Scheme Based on Elliptic Curve Cryptography for Internet of Things

Yasmine Harbi[1(✉)], Zibouda Aliouat[1], Saad Harous[2], and Abdelhak Bentaleb[3]

[1] LRSD Laboratory, Computer Science Department, Ferhat Abbas University Setif1, Setif, Algeria
yasmine.harbi@univ-setif.dz, zaliouat@univ-setif.dz

[2] College of Information Technology, United Arab Emirates University, Al Ain, UAE
harous@uaeu.ac.ae

[3] National University of Singapore, Singapore, Singapore
bentaleb@comp.nus.edu.sg

Abstract. In recent years, Internet of Things (IoT) has made extraordinary progress in human lives from healthcare applications to daily chores. The IoT enables everyday object to be connected to the Internet. These devices are embedded with sensors and actuators in order to collect and share data. However, the transmission of the collected data may face several security and privacy concerns. To overcome this problem, we propose a Secure Data Transmission Scheme (SDTS) that improves communication security in cluster-based Wireless Sensor Networks (WSNs). The SDTS is based on Elliptic Curve Cryptography (ECC) due to its ability to provide high security level with small key size. The proposed method achieves several security requirements including confidentiality, integrity, and authentication. Moreover, it resists different security attacks like brute force attack, replay attack, and sinkhole attack. The performance analysis shows that SDTS is relatively efficient in term of communication cost.

Keywords: Security · Privacy · Wireless Sensor Networks
Security requirements · Security attacks

1 Introduction

The IoT is a system of networked physical objects that bridge physical and virtual worlds. Over the past few years, the IoT has gained an enormous mindshare due to its impact on our way of living [1]. The IoT promises to offer smart environments that make healthcare, transport, homes, and other areas affecting human being and the environment more intelligent [2]. The IoT is increasing rapidly and will encompass billions of connected things in the near future [3].

As IoT devices are equipped with sensors and actuators, WSNs play an integral part in the IoT [4,5]. WSNs are usually composed of a large number of tiny

S. Chikhi et al. (Eds.): MISC 2018, LNNS 64, pp. 34–46, 2019.
https://doi.org/10.1007/978-3-030-05481-6_3

sensor nodes deployed in remote environments. Sensor nodes have inherently limited resources (i.e. power, memory, CPU) and communicate with each other through wireless channels [6]. However, the deployment of WSNs in unattended environments makes the sensor nodes vulnerable to various security attacks [7]. Developing security mechanisms for WSNs is an open challenge [8]. The implementation of traditional security solutions to secure WSNs is a hard task due to the resource-constrained nature of these networks [9].

To address this challenge, this paper presents a lightweight security mechanism named Secure Data Transmission Scheme (SDTS). The SDTS aims to secure data transmission in WSNs and prevents information leakage. The sensors are able to encrypt the sensed data using Elliptic Curve Cryptography (ECC) before data transmission process.

The ECC is suitable for constrained environments since it provides high security level with small key size compared to conventional cryptographic techniques such as RSA (Rivest, Shamir and Adleman) [10,11]. Table 1 shows the comparison between ECC and RSA.

Table 1. ECC vs RSA [12]

ECC (bits)	RSA (bits)
160	1024
256	3024
384	7680
512	16360

From Table 1, an ECC system with 160-bit key size can provide the same security level as 1024-bit RSA.

The rest of this paper is organized as follows. Section 2 introduces the concept of ECC. Section 3 describes the related work. Section 4 discusses our proposed scheme. The security and performance of our scheme are analyzed in Sect. 5 and Sect. 6. A comparative analysis of our work with solutions available in the literature is presented in Sect. 7. Finally, a conclusion is given in Sect. 8.

2 Background

Cryptography is an important concept that provides security in WSNs. It is the process of converting a plain text into an unreadable text or ciphered text using mathematical techniques. There are two main types of cryptography (symmetric and asymmetric) that allow information to be kept secret during transmission [13,14].

In symmetric cryptography, the sender and receiver must agree on a shared key which is used for both encryption and decryption process as showed in Fig. 1. Key distribution and management are major drawbacks of the symmetric key cryptography. These shortcomings were the main reasons that lead to the development of asymmetric cryptography [13,14].

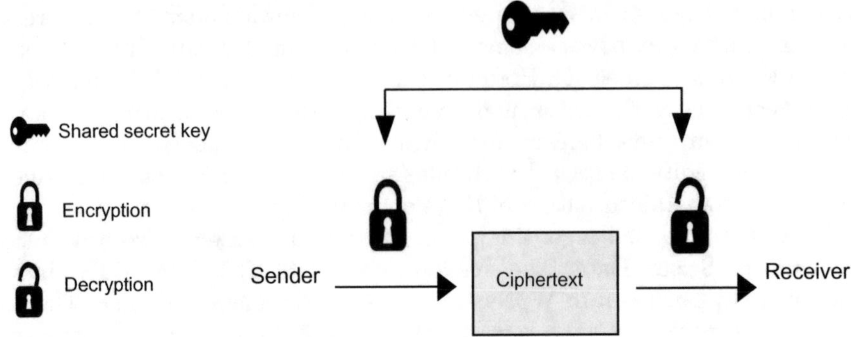

Fig. 1. Symmetric cryptography

In asymmetric key cryptography, two different keys are used for encryption and decryption. Each entity has a key pair consisting of public and private key [13]. Figure 2 illustrates the encryption and decryption using asymmetric cryptography. The public key cryptography eliminates the problems of symmetric key systems [14]. However, it is computationally complex and therefore cannot be applied in WSNs [12].

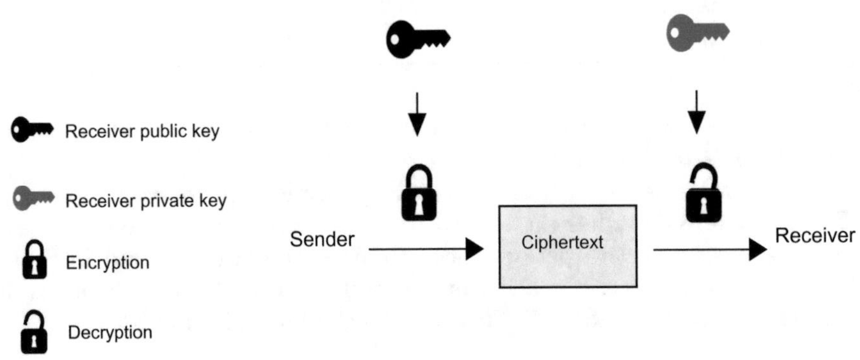

Fig. 2. Asymmetric cryptography

In 1985, Neal Koblitz [15] and Victor Miller [16] introduced separately the use of elliptic curves in cryptography. An elliptic curve is the set of points that satisfy Eq. 1 [14].

$$y^2 = x^3 + ax + b \quad \text{where} \quad 4a^3 + 27b^2 \neq 0 \tag{1}$$

Point addition and point doubling are the basic elliptic curve operations [17]. Multiplication on elliptic curve requires a scalar multiplication operation. This scalar multiplication can be done by a series of addition and doubling operations

[17]. For example, we have an elliptic point P and a positive integer k, the scalar multiplication of kP is defined by Eq. 2.

$$kP = P + P + P + ... + P(k \quad times) \tag{2}$$

When we have two point P and Q, it is difficult to find k such that Q = kP. This operation is known as Elliptic Curve Discrete Logarithm Problem (ECDLP) [14].

In ECC, k is considered as the private key and Q as the public key. The message is mapped to a point of the curve then encrypted to a ciphered point [15,16].

We conclude that ECC is a public key cryptography that offers equivalent security level compared with known public key cryptosystem but with smaller key size. Moreover, it is highly secure since it depends on ECDLP which is computationally hard to solve.

3 Related Work

Recently, many research works based on cryptography have been proposed in order to secure communication in WSNs.

A cryptographic mechanism named ICMDS (Inter-Cluster Multiple key Distribution Scheme) is presented in [18] to secure cluster-based WSNs. The proposed scheme is based on key pre-distribution method where keys are assigned to sensor nodes before network deployment. In ICMDS, the Base Station (BS) generates a timestamp and an asymmetric key pair for each node in the network, then distributes the public keys and timestamps while private keys are kept secrets. Each member node encrypts the sensed data using its public key and sends it to the Cluster Head (CH). This latter aggregates the encrypted data then forwards it to the BS that checks the authenticity of nodes and decrypts the data. The ICMDS is resilient against man-in-the-middle attacks. However, it is not scalable since the number of stored cryptographic keys depends on the number of nodes in the network.

The work proposed in [19] provides end-to-end data security in multi-hops WSNs. The authors use path key pre-distribution to secure data transmitted in the path. The proposed protocol consists of four phases: key pre-distribution, path establishment, path key establishment, and data protection. Firstly, all nodes are pre-loaded with m keys where m is the number of nodes on the path. Secondly, all paths from source node to destination node are identified. Thirdly, the encryption key, decryption key, and authentication key are computed. Finally, the source node encrypts and authenticates the data using encryption key and authentication key respectively. Each intermediate node checks the data authenticity and forwards it to the destination node which performs the decryption process. This approach ensures data confidentiality and authenticity but it requires large memory space to store the pre-distribution keys.

The authors in [20] presented an authentication scheme to secure communication in WSNs. The proposed work used ECC and hash functions to address the security weaknesses of [21]. Their network includes sensors, a gateway, and users. Before data transmission process, the sensors and users need to authenticate on the gateway. The security analysis shows that the proposed work resists various attacks including replay attack, node capture attack, and insider attack. However, the main drawback of this scheme is the inefficiency in terms of communication cost which is high.

In [22], the authors proposed a signcryption scheme that combines the functions of digital signature and encryption. The authors use pairings based on elliptic curve to secure transmission of collected data. Their network model consists of a Public Key Generator (PKG), sensor nodes, and a server. The PKG is responsible for generating the system parameters and nodes key pair. Each sensor node signs and encrypts the data then sends it to the server. This mechanism offers a desired security level but inherently suffers from key escrow problem.

To improve communication security in cluster-based WSNs, Elhoseny et al. proposed in [23] an efficient mechanism based on ECC and homomorphic encryption. The proposed scheme can secure both text and image data. The authors used a genetic algorithm for CHs selection and clusters formation. The EC parameters are pre-loaded in the sensors to generate encryption/decryption key. The cluster members perform XOR, permutation, and concatenation operations to encrypt the sensed data. The CH aggregates the binary encrypted data using homomorphic encryption technique then forwards it to the BS. This scheme is energy efficient and secures data transmission. But, it is not resilient to node capture attack since the public and private keys are embedded in the sensor nodes.

In [24], the authors proposed an enhanced authentication scheme for embedded devices in cloud-based IoT environment. Their work is based on ECC and consists of three phases, namely initialization, registration, login and authentication. Initially, the Cloud Server (CS) sets the system parameters. Each device securely sends its identity and password to register with the CS. This letter generates a pseudo-identity and a cookie and transmits this information to the embedded device. The CS updates the cookie information when it expires, then retransmits the new one. The proposed scheme achieves mutual authentication and session key agreement. However, it does not ensure data integrity.

4 Secure Data Transmission Scheme (SDTS)

4.1 System Model, Assumptions, and Notations

The proposed system model consists of a BS and a large number of tiny and homogenous sensors that monitor a particular physical phenomenon. The sensors are grouped into clusters in order to decrease network overhead and delay. The captured data is transmitted to CH using one-hop communication. The CH aggregates the sensed data and then forwards it directly to the BS or Sink. Figure 3 depicts the proposed system model.

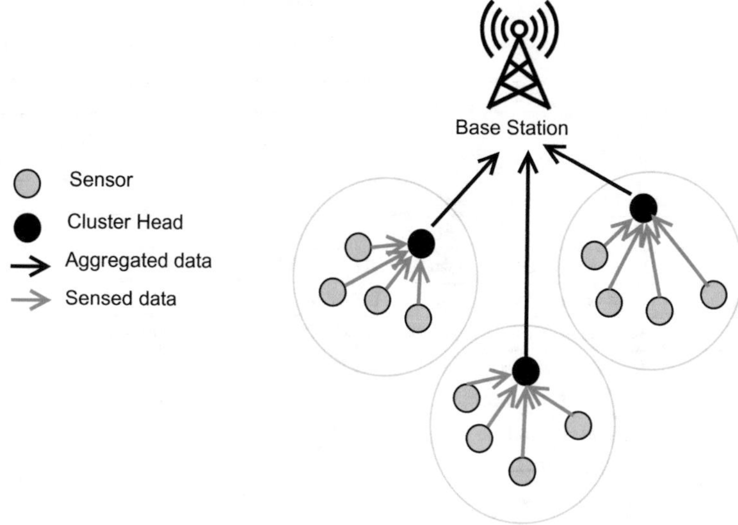

Fig. 3. SDTS's system model

Some assumptions have to be considered in our system model. The first one is that the sensor nodes are stationary and deployed randomly. They periodically capture and send data to the CH. The second assumption is that we use 160-bit ECC to provide the desired security level. Message mapping process is not considered in our work. Finally, we assume that the BS is a powerful computing device and totally secure (cannot be compromised). It acts as a controller or central trust authority.

The main notations used in our proposed scheme are listed in Table 2.

4.2 Proposed Algorithm

In our work, we mainly focus on preventing data transmission from eavesdropping attacks. In order to protect data privacy in WSNs in the context of the IoT, the sensed data needs to be encrypted. The proposed scheme consists of four phases: initialization, key generation, encryption, and decryption. The SDTS is summarized in Fig. 4.

Initialization After the deployment of the network, the BS generates the system parameters by selecting an elliptic curve E, a base point G of order q (where q is a large prime number), and a master secret integer $S_{BS} \in \mathbb{Z}_q$. The BS calculates $P_{BS} = S_{BS}$ * G then publishes the system parameters G, q, P_{BS}.

Table 2. SDTDS's notations

Notation	Description
$\mathbb{Z}_q$	Integer numbers less than q
P_{BS}	Public key of the Base Station
S_{BS}	Private key of the Base Station
P_{Cl}	Public key of a cluster
S_{Cl}	Private key of a cluster
SK	Secret key shared between CH and BS
CK	Ciphered secret key shared between CH and BS
m_i	Sensed data that belongs to node i mapped to a point on the curve
c_i	Ciphered sensed data that belongs to node i
C	Ciphered aggregated data
$\|\|$	Concatenation
$+$	Point addition
$-$	Point subtraction
$*$	Point multiplication

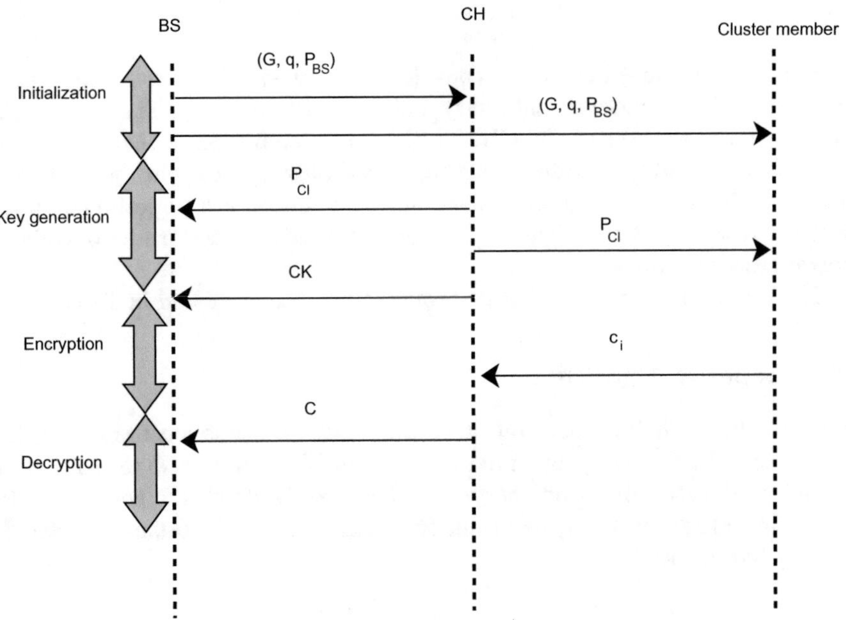

Fig. 4. SDTS's phases

Key Generation Once the CHs are selected in the network, they generate an asymmetric key pair for their clusters and share a secret key with the BS. The asymmetric key pair is used for data encryption/decryption and the symmetric key for signing the encrypted data. The generation of keys follows Algorithm 1.

Algorithm 1 Key generation phase

Input : G, q, P_{BS}
Output : P_{Cl}, S_{Cl}, SK
BEGIN
The CH generates a random integer d $\in \mathbb{Z}_q$
The CH calculates S_{Cl} by adding d to its ID
The CH computes SK = S_{Cl} * P_{BS}
The CH computes $P_{Cl} = S_{Cl}$ * G then sends it to the BS and its member nodes
The BS computes SK = S_{BS} * P_{Cl}
The CH encrypts S_{Cl} with SK (i.e. S_{Cl} + SK = CK) then sends CK to the BS
The BS decrypts S_{Cl} using SK (i.e. CK - SK)
The BS stores the clusters key pair and shared key in a database
END

Encryption Before data transmission process, each cluster node encrypts the sensed data using the clusters public key then sends it to the CH. In order to reduce energy consumption, the CH aggregates the encrypted data of its member nodes without decryption process. The CH creates a signature for each received data using the shared secret key then transmits it to the BS. The encryption phase involves the steps described in Algorithm 2.

Algorithm 2 Encryption phase

Input : m_i, P_{Cl}, SK
Output : c_i, C
BEGIN
for each member node i **do**
  Select a random number $K_i \in \mathbb{Z}_q$
  Compute $V_i = m_i + K_i * P_{Cl}$
  Send $c_i = V_i || K_i$ to CH
end for{upon receiving c_i}
for each c_i **do**
  Create a signature $S_i = V_i + SK$
  Appends S_i to c_i
end for
The CH sends C = $V_1||K_1||S_1, V_2||K_2||S_2, ..., V_n||K_n||S_n$ to the BS
END

Decryption When the encrypted data reaches the BS through the CHs, the BS checks the respective keys on its database and decrypts the ciphered data using Algorithm 3.

The initialization phase is performed one time after the deployment of the network, while the key generation, encryption, and decryption phase are run each round after the CH selection.

Algorithm 3 Decryption phase

Input : C, S_{Cl}, SK
Output : m_i
BEGIN
for each $V_i||K_i||S_i$ **do**
  Calculate S_i - SK
  if S_i - SK = V_i **then**
    Compute $m_i = V_i - K_i S_{Cl} * G$
  else
    Discard the message
  end if
end for
END

5 Security Analysis

In this section, we analyze the security features of our proposed scheme.

- **Confidentiality/eavesdropping resistance**: as the sensed data is encrypted using ECC before forwarding it, an attacker cannot figure out the data without decryption key. Moreover, even if the attacker successfully cracks the asymmetric key pair and secret key for one cluster in one round, he/she cannot get all the keys since these latter are generated every round by the selected CHs. Therefore, message confidentiality is ensured and data privacy is well protected.
- **Integrity and Authentication**: since the CH signs the sensed data of each cluster nodes with the secret shared key, the BS can check that the data have not been altered or modified by verifying the signature of the message. Furthermore, the captured data cannot be signed without having the secret shared key. Hence, the proposed scheme can guarantee data integrity and authentication.
- **Brute force attack resistance**: because we adopt ECC with 160-bit key size, the probability of finding a key through brute force attack is $1/2^{160}$. Hence, the possibility of a successful brute force attack is negligible. Moreover, the generated keys are used to encrypt data for one round. Thus, compromising the keys of one round does not affect the system security.
- **Resilience to sinkhole attack**: as the private keys are stored in the BS which is highly trusted and secured, sinkhole node cannot decrypt the ciphered data even if it attracts all network traffic.
- **Resilience to replay attack**: in our proposed scheme, all transmitted messages include a random number K that disables the adversary to perform a replay attack. If an attacker replays a message without modifying K, the BS or the CH detects immediately the replayed message.

6 Performance Analysis

This section provides an evaluation of the performance of the proposed SDTS in terms of computational complexity and communication overhead. Table 3 shows the operations for computational complexity.

Table 3. Computational complexity's operations

Symbol	Operation
C_s	Time required to generate system parameters
E_a	Time required for asymmetric encryption
G_a	Time required for asymmetric key generation
G_s	Time required for symmetric key generation
E_s	Time required for symmetric encryption
D_s	Time required for symmetric decryption
D_a	Time required for asymmetric decryption

6.1 Computation Costs

We compute the computational complexity for member node, CH, and BS for one data transmission.

On Member Node's Side The cluster member encrypts the sensed data with the clusters public key using ECC. Hence, the sensor node uses E_a to encrypt the data.

On CH's Side In Algorithm 1, the CH generates both symmetric and asymmetric key, which takes $G_s + G_a$. In Algorithm 2, the CH signs the data received from member nodes using the symmetric key which takes E_s. The computation costs on each CH is $G_s + G_a + E_s$.

On BS's Side The BS performs the following computations: system parameters generation, symmetric key generation, signature verification, and data decryption. Thus, the time needed is $C_s + G_s + D_s + D_a$.

In our proposed scheme, we combine symmetric and asymmetric ECC to secure data transmission in WSNs. Most of the computation is performed by the BS to reduce energy consumption and prolong network lifetime.

6.2 Communication Costs

We consider communication overhead between a member node, CH, and BS for one data transmission. Recall that we choose 160-bit ECC (i.e. the elliptic curve point length is 320 bits).

During key generation phase, the CH publishes the clusters public key which is 160 bits long. During encryption phase, the member node sends the ciphered data and a random number; the message size is 320+160 = 480 bits. After receiving and signing the encrypted data, the CH transmits 320+160+320 = 800 bits to the BS. There is no communication during the decryption phase. The total communication cost of one transmission is 1280 bits.

7 Comparative Analysis

This section provides a comparative analysis of our proposed scheme to the schemes [18–20, 22–24]. The comparison of security properties and performance is presented in Table 4.

Table 4. Comparative analysis

	SDTS	[18]	[19]	[20]	[22]	[23]	[24]
Confidentiality	✓	✓	✓	✓	✓	✓	✓
Integrity	✓	×	×	✓	✓	×	×
Authentication	✓	✓	✓	✓	✓	✓	✓
Brute force attack resistance	✓	×	✓	×	×	✓	✓
Replay attack resistance	✓	×	✓	✓	×	×	✓
Sinkhole attack resistance	✓	×	×	×	×	✓	–
Communication cost (bits)	1280	–	–	3680	–	–	1760

From Table 4, we conclude that our proposed solution achieves a desired level of security. Compared to the scheme presented in [20], our algorithm is almost 3 times better in term of communication cost. Hence, it is secure and efficient for constrained IoT environments.

8 Conclusion

In this paper, we proposed a lightweight security mechanism to secure communication in WSNs in the context of IoT. The proposed scheme (SDTS) is based on ECC which is suitable for devices with limited resources. SDTS ensures fundamental security requirements such as confidentiality, integrity, and authentication. Furthermore, it is secure against known security attacks like brute force attack, replay attack, and sinkhole attack. Comparison with recent related works shows that our system is lightweight, secure, and efficient. Thus, it can be implemented in constrained IoT environments.

References

1. Rayes, A., Salam, S.: Internet of Things from Hype to Reality: the Road to Digitization. Springer, Cham (2016)
2. Atzori, L., Iera, A., Morabito, G.: The internet of things: a survey. Comput. Netw. **54**(15), 2787–2805 (2010)
3. Li, S., Da Xu, L., Zhao, S.: The internet of things: a survey. Inf. Syst. Front. **17**(2), 243–259 (2015)
4. Hammoudi, S., Aliouat, Z., Harous, S.: Challenges and research directions for internet of things. Telecommun. Syst. **67**(2), 367–385 (2018)
5. Gubbi, J., Buyya, R., Marusic, S., Palaniswami, M.: Internet of things (IoT): a vision, architectural elements, and future directions. Futur. Gener. Comput. Syst. **29**(7), 1645–1660 (2013)
6. Sohraby, K., Minoli, D., Znati, T.: Wireless Sensor Networks: Technology, Protocols, and Applications. Wiley, New York (2007)
7. Azzabi, T., Farhat, H., Sahli, N.: A survey on wireless sensor networks security issues and military specificities, pp. 66–72. IEEE (2017)
8. Radhappa, H., Pan, L., Xi Zheng, J., Wen, S.: Practical overview of security issues in wireless sensor network applications. Int. J. Comput. Appl. 1–12 (2017)
9. Eisenbarth, T., Kumar, S., Paar, C., Poschmann, A., Uhsadel, L.: A survey of lightweight-cryptography implementations. IEEE Des. Test Comput. **6**, 522–533 (2007)
10. Rivest, R.L., Shamir, A., Adleman, L.: A method for obtaining digital signatures and public-key cryptosystems. Commun. ACM **21**(2), 120–126 (1978)
11. Gura, N., Patel, A., Wander, A., Eberle, H., Shantz, S.C.: Comparing elliptic curve cryptography and RSA on 8-bit CPUs. In: International Workshop on Cryptographic Hardware and Embedded Systems, pp. 119–132. Springer (2004)
12. Malik, M.Y.: Efficient implementation of elliptic curve cryptography using low-power digital signal processor. In: 2010 The 12th International Conference on Advanced Communication Technology (ICACT), vol. 2, pp. 1464–1468. IEEE (2010)
13. Katz, J., Menezes, A.J., Van Oorschot, P.C., Vanstone, S.A.: Handbook of Applied Cryptography. CRC Press, Boca Raton (1996)
14. Hankerson, D., Menezes, A.J., Vanstone, S.: Guide to Elliptic Curve Cryptography. Springer Science & Business Media, New York (2006)
15. Koblitz, N.: Elliptic curve cryptosystems. Math. Comput. **48**(177), 203–209 (1987)
16. Miller, V.S.: Use of elliptic curves in cryptography. In: Conference on the Theory and Application of Cryptographic Techniques, pp. 417–426. Springer, Heidelberg (1985)
17. Silverman, J.H.: The Arithmetic of Elliptic Curves, vol. 106. Springer Science & Business Media, New York (2009)
18. Mehmood, A., Umar, M.M., Song, H.: ICMDS: secure inter-cluster multiple-key distribution scheme for wireless sensor networks. Ad Hoc Netw. **55**, 97–106 (2017)
19. Harn, L., Hsu, C.F., Ruan, O., Zhang, M.Y.: Novel design of secure end-to-end routing protocol in wireless sensor networks. IEEE Sens. J. **16**(6), 1779–1785 (2016)
20. Wu, F., Xu, L., Kumari, S., Li, X.: A privacy-preserving and provable user authentication scheme for wireless sensor networks based on internet of things security. J. Ambient. Intell. Hum. Comput. **8**(1), 101–116 (2017)
21. Hsieh, W.B., Leu, J.S.: A robust ser authentication scheme sing dynamic identity in wireless sensor networks. Wirel. Pers. Commun. **77**(2), 979–989 (2014)

22. Li, F., Zheng, Z., Jin, C.: Secure and efficient data transmission in the internet of things. Telecommun. Syst. **62**(1), 111–122 (2016)
23. Elhoseny, M., Elminir, H., Riad, A., Yuan, X.: A secure data routing schema for wsn using elliptic curve cryptography and homomorphic encryption. J. King Saud Univ. Comput. Inf. Sci. **28**(3), 262–275 (2016)
24. Kumari, S., Karuppiah, M., Das, A.K., Li, X., Wu, F., Kumar, N.: A secure authentication scheme based on elliptic curve cryptography for IoT and cloud servers. J. Supercomput. 1–26 (2017)

QoS Multicast Routing Based on a Quantum Chaotic Dragonfly Algorithm

Mohammed Mahseur[1(✉)], Abdelmadjid Boukra[1], and Yassine Meraihi[2]

[1] Department of Informatics, Faculty of Electronics and Informatics, University of Sciences and Technology Houari Boumediene, El Alia Bab Ezzouar, 16025 Algiers, Algeria
mahseur.mohammed@gmail.com, aboukra@usthb.dz

[2] Automation Department, University of MHamed Bougara Boumerdes, Avenue of Independence, 35000 Boumerdes, Algeria
yassine.meraihi@yahoo.fr

Abstract. Optimizing the quality of service in a multicast routing is a persistent research problem for data transmission in computer networks. It is known to be an NP-hard problem, so several meta-heuristics are applied for an approximate resolution. In this paper, we resolve the quality of service multicast routing problem (QoSMRP) with using a combined approach that uses a newly meta-heuristic called Dragonfly Algorithm (DFA) and Quantum Evolutionary Algorithm (QEA), we adopted a quantum representation of the solutions by a vector of continuous real values which allowed us to use the continuous version of the DFA without discretization, we also use the equation of DFA to calculate $\Delta\theta$ in QEA. The interest of these contributions is to avoid premature convergence, to improve the diversity of solutions, and to increase the efficiency and performance of the proposed algorithm. The experimental results show the feasibility, scalability, and effectiveness of our proposed approach compared to other algorithms such as Genetic Algorithm (GA), Quantum Evolutionary Algorithm (QEA), and Dragonfly Algorithm (DFA).

Keywords: Quantum evolutionary algorithm · Dragonfly algorithm
Quality of Service (QoS) · Multicast routing

1 Introduction

The huge evolution of telecommunication networks has allowed the emergence of new applications that are known by their voracious consumption of different network resources such as online games, video conferencing, online meeting, e-learning, and video streaming.

So, the implementation of routing protocols suitable for this generation of applications is imperative. Multicast routing is a mode of transmission that sends data from a source to a set of destinations by saving the cost of routing

S. Chikhi et al. (Eds.): MISC 2018, LNNS 64, pp. 47–59, 2019.
https://doi.org/10.1007/978-3-030-05481-6_4

and satisfying a set of constraints such as delay, bandwidth, jitter, and loss rate. Finding an optimal multicast tree is an NP-hard problem [1], so the approximate resolution is the best way to deal with this kind of problem. In this stage, several meta-heuristics are adapted and applied with acceptable success.

In recent years, many efficient meta-heuristics have been applied to solve the multicast routing problem with delay constraint DCLC (Delay Constrained Least Cost) or with multiple constraints MCLC.

Koyama et al. [2] present a new approach for QoS multicast routing protocol, in which new genetic operators are used. Ghaboosi and Haghighat [3] invented different movements in the Tabu list and have calculated the movements by a new method. Armaghan and Haghighat [4] proposed a new algorithm based on Tabu Search (TS) with a rule of the list of candidates who can find good solutions than other algorithms based on the traditional TS. Huang et al. [5] presented an approach based on the ant colony system (ACS). Wang et al. [6] proposed a new solution based on ACS, where the algorithm is not seeking a unique destination. It aims to find all destination nodes. This algorithm is based on three operations: tree growth, tree pruning, and pheromone updating. Mahseur and Boukra [7] proposed two algorithms to solve the MCLC Problem, the first one is based on Biogeography Based Optimization (BBO) and the second is based on Bat Algorithm (BA), the experimental results are encouraging in comparison with other existing algorithms.

Meraihi et al. [8] proposed an improved chaotic binary bat algorithm for solving the QoS multicast routing problem. Experimental results reveal the performance and the effectiveness of the proposed algorithm compared with other existing algorithms in the literature.

Mahseur et al. [9] proposed an improved quantum chaotic animal migration optimization algorithm for QoS Multicast Routing Problem, the simulation results show the robustness of their solution.

In this paper, we will propose a hybrid approach that combines between Dragonfly Algorithm (DFA) and Quantum Evolutionary Algorithm (QEA) to benefit from its advantages in order to increase search performance. The experimental results show the feasibility and the effectiveness of our solution.

The rest of this paper is organized as follows: In Sect. 2, we define the mathematical multicast routing problem. In Sect. 3, we present the DFA and the QEA. In the fourth Section, we expose our hybrid algorithm called DFAQEA for QoS multicast routing. Section 5 contains the simulation results and analysis of different tests. Finally, we summarize our conclusion and suggest some perspectives for future works in Sect. 6.

2 Multicast Routing Problem Formulation

Throughout this work, the communication network as defined in our previous works [7,10] is modeled as a directed graph $G = (V, E)$ where V and E are the set of nodes and the set of edges of the graph respectively, and $n = |V|$ is the number of nodes and $l = |E|$ is the number of edges. Each link $e = (i, j) \in E$ that connects node i with node j is associated with edge cost $C(e) : E \rightarrow \mathbb{R}^+$, edge delay $D(e) : E \rightarrow \mathbb{R}^+$, edge bandwidth $B(e) : E \rightarrow \mathbb{R}^+$, and edge packet loss rate $PL(e) : E \rightarrow \mathbb{R}^+$, where $\mathbb{R}^+$ is the set of all nonnegative real numbers. In the general case, the computer network is asymmetric, i.e. the links are bidirectional, so it is often possible that $C(e) \neq C(e'), D(e) \neq D(e'), J(e) \neq J(e'), B(e) \neq B(e'), PL(e) \neq PL(e')$, with $e = (i, j) \in E$ and $e' = (j, i) \in E$ represent the edges that connect the node i with the node j and the node j with the node i respectively. We assume that $s \in V$ represents the source node and $M \subseteq \{V - \{s\}\}$ represents a set of multicast destination nodes such that s and M construct a multicast tree $T(s, M)$.

The multicast tree has the following parameters [10–12]:

The total cost of the multicast tree $T(s, M)$, denoted by $Cost(T(s, M))$, is obtained by summing the costs of all edges in that tree. It can be given by:

$$Cost(T(s, M)) = \sum_{e \in T(s,M)} C(e) \tag{1}$$

The total delay of the path $P_T(s, m)$, denoted by $Delay(P_T(s, m))$, is simply the sum of the delays of all edges along the path:

$$Delay(P_T(s, m)) = \sum_{e \in P_T(s,m)} D(e) \tag{2}$$

where $P_T(s, m)$ represent a path of the multicast tree $T(s, M)$ that connects the source node s with the destination node $m \in M$ (clearly $P_T(s, m)$ is a subset of $T(s, M)$).

The delay jitter of the tree $T(s, M)$ is defined as the average difference of delay on the path from the source to the destination node:

$$Jitter(P_T(s, M)) = \sqrt{\sum_{m \in M} (D(P_T(s, m)) - avg\_delay)^2} \tag{3}$$

where avg_delay represents the paths average delay from source to the destinations.

The bandwidth of the path $P_T(s, m)$, represented by $B(P_T(s, m))$, is defined as the minimum solicited bandwidth at any edge along $P_T(s, m)$:

$$B(P_T(s, m)) = \min_{e \in P_T(s,m)} (B(e)) \tag{4}$$

The Packet loss rate of the path linking the source s with the destination m, denoted by $PL(P_T(s, m))$ is given by:

$$PL(P_T(s, m)) = 1 - \prod_{e \in P_T(s,m)} (1 - PL(e)) \tag{5}$$

The multi-constrained multicast routing problem can be considered as a minimization problem whose goal is to build a multicast tree that minimizes the cost function and satisfies the following constraints [10, 11]:

1. Delay constraint: $Delay(P_T(s,m)) \leq D_{max}, \forall m \in M$ i.e. the allowed delay of a path between the source and the destination must not exceed the delay constraint D_{max};
2. Jitter constraint: $Jitter(P_T(s,M)) \leq J_{max}, \forall m \in M$ i.e. the jitter of the multicast tree must not exceed the jitter constraint J_{max};
3. Bandwidth constraint: $min(B(P_T(s,m))) \geq B_{min}, \forall m \in M$ i.e. the bandwidth of the multicast tree of each edge must be greater or equal to the bandwidth constraint B_{min};
4. Packet loss rate constraint: $PL(P_T(s,m)) \leq PL_{max}, \forall m \in M$ i.e. the packet loss rate of each path linking the source and the destination must not exceed the packet loss rate constraint PL_{max}.

Thus, the problem can be defined as follows:

$$Minimize\ Cost(T(s,M)) \tag{6}$$

Subject to:

$$Delay(P_T(s,m)) \leq D_{max} \tag{7}$$

$$Jitter(P_T(s,M)) \leq J_{max} \tag{8}$$

$$Min(B(P_T(s,m))) \geq B_{min} \tag{9}$$

$$PL(P_T(s,m)) \leq PL_{max} \tag{10}$$

3 Dragonfly Algorithm and Quantum Evolutionary Algorithm

3.1 Dragonfly Algorithm

Dragonfly Algorithm (DA), proposed by Seyedali Mirjalili in 2016 [13], is one of the newly nature-inspired algorithms used to solve various optimization problems. This optimization algorithm is inspired by the static (feeding) and the dynamic (migratory) swarming behaviors in nature. These two swarming behaviors constitute the exploration and exploitation phases of the DA. In static swarms, dragonflies make small groups and fly back and forth over a small area for hunting other flying preys. In dynamic swarms, however, a massive number of dragonflies make the swarms migrate in one direction over long distances. The swarming behavior of dragonflies follows five primitive principles [13]:

1. Separation is the static collision avoidance of the individuals from other individuals in the neighborhood. It is calculated as follows:

$$S_i = -\sum_{j-1}^{N} X - X_j \tag{11}$$

Where X denotes the position of the current individual, X_j the position of the j-th neighboring individual, and N the number of neighboring individuals.

2. Alignment represents the individual's velocity matching according to other neighborhood individuals. It is calculated as follows:

$$A_i = \frac{\sum_{j-1}^{N} V_j}{N} \tag{12}$$

Where V_j represents the velocity of the j-th individual.

3. Cohesion refers to the tendency of individuals towards the center of the mass of the neighborhood. It is mathematically modeled as follows:

$$C_i = \frac{\sum_{j-1}^{N} X_j}{N} - X \tag{13}$$

Where X represents the position of the current individual, X_j shows the position of the j-th neighboring individual and N the number of neighboring individuals.

4. Attraction towards the food source (F) is mathematically modeled by:

$$F_i = X^+ - X \tag{14}$$

Where X denotes the position of the current position and X^+ is the position of the food source.

5. Distraction from the enemies is calculated as follows:

$$E_i = X^- + X \tag{15}$$

Where X denotes the position of the current position and X^- denotes the enemy's position.

Step vector (ΔX) and position vector (X) are considered to update the position of artificial dragonflies in a search space. The step vector is analogous to the velocity vector in PSO. The step vector is defined and updated as follows:

$$\Delta X_i^{t+1} = (sS_i + aA_i + cC_i + fF_i + eE_i) + \omega \Delta X_i^t \tag{16}$$

where s indicates the separation weight, S_i shows the separation of the i-th individual, a is the alignment weight, A_i is the alignment of i-th individual, c represents the cohesion weight, C_i is the cohesion of the i-th individual, f is the food factor, F_i is the food source of the i-th individual, e is the enemy factor, E_i is the position of enemy of the i-th individual, ω is the inertia weight, and t is the iteration number. Then, the position of the i-th dragonfly at iteration $t+1$ is updated as follows:

$$X_i^{t+1} = X_i^t + \Delta X_i^{t+1} \tag{17}$$

The randomness, stochastic behavior, and exploration of the artificial dragonflies can be improved by introducing random walk (Levy flight) when there are no neighboring solutions. In this case, the position of the i-th dragonfly at iteration $t+1$ is updated as follows:

$$X_i^{t+1} = X_i^t + Levy(d) \times X_i^t \tag{18}$$

where d is the dimension of the position vectors. The pseudo-code of the dragonfly algorithm is illustrated in Algorithm 1 [13].

Algorithm 1. The pseudo-code of the Dragonfly Algorithm

```
Initialize the dragonflies population X_i(i = 1, 2, ..., n) randomly
Initialize the step vectors ΔX_i(i = 1, 2, ..., n)
while (not terminate-condition) do
    Calculate the fitness of each dragonfly f(X_i)
    Update the source food and enemy
    Update w, s, a, c, f, and e
    Calculate S, A, C, F, and E using equations 11 to 15
    Update neighboring radius
    if a dragonfly has at least one neighboring dragonfly then
        Update velocity vector using equation 16
        Update position vector using equation 17
    else
        Update position vector using equation 18
    end if
    Check and correct the new positions based on the boundaries of variables
end while
```

3.2 Quantum Evolutionary Algorithm

The Quantum evolutionary algorithm (QEA) proposed firstly by Kuk Hyuan Han in 2002 [14] is a new probabilistic searching optimization algorithm based on the concept of quantum computation theory [15]. It uses the quantum bit(Q-bit) to represent the probabilistic state of individuals. The Q-bit is the smallest unit of information stored in a two-state quantum numbers [14,16,17]. The state of a Q-bit is described as [18]:

$$\Psi\rangle = \alpha|0\rangle + \beta|1\rangle \tag{19}$$

Such that α and β represent the terms of a complex number, $\alpha = \sin(\theta)$, $\beta = \cos(\theta)$, and the Q-bit takes '0' with probability $|\alpha|^2$, or '1' with probability $|\beta|^2$; $(|\alpha|^2 + |\beta|^2 = 1)$. In QEA, an individual can be represented as a string of m Q-bits that is able to represent 2^m states. The *i-th* individual Q_i^t is given as follow:

$$Q_i^t = \begin{bmatrix} \alpha_{i1}^t & \alpha_{i2}^t & \dots & \alpha_{im}^t \\ \beta_{i1}^t & \beta_{i2}^t & \dots & \beta_{im}^t \end{bmatrix} \tag{20}$$

where m is the number of Q-bits, i.e. the dimension of the Q-bit individual.

An operator called Q-gate, that makes a rotation on the angle θ for performs the development of members of the population, it is calculated using Eq. 21:

$$U(\Delta\theta) = \begin{bmatrix} \cos(\Delta\theta) & -\sin(\Delta\theta) \\ \sin(\Delta\theta) & \cos(\Delta\theta) \end{bmatrix} \tag{21}$$

A Q-bit (j) of the member (i) is updated using Eq. 22:

$$\begin{bmatrix} \alpha_{ij}^{t+1} \\ \beta_{ij}^{t+1} \end{bmatrix} = U(\Delta\theta_{ij}^{t}) \begin{bmatrix} \alpha_{ij}^{t} \\ \beta_{ij}^{t} \end{bmatrix} = \begin{bmatrix} \cos(\Delta\theta_{ij}^{t}) & -\sin(\Delta\theta_{ij}^{t}) \\ \sin(\Delta\theta_{ij}^{t}) & \cos(\Delta\theta_{ij}^{t}) \end{bmatrix} \begin{bmatrix} \alpha_{ij}^{t} \\ \beta_{ij}^{t} \end{bmatrix} \quad (22)$$

The original QEA is defined as follows:

Algorithm 2. The pseudo-code of the original QEA

$t = 0$
Initialise $Q(t)$
Make $P(t)$ by observing the states of $Q(t)$
Evaluate $P(t)$
Store the best individual (solution) among $P(t)$ into $B(t)$
while ($\textit{not terminate} - \textit{condition}$) **do**
    $t = t + 1$
    Make $P(t)$ by observing the states of $Q(t-1)$
    Evaluate $P(t)$
    Update $Q(t-1)$ using suitable quantum gates $U(t)$
    Store the best individual (solution)
end while

4 Proposed Algorithm Based on the Hybridization of DFA and QEA for QoS Multicast Routing

This solution is based on the QEA algorithm, we found that the parameter $\Delta\theta$ that represents the displacement step is a fixed parameter in the original version of QEA, which slows down the search process. For this, we have proposed a new method to calculate this parameter following the behavior of the dragonflies, the parameter $\Delta\theta$ will be calculated as follows:

$$\Delta\theta_i^{t+1} = (sS_i + aA_i + cC_i + fF_i + eE_i) + \omega\Delta\theta_i^t \quad (23)$$

The new k^{th} component q_k^{t+1} of the solution i is calculated as follows:

$$q_{ik}^{t+1} = q_{ik}^{t} \mp \Delta\theta_i \quad (24)$$

The sign of $\Delta\theta_i$ is defined based on the Table 1:

Where b_{ik} and b_{jk} represent the kth component of the binary solution B_i and B_j respectively.

To calculate the parameters w, s, a, c, f, and e, we use a sinusoidal map [19] presented by Eq. 25:

$$X_{k+1} = a \cdot X_k^2 \sin(\pi X_k) \tag{25}$$

In a particular case where $a = 2.3$ and $X_0 = 0.7$, it can be simplified as follows:

Table 1. Determination of $\Delta\theta_i$

$Cost(B_i) \geq Cost(B_j)$	b_{ik}	b_{jk}	$\Delta\theta_i$
True	X	X	0
False	0	0	0
False	0	1	$+\Delta\theta_i$
False	1	0	$-\Delta\theta_i$
False	1	1	0

$$X_{k+1} = sin(\pi X_k) \tag{26}$$

The pseudo-code of DFAQEA is presented in Algorithm 3.

Algorithm 3. The pseudo-code of DFAQEA

$t = 0$
Initialise $Q(t)$
Make $P(t)$ by observing the states of $Q(t)$
Evaluate $P(t)$
Store the best individual among $P(t)$ into $B(t)$
while (*not terminate* − *condition*) **do**
    $t = t + 1$
    Make $P(t)$ by observing the states of $Q(t-1)$
    Repair solutions
    Evaluate $P(t)$
    Calculate w, s, a, c, f, and e using sinusoidal map (Equation 26)
    Calculate S, A, C, F, and E using equations 11 to 15
    for $i = 1$ to n(n is the number of solutions) **do**
        for $j = 1$ to n **do**
            if $(cost(X_i^t) > cost(x_j^t))$ **then**
                Calculate $\Delta\theta_i^{t+1}$ using equation 23
                Update X_i^{t+1} using equation 24
            end if
        end for
    end for
    Store the best individual
end while

5 Simulation Results

In order to evaluate the performance of the proposed algorithm DFAQEA, we have conducted a set of simulations in which we compared our algorithm with GA, QEA, and DFA. The positions of the nodes are randomly distributed in a rectangle of 4000 Km × 2400 Km. We used the WAXMAN topology model [20] in the experiments to generate the graph. In this model, the links are created between the pairs of nodes i and j by using a probability given by:

$$P(i,j) = \beta \cdot \exp(-l(i,j)/\alpha L) \tag{27}$$

where $l(i,j)$ is the Euclidean distance between node i and node j, L is the maximum distance between any two nodes in the graph. The parameter α controls the number of short links in the graph. The parameter β controls the number of links in the graph. In our simulation, we fixed the value of α and β to 0.8 and 0.7 respectively. The cost, delay, bandwidth, and loss rate of each link are distributed in the range [1, 100], [1, 30], [2, 10] and [0.0001, 0.01] respectively. D_{max}, J_{max}, and PL_{max} are fixed to 120 ms, 60 ms, and 0.05 respectively. The bandwidth B_{min} required by the flow is randomly generated in the interval [2 mbps, 10]. The source node and the set of destinations are selected randomly from the graph. The average cost of the multicast tree is obtained by ten executions of the algorithms.
We studied the performance of our algorithm in two scenarios.

5.1 First Scenario

In the first scenario, we varied the number of nodes from 20 to 120 with 20% nodes as destinations in order to evaluate the performance of our algorithm in terms of multicast tree cost, delay and jitter. The results are given in Figs. 1,2 and 3.

Figure 1 presents the multicast tree cost of GA, QEA, DFA, and DFAQEA, according to the network size. First, we remark that the multicast tree cost

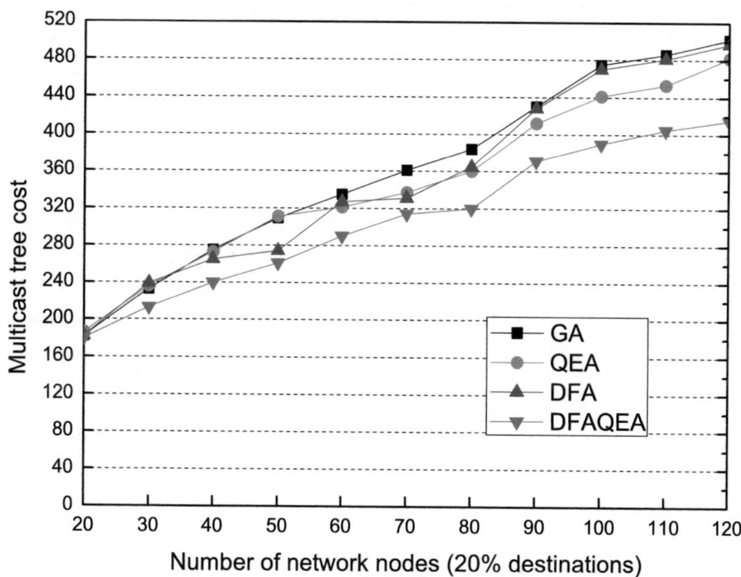

Fig. 1. The cost of the multicast tree while varying the number of nodes in the network

increases with the increase in the network size. Second, we observe that the multicast tree cost generated by DFAQEA is less than the others multicast tree cost.

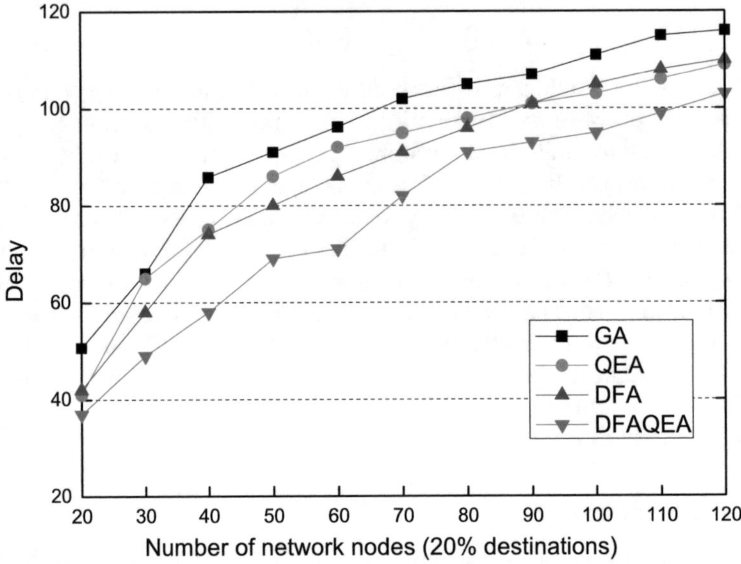

Fig. 2. The delay of the multicast tree while varying the number of nodes in the network

Figure 2 presents the multicast tree delays of four algorithms, according to the network size, we observe that the delay consumed by the proposed algorithm is the lowest compared to the other algorithms evaluated.

Figure 3 presents the multicast tree jitters of algorithms while varying the number of network nodes. We observe that the jitter of the multicast routing tree computed by our algorithm is the most reduced.

5.2 Second Scenario

In the second scenario, we varied the size of the set of destinations from 10% to 80% with an increment of 5 nodes, in a network of 120 nodes, in the aim to study the performance of our algorithm. Results are shown in Fig. 4.

Figure 4 shows the cost of the multicast tree generated by each algorithm while varying the number of destination nodes. It is shown that the cost of the multicast tree for each algorithm increases with increasing the number of destination nodes. It is also observed that the multicast tree generated by our algorithm have the lowest costs as compared to DFA, QEA, and GA.

6 Conclusion

In this paper, we proposed a new hybrid approach based on the hybridization of DFA with QEA for solving the Quality of service multicast routing problem

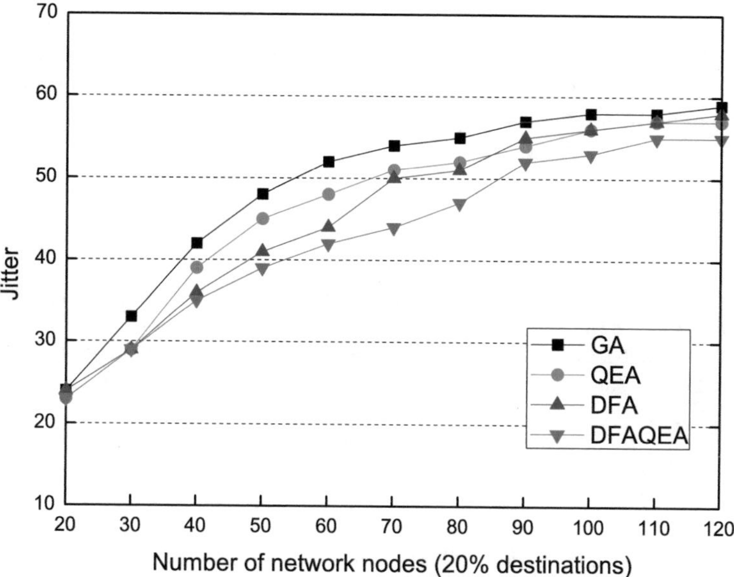

Fig. 3. The jitter of the multicast tree while varying the number of nodes in the network

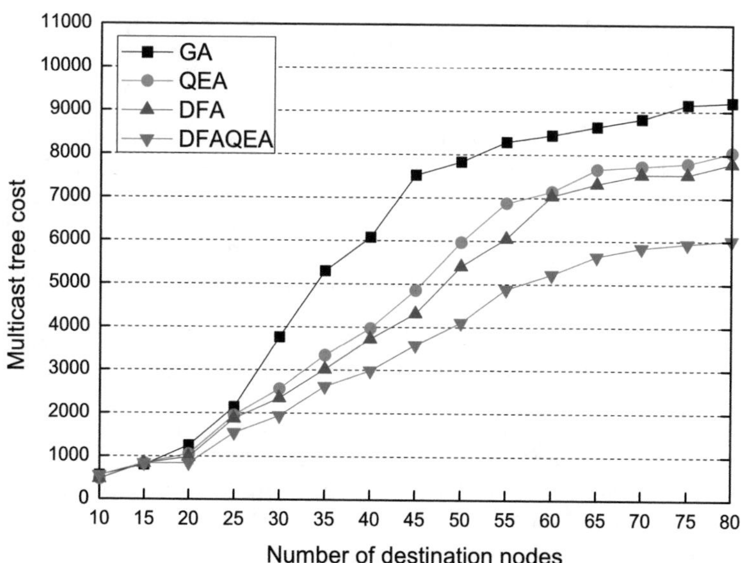

Fig. 4. The cost of the multicast tree while varying the number of nodes in the network

under multiple constraints. In this approach, called DFAQEA, the evolutionary equation of DFA is used to calculate $\Delta\theta$ in QEA. To assert the performance of the proposed algorithms, simulations were carried out while varying the number of nodes in the network and also varying the size of the set of destinations. Our algorithm and the existing algorithms GA, QEA and DFA have been experimented with and evaluated. The simulation results clearly prove the superiority and the efficiency of the proposed algorithm over GA, QEA, and DFA. Future work will be to study the same problem with the integration of the evolutionary equation of QEA in DFA.

References

1. Wang, Z., Crowcroft, J.: Quality-of-service routing for supporting multimedia applications. IEEE J. Sel. Areas Commun. **14**(7), 1228–1234 (1996)
2. Koyama, A., Nishie, T., Arai, J., Barolli, L.: A GA-based QoS multicast routing algorithm for large-scale networks. Int. J. High Perform. Comput. Netw. **5**(5–6), 381–387 (2008)
3. Haghighat, A.T., Nejla Ghaboosi and: Tabu search based algorithms for bandwidth-delay-constrained least-cost multicast routing. Telecommun. Syst. **34**(3–4), 147–166 (2007)
4. Armaghan, M., Haghighat, A.T.: QoS Multicast Routing Algorithms Based on Tabu Search with Hybrid Candidate List, pp. 285–294. Springer, Heidelberg (2009)
5. Huang, L., Han, H., Hou, J.: Multicast routing based on the ant system. Appl. Math. Sci. **1**(57), 2827–2838 (2007)
6. Wang, H., Hong, X., Yi, S., Shi, Z.: A tree-growth based ant colony algorithm for QoS multicast routing problem. Expert. Syst. Appl. **38**(9), 11787–11795 (2011)
7. Mahseur, M., Boukra, A.: Using bio-inspired approaches to improve the quality of service in a multicast routing. Int. J. Commun. Netw. Distrib. Syst. **19**(2), 186–213 (2017)
8. Meraihi, Y., Acheli, D., Ramdane-Cherif, A.: An improved chaotic binary bat algorithm for QoS multicast routing. Int. J. Artif. Intell. Tools **25**(04), 1650025 (2016)
9. Mahseur, M., Boukra, A., Meraihi, Y.: Improved quantum chaotic animal migration optimization algorithm for qos multicast routing problem. In: 6th IFIP TC 5 International Conference on Computational Intelligence and Its Applications, CIIA 2018, Oran, Algeria, May 8–10, 2018, Proceedings 6, pp. 128–139. Springer (2018)
10. Meraihi, Y., Acheli, D., Ramdane-Cherif, A., Mahseur, M.: A quantum-inspired binary firefly algorithm for QoS multicast routing. Int. J. Metaheuristics **6**(4), 309–333 (2017)
11. Abdel-Kader, R.F.: Hybrid discrete PSO with GA operators for efficient QoS-multicast routing. Ain Shams Eng. J. **2**(1), 21–31 (2011)
12. Meraihi, Y., Acheli, D., Ramdane-Cherif, A.: QoS multicast routing for wireless mesh network based on a modified binary bat algorithm. Neural Comput. Appl. 1–17 (2017)
13. Mirjalili, S.: Dragonfly algorithm: a new meta-heuristic optimization technique for solving single-objective, discrete, and multi-objective problems. Neural Comput. Appl. **27**(4), 1053–1073 (2016)
14. Han, K.-H., Kim, J.-H.: Quantum-inspired evolutionary algorithm for a class of combinatorial optimization. IEEE Trans. Evol. Comput. **6**(6), 580–593 (2002)
15. Benioff, P.: The computer as a physical system: a microscopic quantum mechanical hamiltonian model of computers as represented by turing machines. J. Stat. Phys. **22**(5), 563–591 (1980)

16. Han, K.H., Kim, J.H.: Genetic quantum algorithm and its application to combinatorial optimization problem, vol. 2, pp. 1354–1360. IEEE (2000)
17. Ying, M.: Quantum computation, quantum theory and AI. Artif. Intell. **174**(2), 162–176 (2010)
18. Hey, T.: Quantum computing: an introduction. Comput. Control. Eng. J. **10**(3), 105–112 (1999)
19. Li, Y., Deng, S., Xiao, D.: A novel hash algorithm construction based on chaotic neural network. Neural Comput. Appl. **20**(1), 133–141 (2011)
20. Waxman, B.M.: Routing of multipoint connections. IEEE J. Sel. Areas Commun. **6**(9), 1617–1622 (1988)

Service Discovery in the Internet of Things: A Survey

Sami Abdellatif[✉], Okba Tibermacine, and Abdelmalik Bachir

Computer Science Department, Biskra University, Biskra, Algeria
{sami.abdellatif,o.tibermacine,a.bachir}@univ-biskra.dz

Abstract. One of the main elements that enables us to benefit from the potential of Internet of Things, is an efficient service discovery mechanism. Many research efforts have been put into the design and implementation of service discovery approaches and tools, though there are still some issues that need further investigation. In this paper, we provide a survey of the most recent works on service discovery in the Internet of Things. We propose a novel classification of existing approaches, based on their architecture, mode of operation, search method and the dependency to Internet protocols. The analysis of studied approaches, demonstrates their advantages and drawbacks, and help reveal future research directions.

Keywords: Internet of things · Service discovery · Classification Survey

1 Introduction

The Internet of Things (IoT), is an emerging technology intended to close the gap between the physical and virtual worlds, by embedding small intelligent devices (e.g. sensors/actuators) into the physical environment. Each device offers its hosted functionalities as services, accessible by human users or even other IoT devices. By 2020, there will be from 50 to 100 billion devices connected to the Internet [1], each device will offer its services to both public and private requesters. The discovery and selection of appropriate IoT devices is a primordial task, for an effective use of IoT devices in open online cyber-physical systems. Generally speaking, discovery and selection mechanisms in IoT environments, should support locating resources and services, and facilitate for users the identification of relevant devices/services, that matche their functional and extra functional (e.g. Quality of Services) requirements. Unfortunately, traditional web service discovery solutions are not suitable in this context, where services/resources are enabled by devices located in widely distributed, and often heterogeneous constrained information systems [2].

S. Chikhi et al. (Eds.): MISC 2018, LNNS 64, pp. 60–74, 2019.
https://doi.org/10.1007/978-3-030-05481-6_5

In fact, designing an effective and efficient service discovery (SD) mechanism, has been the topic of many research papers in the literature. Many surveys have been conducted on the service discovery approaches, where a general classification was proposed based on the used architecture [3,4], or the generated network overhead [2]. However, none of the above surveys discusses the search interface employed by users to locate services, nor do they investigate the dependency level of the existing solutions on the Internet network. Thus, this paper focuses on:

- Defining a new classification scheme that takes into account different search methods, architectures and Internet dependency.
- Providing a comprehensive review on recent IoT service discovery mechanisms, then comparing and analyzing surveyed solutions based on the proposed classification scheme.
- Highlighting guidelines and future research directions, on how to design an efficient IoT service discovery solution and avoiding existing deficiencies.

The rest of the paper is organized as follows: Sect. 2 presents a novel classification scheme for IoT service discovery approaches. Section 3 outlines recent solutions highlighting functional features and implementation issues. Findings of the survey are provided in Sect. 4, and the paper is concluded in Sect. 5.

2 Classification Scheme

One of IoT's main requirements is the dynamic discovery of services, provided by distributed objects [5]. Service Discovery is the process of locating available network services, that can meet a requester need. It involves: locating requested services (service location); matching service descriptions to clients requests (description and matchmaking); and selecting the most relevant services (service selection) [3]. In the literature, many notable contributions have been proposed in the field of IoT service discovery. To study their advantages and drawbacks, we classify and analyze them based on the classification scheme depicted on Fig. 1. As it is shown in the figure, the proposed classification is represented by six layers, namely search scope (local or remote), search method, network (Internet dependency), architecture (centralized, distributed or hierarchical), usability operation mode and finally description type. Each layer represents an important aspect that affects the nature and the quality of the proposed IoT discovery solution. These layers represent the criteria by which we compare between existing works in this field. Details on each criterion is covered in the following subsections.

2.1 Description

Most of SD approaches relay on the description method. It could be either a syntactic based or a semantic based description. Syntactic description methods use

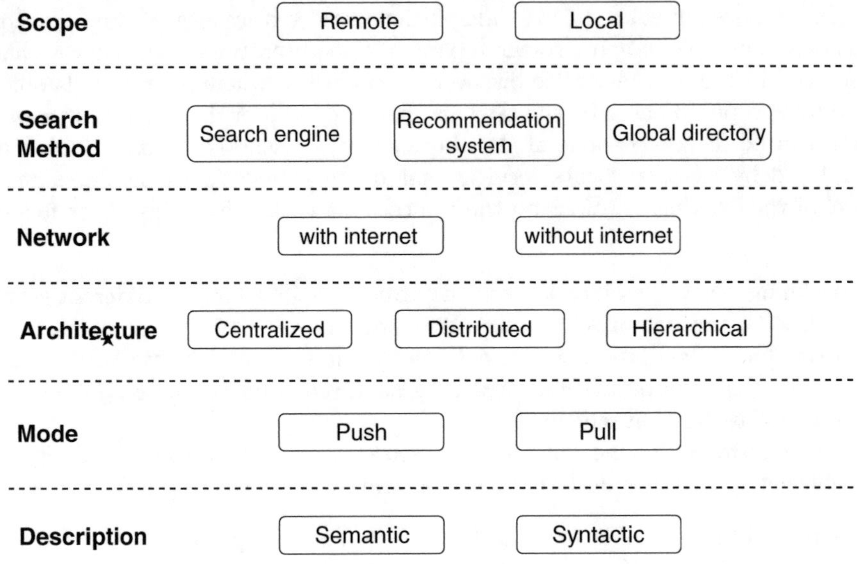

Fig. 1. Classification scheme for service discovery

a set of attribute-value pairs to describe services. For instance, a CoAP (Constrained Application Protocol) path to a temperature service could be described as: *Path* = "*coap://[2001::12]:5683/sensor/temp*".

Semantic-based description methods adopt SOA (Service-Oriented Architecture) principals, and consider Things as a semantic service. In most cases, these methods utilize ontologies for providing semantic annotations of services. In matter of expressiveness these approaches are by far the best, but they usually require more processing power, and generate high network overhead [6].

2.2 Architecture

Generally, discovery mechanisms adopt either a directory-based, or a directory-less solution. The later eliminates the need for a central registry, and instead most devices maintain a cache of already known services, and use multicast transmission to transmit queries/advertisements. They require less human intervention (zeroConf), and generate low network overhead. However, multicast transmission is limited to the local scope only, making this kind of discovery suitable only for local networks.

In directory based architecture (i.e. **Centralized architecture**), a common entry point for users and service providers is introduced. Its main functionality is the registration and selection of services. This type of solutions often suffers from a lack of scalability, and fault tolerance. Researchers addressed this problem by distributing the workload on a number of interconnected cooperating directories (i.e. **Distributed architecture**), which guarantee a better performance. A wide range of solutions make use of P2P connection, between these distributed directories, while other solutions maintain a master-slave relation, forming a hierarchy of directories (i.e. **Hierarchical Architecture**), governed by a root element usually referred to as global directory.

2.3 Search Method

To allow a global access of the IoT distributed things, a sophisticated search interface is usually presented to users. Where this search tool must support an endless number of devices and respond to user's request with most relevant services in real time. We can identify three types of search interfaces including: (i) IoT **Search engines** like "ThingSeek" [7], (ii) **Recommendation system** like the work proposed in [8], and (iii) **Global directory** accessible only by proprietary application (e.g. [9]).

2.4 Internet Dependency

The IoT is foreseen as an extension of the actual Internet, with the addition of smart embedded devices capable of integrating the physical environment, through the data provided by sensors and actions performed by actuators. Thus, service discovery solutions should be based on Internet, in order to achieve their goal. Nonetheless, the level of dependency between the discovery method and the Internet varies. Some approaches adopt the Internet network, and its infrastructure, taking advantage of the wide spread network and making it inter-operable with existing systems and easier to deploy. However, the absence of internet implies the non-availability of IoT services. In some cases the availability of services is crucial, that led researchers to think of a standalone infrastructure, that takes robustness as the most important criterion. The resulting solution depends only on Internet protocols such as "TCP", "HTTP", "DNS".

2.5 Push/Pull Modes

Service discovery process consists basically of query response messages, being sent over the network in order to locate a desired service. Also, it allows devices to announce their services periodically (i.e. the **push method** sends service status to other devices), so they can be discovered and used. This periodic multicast of advertisement messages tends to be energy consuming. On the contrary, the **pull method** uses query messages sent over multicast in case of decentralized architecture, or unicast in the presence of a directory.

2.6 Scope

A Service discovery protocol is characterized by its range, based on which it operates in. Generally, there are SD mechanisms that focus on **Local Discovery**, offering to the user the ability to monitor, and interact with its immediate environment. Its main objective is to minimize network overhead, and energy consumption. Whereas, other SD mechanisms focus on **Remote Discovery** of services, throughout a wide area and over either internet or a multi-hop network. Providing a user with remote access to distant services and devices. The main goal here is the accurate and rapid selection of appropriate services.

3 Service Discovery in IoT

In this section, we review some recent contributions in the field of IoT service discovery. We classify and list papers based on the scope (local discovery 3.1, Remote discovery 3.2) and the search method (Search engine 3.3, Recommendation systems 3.4). Other criteria mentioned above in the classification scheme are used to analyze and compare between SD solutions in each category. Refer to Table 1 for a comparative overview of the reviewed contributions.

3.1 Local Discovery

A local service discovery is intended to provide a client with services residing within the local network. It reveals to a user the presence of smart objects that are embedded in his surrounding environment and allows him to consume their offered services. The user can then interact with and monitor his surroundings seamlessly. Based on the way discovery messages are transmitted we distinguish two types of IoT local discovery protocols:

Unicast based discovery. Usually a centralized node called service directory is used. The directory is deployed and made available on local network for users to access it. It is responsible for registering services descriptions, collected either by constantly sending "Who is there" messages or by listening to advertisement messages coming from devices. To discovery available services a user can send a query to the local directory through a unicast transmission asking for services (a user pulls services from the directory) or he/she can subscribe to the directory and as soon as services are available or updated the user gets notified (the directory pushes services to end users). Lee et al. [10] suggest to deploy a DNS server as a local directory. They also propose a framework called "DINAS" that helps IoT devices to automatically generate their own names and register their available service in the local DNS server. The DNS server stores available services as DNS RR record, but before registering any device or service, a physical-contact-based authentication must take place using users smart phones over near field communication (NFC). The proposed framework focuses on reducing communication overhead by avoiding multicast transmission and offering security by registering authenticated devices only. However, the framework supports only devices equipped with NFC and does not offer a solution for non NFC devices. The Proposed solution also Requires the deployment of routers or access points and authentication server.

Abdelfadeel et al. [9] introduced a lightweight application-layer service discovery protocol, called "6LoWDIS". The protocol is built on top of "CoAP/HTTP" and exploits the capabilities of RESTful architecture. "6LoWDIS" divides the service discovery (SD) into two different stages: Local SD on the "6LoWPAN" cloud and Global SD on the Internet cloud by introducing an IoT gatways (IoTGW). These IoTGW are considered as directories. A key feature of "6lowdis" is that it addresses the heterogeneity problem between HTTP enabled devices and CoAP enabled devices by supporting/implementing both on the gateway level. However, it requires the presence of IoTGW in order to become fully operational.

The use of central directories offers more control over the discovery process and generates less network overhead. Also it has the ability to evolve into a remote discovery by connecting multiple directories. However, it requires the deployment of specialized hardware that can handle directory tasks. In addition, it makes the unicast based SD prone to single point of failure and less suitable for large scale and dynamic networks.

Multicast based discovery. Directory based SD protocols suffers from single point of failure problem and lack of scalability. Due to these issues, most of IoT service discovery protocols that are designed for local networks use multicast transmission to search/advertise a service. For instance, [11] proposes a local service discovery mechanism that supports both pull and push modes using link-layer broadcast messages. The proposed discovery mechanism considers three types of applications hosted at each device: (1) a Discovery Daemon, (2) Services and (3) Clients. The Discovery Daemon is responsible for service discovery and announcement processes by exchanging information within the local network using broadcast transmissions. It is responsible of registering/unregistering of local services after receiving a request from local service application. Additionally, It can query for Services, and consequently receive Responses from other Discovery Daemons. The Discovery Daemon can also operate proactively, by sending periodic advertisements (push), containing the information regarding the services being provided at the node. A major challenge for multicast based SD is the high network overhead generated by the broadcast/multicast packets and the occurrence of transmission loops. [12] addresses the multicast transmission loop by storing already processed request in a cache. Later bloom filters are used to reduce cache size so that it can fit to constrained devices. [13] propose "uBonjour", a lightweight version of apple's famous mDNS discovery protocol "Bonjour". "uBonjour" is designed to fit into small things while it can still work with other non constrained versions without the need for an application layer gateway. To achieve this, authors focused on reducing memory usage and code size. Similarly to [12], "uBonjour" addresses the transmission loop problem by dropping received requests that are directed to other nodes.

3.2 Remote Discovery

Remote discovery is the process of selecting and locating a distant IoT device and its hosted service among a huge number of other heterogeneous devices throughout a wide area. A remote service discovery mechanisms offers to a client the possibility to request and interact with a device/service even if the two (device/service and the client) are not in the same local network. Existing IoT Remote SD mechanisms relay on many directories deployed across a wide area. These distributed directories collect, store and aggregate services description then cooperate to resolve a service query receive from a client. Existing remote SD mechanisms differ from the way distributed directories are interconnected. Based on our literature review, the existing solutions can be classified into four subcategories:

Unstructured Distributed directories. SD mechanisms falling under this category are fully decentralized and do not follow any structure when connecting distributed directories. However, some approaches introduce a set of rules that determines which directory to query or forward a query to. In addition, these rules are used to determine if a pushed service description can be stored in local directory or not. An example of these rules is Social based rules. For instance, [14,15] both propose a fully decentralized and distributed service discovery framework, that takes inspiration from some well established social interaction theories. [15] uses a trust based rules and semantic descriptions to locate services based on requester quality of service demands and changing context requirements. Moreover, it integrates an efficient trust based propagation mechanism to the service discovery. Whereas "SESD" [14], uses two kinds of bi-dimensional lists: A white list in which it stores friends that were able to answer a query for a given service, and a black list where it stores peers that were unable to answer queries. Both lists represent the knowledge base of requests, giving benefit to previous searches and allowing not only simple queries but also complex ones to be efficiently processed. In addition, "SESD" uses a routing algorithm to determine which nodes will receive a query message next, based on their capability to answer the query or on a correlation degree, that represents how much does a node know of the query area. This approach uses an adaptive number of receivers to avoid flooding the network.

Other solutions take inspiration from some biological phenomena like in [16], where Rapti et al. suggest a bio-inspired decentralized solution, for discovery of services in the Internet of Things. This approach was inspired from the response threshold model, offering scalability, fault tolerance and flexibility. The SD problem was modelled as a multi-agent system, an agent requester chooses a promising neighbour, to which it forwards the remaining of the request. The selection is performed by using a probability value, calculated based on a stimulus that is the potential of a node to fulfill the request. Another bio-inspired SD is introduced in [17] based on swarm intelligence. In general, a number of directories collect and store services and content. Each service/content in these directories are mapped to meta-data using a locality preserving hash function. Then the content of these directories is reorganize and redistribute in such a way that similar services/content are stored in the same directory. To achieve this goal mobile agents equipped with two main functions are deployed (on load and on leave). The on load function is called to inject services carried by the agent into the current server, and on leave function collects services from the current server when departing. Agents decide which service to collect or which to inject based a similarity value calculated using a hamming function. A discovery process start by sending a query to a local server. The local server starts by calculating the similarity between the query and other stored meta-data. If a neighbouring server has a higher similarity value, then the query is forwarded to this server else the server returns all services similar to the issued query. The search process supports both simple query and ranged query. The main advantages of this approach is that it reorganizes the content of servers and reduces the load on a single server by distributing the load on a number of servers. However, relying only on similarity could result in an unbalanced load among servers. Besides, performance relays heavily on the type of description used and meta-data vector size.

Structured P2P directories. The scalable nature of P2P networks made it the center of many propositions, where in [18,19] a peer-to-peer (p2p) service discovery algorithm in an IoT network is presented. The algorithm is based on routing information caching (RIC). Cache size is bounded to meet the requirement of IoT constrained nodes, and three Maintenance phases were introduced to maximize the usefulness of the RIC. First, it builds an ontology of IoT device capabilities and designs a hierarchical Bloom filter based Indexing (HBFI) to index the capabilities in the ontology. Second, a cache update mechanism is used to maintain the freshness and usefulness of entries. It considers the reference time of the entries, the mobility of the IoT nodes, and the inclusiveness of the capability concept. Third, a Service discovery process that relies on the (HBFI) is used to answer received query faster. Unlike [18–20] propose to a use "Skipnet" overlay instead of DHT to ensure a better control over the location of services. This enhances the security and integrity of an organization by supporting content and path locality. In [21] authors suggest to connect gateways through two kinds of overlays, GDT and DLS to support both geographic based search (GDT), and service type based search using (DLS). Local service discovery uses mDNS protocol in both push and poll modes.

[22] propose an Edge-centric distributed architecture to provide IoT applications with a common service for remote and global discovery. This service is realized by interconnecting distributed IoT Gateways using a structured Peer-to-Peer (P2P) overlay, implemented using a Distributed Hash Table (DHT). The proposed solution uses CoAP protocol to implement the standard interface exposed to applications. A gateway is responsible for discovering, registering services and also resolving and proxiying received queries. The proposed solution supports push and pull modes, and uses key value descriptions to represent services stored at the gateway level.

Hierarchical directories. In contrast to other remote SD, Hierarchical directories based SD mechanisms impose a parent-child relation between different directories. Parent directories federates many low-tier directories and collects aggregated services description by pulling its following directories. A low-tier directory covers its own local area and tries to resolve service queries from within its local network. When it fails it forwards the service query to its parent directory. Authors in [23] uses a cluster head that could be any sensor or smart object in the network, it takes the responsibility of a directory, in registering other devices services and satisfying requests, from only 1-hop neighbours. Cluster heads (CHs) can communicate with each other, to extend the boundaries of the SD region, they are relayed to a higher level of the hierarchy called area routers. These area routers aggregate the services offered at the CHs, and guaranties a considerable success rate of answering given requests. In addition, the proposed approach utilizes smart-phones density in the network, to properly enable a dynamic sleep schedule, leading to further improvements in energy efficiency. The authors claim that the maturity of the smart object, and recent advancements in the sensors hardware capacity, enables it to fulfill the duties of a cluster head. But managing an entire network, in case of an area router, or even a sub network in case of cluster head, is too much of a taxing task, on both computational and energy units. The proposed approach is limited only to local discovery. [24] In this paper, semantic annotation of services is supported through ontologies defined for API definition languages such as Swagger and RAML. CoAP and linked-data serialization format (JSON-LD) are used to represent services, entities, and properties in a semantic-aware framework. This enables intelligent discovery of services. This approach combines the power of semantic-aware service discovery with modern and standard service definition methods. Taking into account the special requirements of IoT such as limited and scattered resources. The proposed system uses API definition languages and semantic tags to define and describe services and discovery queries. Both CoAP and HTTP are supported under pull mode.

Centralized directory. To further extend the search scope, global centralized directories are often used such as "Trendy" [25]. In this work, authors propose a RESTful web services based Service Discovery mechanism, that depends on a root directory called DA, and a number of group leader (GL) with the responsibility of a proxy, aggregation and even caching. "Trendy" relays on pull mode to discover and collect service. In addition, it uses a demand based adaptive timer to represent the lifetime of a service description in the registry. Unlike ordinary TTL, this timer is not fixed, rather it changes depending on the popularity of the service (i.e. how many times the service is searched). This keeps the popular services fresh and up to date. Consequently, it minimizes the communication overhead by only updating and requesting most searched services. "Trendy" uses context information like the location and energy level, to provide optimal service selection.

3.3 IoT Search Engine

Perhaps one of the issues that still needs further investigation, is the search interface used to initiate the discovery process. Many of the suggested search engines for IoT are concerned with the constrained nature of devices and try to minimize network overhead and energy consumption. For instance, "Snoogle" [26] an information retrieval system (search engine) that follows a two tier architecture. It uses a description compression technique in local registry based on bloom filters and distributed top-k query algorithm to reduce the network overhead. Also, It offers a security and privacy framework for users by leveraging access rights and PKC (public key cryptography). Additionally to low overhead, [27] introduced precision prediction models to improve the service search efficiency. Authors, assume that smart objects periodically push their data to a local gateway, which then indexes received data and generates prediction models. These prediction models are used to infer the state of the smart object in between two push periods. A client is presented with a search interface connected directly to a global directory. When receiving a query, the global directory forwards the query to the appropriate local gateway capable of resolving it. Three methods to optimize the search process were introduced: First, an approximation method based on least squares polynomial approximation, that formulate the sensors values into a polynomial equation. Second, a multi step prediction method is proposed to accurately estimate the state of a sensor between push periods using the polynomial approximation and a variant of SVM. Third, a ranking method is used to select and rank most appropriate sensors that can answer a given query based on their predicted value at query time. Less computational overhead was achieved by using a lightweight prediction model in the third method.

Other search engines are more concerned by real time data by claiming that data related to IoT is dynamic and possesses a short life span. In [28], a real time search engine that addresses the key challenge of scalability is proposed. The engine itself follows a centralized architecture and fetches sensors description following a pull method. Prediction model gets generated from fetched web pages made online by IoT entities (a room or sensors). The fetch process is executed over two crawling methods and occurs at regular intervals. First, Google is used to search and crawl sensors pages. Second depends on gateways linked to the "Dyser" search engine. After a query arrives (described in a key value based format), the indexer ranks the probabilistic models (a set of probabilities on the state of an entity or a sensor in a certain time) based on their values. Then it chooses "K" best results. Afterwards, it fetches their real values in real time to finally rank the results by relevance and send them to the user. "Dyser" assumes that the URL to the data source (web pages, gateways) are all available and publicly shared by the service owner. In [29], authors propose a centralized framework with a search engine on top named "digcovery". The later is suitable for global discovery of devices and sensors. The framework allows user to register their own devices into the infrastructure and discovers available resources through mobile phones. The primary focus of this work is on the discovery of devices based on context awareness and geo-location information. The implemented solution is scalable and supports a variety of devices and domains. It also supports the integration of heterogeneous technologies which include RFID, sensor networks, Bluetooth, and WiFi. The novelty of this work is that the application uses sensors values, geo-location data and context awareness capabilities through the deployed back-end or through RFID/QR codes to discover near devices. Besides, the solution is able to integrate and support legacy technologies.

[30] Presents "ForwarDS-IoT", a decentralized resource discovery structured upon a federated architecture. "ForwarDS-IoT" uses an ontology-based information model with a consistent vocabulary of concepts related to IoT resources, services, and relationships. The proposed architecture comprises four layers and exposes a web interface that forwards SPARQL queries to ForwarDS-IoT. This search engine supports both push and pull modes and simple and ranged queries.

3.4 IoT Recommendation System

[8,34] are two web service clustering based recommendation systems, with the objective of reducing the search space. Ben Fredj et al. [34] use a hierarchy of distributed directories deployed on gateways and incremental clustering mechanisms to offer efficient cluster management in a dynamic environment. In contrast, Chirila et al. [8] use a central directory storing semantic description of services, and an agglomerative clustering technique. Two semantic similarity measures where used; the NGD (normalized Google distance) for measuring the similarity between two web service's simple features like names. Complex web services are evaluated by a weighted average of WSDL elements. The clustering technique creates a number of service clusters with a centroid that represents the service with a minimal distance to all cluster members. The discovery process begins by calculating the distance between the query and all centroids. The centroid with a minimal distance is expected to be the closest to the desired service. Therefore, the system then recommends to the user the most relevant services that are in the same cluster as the centroid. This approach use semantic description excessively, which in case of constrained network could be unsuitable. Also, the clustering technique requires constant updates.

4 Evaluation

After reviewing a certain number of existing service discovery approaches, we noticed that most papers focus on, either local discovery or global discovery. Only a few of the reviewed papers treated both cases. In local discovery, directory-less approaches seem more suitable than directory approaches because they ensure scalability and robustness. However, due to the excessive use of multicast transmissions, this kind of solutions can operate only on a limited scope, and often requires advanced flooding and forwarding techniques. Local discovery operates under two kinds of modes push and pull. The pull mode can be expected to generate smaller communication volume, still pulling all sensors upon each query

Table 1. Comparative table between the surveyed solutions

Approach	Scope	Search method	Internet	Architecture	Push/Pull	Description	Main focus
[8]	remote	Recommendation system	✓	Hierarchical	Push	Semantic	Precision and context awareness
[9]	Local and remote	Global directory	✓	Centralized	hybrid	Syntactic	Overhead and Energy
[10]	Local and remote	N/A	✗	Centralized	Pull	Semantic	Large scale
[11]	Local	N/A	✗	Distributed	Hybrid	N/A	Large scale
[12]	Local and remote	N/A	Protocols only	Decentralized	Push	Syntactic	Overhead and energy
[13]	Local	N/A	protocols only	Decentralized	Pull	Syntactic	Overhead and energy
[14]	Remote	N/A	✗	Decentralized	Pull	Semantic	precision
[15]	Remote	N/A	N/A	Distributed P2P	Pull	Semantic	Precision and context awareness
[16]	Remote	N/A	✗	Decentralized P2P	Pull	Semantic	Large scale and low overhead
[17]	Remote	N/A	✗	Distributed	Pull	N/A	Heterogeneity
[19]	Remote	N/A	✗	Distributed	Pull	Semantic	Overhead and energy
[20]	Remote	N/A	Protocols only	Decentralized P2P	Push	Semantic	Precision and context awareness
[21]	Local and remote	N/A	Protocols only	Distributed P2P	Hybrid	Syntactic	Large scale
[23]	Local	Search engine	✗	Hierarchical	pull	Syntactic	Overhead and energy
[24]	Remote	Search engine	✓	Hierarchical	Push	Semantic	Precision and Context awareness
[25]	Remote	N/A	✗	Hierarchical	Push	Semantic	Precision and context awareness
[26]	Local and remote	Search engine	✗	Hierarchical	Push	Syntactic	Overhead and energy
[27]	Remote	Search engine	✓	Hierarchical	Push	Syntactic	Overhead
[28]	Remote	Search engine	✓	Centralized	Pull	Semantic	Large scale and heterogeneity
[29]	remote	Search engine	✓	Centralized	Push	Semantic	Heterogeneity
[30]	Remote	Search engine	✓	Distributed	Hybrid	Semantic	Large scale/Precision and context awareness
[31]	Local	N/A	✓	Centralized	hybrid	Semantic	Overhead and energy
[32]	Remote	N/A	✓	Centralized	Push	Semantic	Heterogeneity
[33]	Local and remote	N/A	✗	Distributed P2P	hybrid	Syntactic	Overhead and energy
[34]	Remote	Recommendation system	✓	Hierarchical	Push	Semantic	Precision and context awareness

would not scale. Whereas push mode offers users with real time notification, but it generates high network traffic. For Global discovery, gateways are used in most cases. The main difference between studied approaches is the way in which gateways are connected. In a number of SD solutions a hierarchy of gateways is used with a root node on top, whereas others adopt a P2P architecture to add more robustness to the systems. To facilitate the SD process for end user, a search engine is usually proposed on top of the adopted architecture. One of the main responsibility of the search engine is to provide a common entry point, featuring advanced search and indexing techniques. IoT search engines defer from traditional web search engines where the former has to deal with the endless amount of data and their dynamic nature, making most of IoT search engines use prediction models instead of constantly fetching data from smart objects.

5 Conclusion and Future Research Directions

The Internet of Things is an emerging technology intended to close the gap between physical and virtual world. One of the challenging topics in IoT is service discovery. In this article we surveyed a number of papers and we proposed a new classification scheme based on which we compared these solutions. We focused more on the advantages and drawbacks of the studied approaches. As we have seen, most of global discovery mechanisms utilize a central directory maintained by a gateway to allow a remote discovery. Yet this gateway represents a single point of failure that could threaten the availability of services. Existing approaches that do not require a persistent gateway are limited only to the local scope. We conclude that it is missing an efficient mechanism that could achieve global discovery without fully relying on gateways or a central directory.

The huge number of Device and service envisioned by IoT is one of the most challenging problems that a remote service discovery mechanism must overcome. One Promising solution is the use of decentralized architecture and P2P networks. Another IoT most challenging problem is the heterogeneity of devices. Although Service oriented Architecture and semantic based service discovery help overcome a part of this problem by abstracting the software layer and allowing for interoperability between application layer protocols. They do not solve the heterogeneity of hardware and communication standards and also the heterogeneity of operational modes.

References

1. Sundmaeker, H., Guillemin, P., Friess, P., Woelfflé, S.: Vision and challenges for realising the internet of things. Clust. Eur. Res. Proj. Internet Things, Eur. Commision **3**(3), 34–36 (2010). https://doi.org/10.2759/26127
2. Villaverde, B.C., de Paz Alberola, R., Jara, A.J., Fedor, S., Das, S.K., Pesch, D.: Service discovery protocols for constrained machine-to-machine communications. IEEE Commun. Surv. Tutor. **16**(1), 41–60 (2014)

3. Djamaa, B., Richardson, M., Aouf, N., Walters, B.: Service discovery in 6LoWPANs: classification and challenges. In: Proceedings of the IEEE 8th International Symposium on Service Oriented System Engineering, SOSE 2014, pp. 160–161 (2014). https://doi.org/10.1109/SOSE.2014.67
4. Balakrishnan, S.M., Sangaiah, A.K.: Aspect oriented middleware for internet of things: a state-of-the art survey of service discovery approaches. Int. J. Intell. Eng. Syst. **8**(4), 16–28 (2015)
5. Nitti, M., Pilloni, V., Colistra, G., Atzori, L.: The virtual object as a major element of the internet of things: a survey. IEEE Commun. Surv. Tutor. **18**(2), 1228–1240 (2016). https://doi.org/10.1109/COMST.2015.2498304
6. Talal, B.K., Rachid, M.: Service discovery–a survey and comparison. arXiv preprint arXiv:1308.2912 (2013)
7. Shemshadi, A., Sheng, Q.Z., Qin, Y.: ThingSeek. In: Proceedings of the 39th International ACM SIGIR Conference on Research and Development in Information Retrieval - SIGIR 2016, pp. 1149–1152 (2016). https://doi.org/10.1145/2911451.2911471, http://dl.acm.org/citation.cfm?doid=2911451.2911471
8. Chirila, S., Lemnaru, C., Dinsoreanu, M.: Semantic-based IoT device discovery and recommendation mechanism. In: 2016 IEEE 12th International Conference on Intelligent Computer Communication and Processing (ICCP), pp. 111–116. IEEE (2016)
9. Abdelfadeel, K.Q., Elsayed, K.: 6LoWDIS: a lightweight service discovery protocol for 6LoWPAN. In: 2016 IEEE International Conference on Communications Workshops, ICC 2016, pp. 284–289 (2016). https://doi.org/10.1109/ICCW.2016.7503801
10. Lee, K., Kim, S., Jeong, J.P., Lee, S., Kim, H., Park, J.S.: A framework for DNS naming services for internet-of-things devices. Futur. Gener. Comput. Syst. (2018)
11. Quevedo, J., Guimarães, C., Ferreira, R., Corujo, D., Aguiar, R.L.: Icn as network infrastructure for multi-sensory devices: Local domain service discovery for icn-based iot environments. Wirel. Pers. Commun. **95**(1), 7–26 (2017)
12. Antonini, M., Cirani, S., Ferrari, G., Medagliani, P., Picone, M., Veltri, L.: Lightweight multicast forwarding for service discovery in low-power IoT networks. In: 2014 22nd International Conference on Software, Telecommunications and Computer Networks, SoftCOM 2014, pp. 133–138 (2014). https://doi.org/10.1109/SOFTCOM.2014.7039103
13. Klauck, R., Kirsche, M.: Bonjour contiki: A case study of a DNS-based discovery service for the internet of things. Ad-Hoc Mob. Wirel. Netw. 316–329 (2012)
14. Liu, L., Antonopoulos, N., Zheng, M., Zhan, Y., Ding, Z.: A socioecological model for advanced service discovery in machine-to-machine communication networks. ACM Trans. Embed. Comput. Syst. (TECS) **15**(2), 38 (2016)
15. Li, J., Bai, Y., Zaman, N., Leung, V.C.: A decentralized trustworthy context and QoS-aware service discovery framework for the internet of things. IEEE Access **5**, 19154–19166 (2017)
16. Rapti, E., Houstis, C., Houstis, E., Karageorgos, A.: A bio-inspired service discovery and selection approach for IoT applications. In: 2016 IEEE International Conference on Services Computing (SCC), pp. 868–871 (2016)
17. Forestiero, A.: A smart discovery service in internet of things using swarm intelligence, pp. 75–86. Springer, Cham (2017)
18. Moeini, H., Yen, I.L., Bastani, F.: Efficient caching for peer-to-peer service discovery in internet of things, pp. 196–203. IEEE (2017)

19. Moeini, H., Yen, I.L., Bastani, F.: Routing in IoT network for dynamic service discovery. In: 2017 IEEE 23rd International Conference on Parallel and Distributed Systems (ICPADS), pp. 360–367. IEEE (2017)
20. Li, J., Zaman, N., Li, H.: A decentralized locality-preserving context-aware service discovery framework for internet of things. In: Proceedings of the 2015 IEEE International Conference on Services Computing, SCC 2015, pp. 317–323 (2015). https://doi.org/10.1109/SCC.2015.51
21. Cirani, S., Davoli, L., Ferrari, G., Leone, R., Medagliani, P., Picone, M., Veltri, L.: A scalable and self-configuring architecture for service discovery in the internet of things. IEEE Internet Things J. **1**(5), 508–521 (2014). https://doi.org/10.1109/JIOT.2014.2358296
22. Tanganelli, G., Vallati, C., Mingozzi, E.: Edge-centric distributed discovery and access in the internet of things. IEEE Internet Things J. **5**(1), 425–438 (2018)
23. Helal, R., ElMougy, A.: An energy-efficient Service Discovery protocol for the IoT based on a multi-tier WSN architecture. In: Proceedings of the Conference on Local Computer Networks, LCN 2015-Decem, pp. 862–869 (2015). https://doi.org/10.1109/LCNW.2015.7365939
24. Khodadadi, F., Sinnott, R.O.: A semantic-aware framework for service definition and discovery in the internet of things using CoAP. Procedia Comput. Sci. **113**, 146–153 (2017)
25. Butt, T.A., Phillips, I., Guan, L., Oikonomou, G.: Adaptive and context-aware service discovery for the internet of things. In: Internet of Things, Smart Spaces, and Next Generation Networking, pp. 36–47. Springer, Heidelberg (2013)
26. Wang, H., Tan, C.C., Li, Q.: Snoogle: a search engine for pervasive environments. IEEE Trans. Parallel Distrib. Syst. **21**(8), 1188–1202 (2010)
27. Zhang, P., Liu, Y., Wu, F., Liu, S., Tang, B.: Low-overhead and high-precision prediction model for content-based sensor search in the internet of things. IEEE Commun. Lett. **20**(4), 720–723 (2016). https://doi.org/10.1109/LCOMM.2016.2521735
28. Ostermaier, B., Römer, K., Mattern, F., Fahrmair, M., Kellerer, W.: A real-time search engine for the web of things. Internet Things (IOT) **2010**, 1–8 (2010). https://doi.org/10.1109/IOT.2010.5678450
29. Jara, A.J., Lopez, P., Fernandez, D., Castillo, J.F., Zamora, M.A., Skarmeta, A.F.: Mobile digcovery: discovering and interacting with the world through the internet of things. Pers. Ubiquitous Comput. **18**(2), 323–338 (2014). https://doi.org/10.1007/s00779-013-0648-0
30. Gomes, P., Cavalcante, E., Rodrigues, T., Batista, T., Delicato, F.C., Pires, P.F.: A federated discovery service for the internet of things, pp. 25–30. ACM (2015)
31. Djamaa, B., Richardson, M., Aouf, N., Walters, B.: Towards efficient distributed service discovery in low-power and lossy networks. Wireless Netw. **20**(8), 2437–2453 (2014). https://doi.org/10.1007/s11276-014-0749-3
32. Jung, Y., Peradilla, M., Saini, A.: Software-defined naming, discovery and session control for IoT devices and smart phones in the constraint networks. Procedia Comput. Sci. **110**, 290–296 (2017)
33. Amoretti, M., Alphand, O., Ferrari, G., Rousseau, F., Duda, A.: DINAS: a lightweight and efficient distributed naming service for All-IP wireless sensor networks. IEEE Internet Things J. **4**(3), 670–684 (2017)
34. Ben Fredj, S., Boussard, M., Kofman, D., Noirie, L.: Efficient semantic-based IoT service discovery mechanism for dynamic environments. In: IEEE International Symposium on Personal, Indoor and Mobile Radio Communications, PIMRC 2015-June, pp. 2088–2092 (2015). https://doi.org/10.1109/PIMRC.2014.7136516

Spectral Band Selection Using Binary Gray Wolf Optimizer and Signal to Noise Ration Measure

Seyyid Ahmed Medjahed[1(✉)] and Mohammed Ouali[2]

[1] Centre Universitaire Ahmed Zabana, Relizane, Algérie
sa.medjahed@gmail.com, seyyid.ahmed@univ-usto.dz
[2] Thales Canada Inc., 105 Moatfield Drive, North York, ON M3B 0A4, Canada
mohammed.ouali@usherbrooke.ca

Abstract. In remote sensing, spectral band selection has been a primordial step to improve the classification of hyperspectral images. It aims at finding the most important information from a set of bands by eliminating the irrelevant, noisy, and highly correlated bands. In this paper, the band selection problem is regarded as a combinatorial optimization problem. We propose a new band selection approach for hyperspectral image classification based on the Gray Wolf Optimizer (GWO) which is a new meta-heuristic that simulate the hunting process of gray wolf in nature. A new binary version of GWO based on transfer function is proposed. In addition, a new fitness function is designed using two terms: the first term is the SVM classifier and the second term of the fitness function is SNR measure (Signal to Noise Ration) which measures the capacity of discrimination. The proposed approach is benchmarked on three hyperspectral images widely used in band selection and hyperspectral images classification. The experimental results show that this approach is suitable to the challenging problem of spectral band selection and provides a higher classification accuracy rate compared to the other band selection methods.

Keywords: Binary Gray Wolf Optimizer · Support vector machine
Spectral band selection · Hyperspectral image classification
Feature selection

1 Introduction

Feature selection is an active research field in various applications such as bioinformatics, image processing, remote sensing, and several data analysis field [1–5]. The process of feature selection technique aims at reducing the dimensionality by selecting the relevant features of the data to be processed without degrading

The original version of this chapter has been revised. The author "Mohammed Ouali's affiliation has been updated. The correction to this chapter can be found at https://doi.org/10.1007/978-3-030-05481-6_26

S. Chikhi et al. (Eds.): MISC 2018, LNNS 64, pp. 75–89, 2019.
https://doi.org/10.1007/978-3-030-05481-6_6

the classification accuracy rate. The performance of classifier systems depends largely on the quality of the data used to build the classification model [6]. The most important advantage of feature selection approach resides in the fact that it decreases the computational time and the complexity of the classifier system [7–10]. Another benefit is the improvement of the classification accuracy rate and the capacity to reduce the risk of the overfitting. Feature selection process contains three main criteria: search strategy, objective function and stopping criterion. Feature selection methods can be classified as filter and wrapper methods. The filter methods are based on the utilization of statistical techniques or mutual information measures to validate the quality of the selected features. These techniques are independent of any classifier but less computationally intensive. The most popular approaches are: principal component analysis (PCA), factor analysis (FA), independent component analysis (ICA), mutual information maximization (MIM) [11], mutual information feature selection (MIFS) [12], joint mutual information (JMI) [13], minimum redundancy maximum relevance (mRmR) [14], conditional informax feature extraction (CIFE), fast correlation-based filter (FCBF), Relief, etc. [15]. Another category of feature selection method is the wrapper methods which are based on classifier systems. The wrapper methods use the classification accuracy rate as the performance measure. The basic idea is tantamount to select the smallest subset of features that minimizes the error rate or maximizes the accuracy rate. This category provides high computational complexity but they select the most relevant feature subset. To search the optimal feature subset, the wrapper methods use generally a meta-heuristic to minimize the classification error rate. Many wrappers approaches based on meta-heuristics have been developed recently: In [16], the authors have proposed a new approach based on binary bat algorithm to select the relevant feature. Amir Rajabi et al. [17], developed a feature subset selection technique by using the binary gravitational search algorithm for intrusion detection system. The simulated annealing algorithm has been also used in [18] for feature selection and parameter determination in SVM. In [19], a binary version of particle swarm optimization has been used for feature selection in detection on infants with hypothyroidism.

In remote sensing and specifically hyperspectral images classification, feature selection or spectral band selection is an important process using to select a minimal and effective subset of bands. The problem resides in the fact that hyperspectral images are high dimensional which causes the problem of Huges phenomenon. To avoid this problem, the goal is to remove parts of bands such as adjacent bands which are highly correlated, noisy bands, irrelevant bands without affecting the classification accuracy rate.

In this paper, we propose a new binary version of Gray Wolf Optimizer for spectral band selection namely BGWO: Binary Gray Wolf Optimizer and based on the transfer function. Gray wolf optimizer (GWO) is a new meta-heuristic method inspired from Gray wolves behavior and used for continuous optimization problem. This method is based on the hunting mechanism of Gray wolves in nature. Recent studies demonstrate that the GWO algorithm is able to provide good results compared to Particle Swarm Optimization (PSO), Gravitational

Search Algorithm (GSA), Differential Evolution (DE), Evolutionary Programming (EP), and Evolution Strategy (ES) [20]. The problem of spectral band selection is reformulated as a combinatorial problem and we attempt to optimize a new fitness function based on SNR (Signal to Noise Ration) and SVM classifier. The experimentations are conducted on three hyperspectral images largely used in remote sensing. In order to evaluate the robustness of the proposed approach, we compared it with several feature selection methods. The rest of paper is organized as follows: in Sect. 2, we present and describe the proposed approach for spectral band selection. In Sect. 3, we discuss the experimental results. Finally, conclusions are drawn in Sect. 4.

2 Spectral Band Selection Using Binary Gray Wolf Optimizer

2.1 Gray Wolf Optimizer

The Gray wolf is also called canis lupus and it is a canid native to the wilderness. The Gray wolves live in a group with very special social hierarchy. We distinguish four types of Gray wolves: the leaders are a male and a female called alpha. The alpha (dominant wolf) is the responsible and makes all the decisions to the group. The subordinate is called beta, it can be male or female and help the alpha to make the decisions. The third level in the hierarchy of Gray wolves is omega. The omegas wolves have the same role as a scapegoat and have to submit to all the other dominant wolves. The last level is delta wolves and is divided into five categories: scouts, sentinels, elders, hunters, and caretakers [20]. The hunting process of Gray wolves (tracking, encircling and attacking prey) can be seen as an optimization phases defined as follows:

1. the process of hunting is the optimization;
2. the prey represents the optimum to find;
3. the alpha, beta, and gamma are the fittest solutions and alpha the best solution.

The first step of hunting process is the encircling of the prey. The Gray wolves can detect the location of preys and encircle them. This phase can be defined mathematically as:

$$D = |C \cdot X_p(t) - X(t)| \tag{1}$$

$$X(t+1) = X_p(t) - A \cdot D \tag{2}$$

The vectors $A$and C are coefficient vectors and are given as follows:

$$A = 2a.r_1 - a \tag{3}$$

$$C = 2.r_2 \tag{4}$$

X is the position vector of the Gray wolf and X_p represents the position of the prey in the current iteration t. The Gray wolf updates its position according to the position of the prey by using the Eq. 2. r_1 and r_2 are random vectors in $[0, 1]$ which allow to the Gray wolf position to be randomly around the prey. a is initialized to 2 and during the iteration of the algorithm, it decreases linearly from 2 to 0 [20].

The second step is the hunting process and it is making by the alpha. This is why; we consider that the potential position of the prey is known by alpha, beta and delta. So, we must save the solutions of alpha, beta and delta to update the solution of other Gray wolves. This simulation can be described by the following equations [20]:

$$D_\alpha = |C_1 \cdot X_\alpha - X|\,,\ D_\beta = |C_2 \cdot X_\beta - X|\,, \tag{5}$$

$$D_\delta = |C_3 \cdot X_\delta - X| \tag{6}$$

$$X_1 = X_\alpha - A_1 \cdot (D_\alpha),\ X_2 = X_\beta - A_2 \cdot (D_\beta), \tag{7}$$

$$X_3 = X_\delta - A_3 \cdot (D_\delta) \tag{8}$$

$$X(t+1) = \frac{X_1 + X_2 + X_3}{3} \tag{9}$$

The last step is to attack the prey when it stops moving. This is modeled by decreased the value of a.

The pseudo code the GWO algorithm can be defined as follows [20]:

Algorithm 1. Gray Wolf Optimizer

Initialize a, A, C and the Gray wolves position X_i, $i = 1, \ldots, m$
Calculate the objective function of each search agent
for $t = 0$ to *max iteration* **do**
  for *each search agent* **do**
    Update the position of the current search agent
  end for
  Update a, A and C
  Calculate the objective function of each agent
  Update X_α, X_β and X_δ
end for
Return X_α

2.2 BGWO: Binary Gray Wolf Optimizer for Spectral Band Selection

The GWO algorithm is based on moving to Gray wolves in the search space solutions towards continuous valued position. The spectral band selection problem is a binary problem which represent each band X_i by a binary value v_i. The problem is to select or not a given band, if $v_i = 1$ the band is selected and used to build the classification model, $v_i = 0$ the band is not selection. The Gray wolves positions is given by a binary vectors.

We propose a binary version of Gray wolf optimizer algorithm based on the transfer function defined in [21]:

$$g(X_i) = \left| \frac{2}{\pi} \arctan\left(\frac{2}{\pi} X_i \right) \right| \tag{10}$$

$$v_i = \begin{cases} 1 \text{ if } g(X_i) > \sigma \\ 0 \text{ otherwise} \end{cases} \tag{11}$$

where σ is a random variable within interval $[0, 1]$. The position of Gray wolf represents the presence or absence of the spectral band.

The proposed BGWO for spectral band selection problem is detailed in the following algorithm:

Lines 1–2 represent the data initialization. We divide the data into three subsets: T_1, T_2 and T_3. T_1 is the training set and it used to train the classifier algorithm. T_2 is the test set used to evaluate the classification by computing the error rate. The last subset T_3 is used to validate the final results provided by the algorithm, in other terms, T_3 is not considered in the algorithm and we used this subset to compute the error rate over the selected bands provided by the algorithm in line 44.

The first loop in lines 3–7 randomly initializes the position of the Gray wolves population X_i. The loop in lines $8-10$ initializes the position of the Gray wolves alpha, beta, and delta wolves to 0.

The main loop starts in 12, the loop in lines 13–21 is used to compute the objective function and update the value of X_α, X_β and X_δ.

In line 14, we randomly create a training set $T_1^{'}$ from T_1 and a test set $T_2^{'}$ from T_2 by using the hold out method. The line 15 the algorithm removes the non-selected bands (bands which have $X_{i,j} = 0$) from $(T_1^{'}, T_2^{'})$ and lines 15–16 compute the fitness function.

From line 18 to 20, the algorithm stores the top three positions in X_α, X_β, X_δ and the top values of objective function in f_α, f_β, f_δ. Line 22 decreases linearly the value of a.

The final loop in lines 23–42 updates the position of the Gray wolves by using the Eqs. 9 and 11.

Finally, in line 44, the output vector X_α contains the selected features and in line 45 the classification error rate is stored in $global\_f$.

Algorithm 2. Binary Gray Wolf Optimizer for Spectral Band Selection

Training set with labels T_1 and Test set without labels T_2
$a \leftarrow 2$
for $i = 1$ to *number of search agents* do **do**
    for $j = 1$ to *number of bands* do **do**
        $X_{i,j} \leftarrow Random\{0,1\}$
    end for
end for
for $j = 1$ to *number of bands* do **do**
    $X_{\alpha,j} \leftarrow 0$; $X_{\beta,j} \leftarrow 0$; $X_{\delta,j} \leftarrow 0$
end for
$f_\alpha \leftarrow +\infty$; $f_\beta \leftarrow +\infty$; $f_\delta \leftarrow +\infty$; $global\_f \leftarrow +\infty$
for $k = 1$ to *max iteration* do **do**
    for $i = 1$ to *number of search agents* do **do**
        Create $T_1^{'}$ and $T_2^{'}$ from T_1 et T_2 respectively by using the hold out method
        $T_1^{'} \leftarrow T_1^{'} - \{bands\ with\ X_{i,j} = 0\}$; $T_2^{'} \leftarrow T_2^{'} - \{bands\ with\ X_{i,j} = 0\}$
        Train classifier over $T_1^{'}$ and evaluate its over $T_2^{'}$
        $err \leftarrow$ store the fitness function
        $X_{\alpha,j} \leftarrow the\ first\ best\ solution$; $f_\alpha \leftarrow the\ first\ best\ function\ value$
        $X_{\beta,j} \leftarrow the\ second\ best\ solution$; $f_\beta \leftarrow the\ second\ best\ function\ value$
        $X_{\delta,j} \leftarrow the\ third\ best\ solution$; $f_\delta \leftarrow the\ third\ best\ function\ value$
    end for
    $a \leftarrow a - \frac{2}{maxiteration}$
    for $i = 1$ to *number of search agents* do **do**
        for $j = 1$ to *number of bands* do **do**
            $r_{1\alpha} \leftarrow Random\{0,1\}$; $r_{2\alpha} \leftarrow Random\{0,1\}$
            $A_\alpha \leftarrow 2a.r_{1\alpha} - a$; $C_\alpha \leftarrow 2.r_{2\alpha}$; $D_\alpha \leftarrow C \cdot X_{\alpha,j} - X_{i,j}$
            $X_1 \leftarrow X_{\alpha,j} - A_\alpha \cdot (D_\alpha)$
            $r_{1\beta} \leftarrow Random\{0,1\}$; $r_{2\beta} \leftarrow Random\{0,1\}$
            $A_\beta \leftarrow 2a.r_{1\beta} - a$; $C_\beta \leftarrow 2.r_{2\beta}$; $D_\beta \leftarrow C_\beta \cdot X_{\beta,j} - X_{i,j}$
            $X_2 \leftarrow X_{\beta,j} - A_\beta \cdot (D_\beta)$
            $r_{1\delta} \leftarrow Random\{0,1\}$; $r_{2\delta} \leftarrow Random\{0,1\}$
            $A_\delta \leftarrow 2a.r_{1\delta} - a$; $C_\delta \leftarrow 2.r_{2\delta}$; $D_\delta \leftarrow C_\delta \cdot X_{\delta,j} - X_{i,j}$
            $X_3 \leftarrow X_{\delta,j} - A_\delta \cdot (D_\delta)$
            $X_{123} \leftarrow \frac{X_1+X_2+X_3}{3}$
            $\sigma \leftarrow Random\,[0,1]$
            if $\left|\frac{2}{\pi}\arctan\left(\frac{2}{\pi}X_i\right)\right| > \sigma$ **then**
                $X_{i,j} \leftarrow 1$
            else
                $X_{i,j} \leftarrow 0$
            end if
        end for
    end for
end for
$X_\alpha : \{binary\ vector\ of\ selected\ bands\}$
$global\_f \leftarrow f_\alpha$

2.3 Fitness Function

The fitness function proposed in this study is composed of two important terms: the first one is the classification accuracy rate calculated by using a classifier system. We use the support vector machine to compute the first term fitness function. We note that the BGWO can use any classifier system.

The second term of the fitness function is the class discrimination measure based on SNR (Signal to Noise Ration).

Classification Accuracy Criterion We use the equation defined in [2] for support vector machine. This equation proposes that the classification accuracy rate is measuring according to:

$$\begin{cases} J_1(X) = \dfrac{\sum_{i=1}^{|Pix|} assess(p_x)}{|Pix|}, p_x \in Pix \\ assess(p_x) = \begin{cases} 1 \; if \; classify(p_x) = c \\ \quad 0 \; otherwise \end{cases} \end{cases} \tag{12}$$

where $J_1(X)$ computes the classification accuracy rate using the selected spectral bands X_i (with $vi = 1$). $assess(p_x)$ is equal to 1 if the pixel is correctly classified else is equal to 0.

X is the set of bands with $X = \{X_1, ..., X_N\}$, N is the total number of bands. Pix is the set of instances which represents the set of pixels. $|Pix|$ represents the total number of pixels and p_x is a pixel.

Class Discrimination Criterion The second term of the fitness function is based on SNR (Signal to Noise Ration) which is a score used to determine the discrimination power of a variable. The SNR formula for two classes is given by:

$$SNR(c_1, c_2) = \frac{2 \times \left|\mu^x_{c_1} - \mu^x_{c_2}\right|}{(\sigma^x_{c_1} - \sigma^x_{c_2})} \tag{13}$$

where $\mu^x_{c_1}$ and $\mu^x_{c_2}$ are the mean values for the instances of class 1 and class 2 respectively. $\sigma^x_{c_1}$ and $\sigma^x_{c_2}$ are the standard deviations for the instances of class 1 and class 2 respectively.

For multiclass problem, SNR can be written as follows:

$$S_{X_i} = \frac{1}{n \times (n-1)} \sum_{i=1}^{n-1} \sum_{j=i+1}^{n} SNR(c_i, c_j) \tag{14}$$

The function of SNR is given by:

$$J_2(X) = \frac{\sum_{i=1}^{N} v_i \times S_{X_i}}{\sum_{i=1}^{N} S_{X_i}} \tag{15}$$

The fitness function is defined as follows:

$$J(X) = \alpha \times J_1(X) + \beta \times J_2(X) \tag{16}$$

where α and β are the weight coefficients allowing to balance between the terms or to give importance to the term regarding to the second.

3 Experimental Results

3.1 Hyperspectral Datasets

In this section, we present the hyperspectral dataset used to evaluate the proposed approach. We propose to conduct the experimentation on three hyperspectral images widely used to demonstrate the performance of the band selection methods in hyperspectral image classification.

Pavia University Firstly, we conduct the experimentation on Pavia University hyperspectral image which was taken by the ROSIS over urban area of Pavia University. The size of the image is 610×340 pixels for 103 spectral bands taken under the wavelength from 0.4 to 0.86 µm. This image is composed of 9 classes, namely: Asphalt, Meadows, Gravel, Trees, Painted Metal Sheets, Bare Soil, Bitumen, Self-Bloking Bricks, and Shadows. Figure 1 illustrates the Pavia University hyperspectral image.

Indian Pines Secondly, the experimentation is conducted on the Indian Pines hyperspectral image. This scene represents a segment of AVIRIS (Airborne Visible InfraRed Imaging Spectrometer) which is acquired over the agricultural area of Northwestern Indiana, USA. The size of this image is 145×145 pixels and it composed of 220 bands. The wavelength is from 0.5 to 2.5µm. This image is composed of 16 classes, namely: Alfalfa, Corn-notill, Corn-mintill, Corn, Grass-pasture, Grass-trees, Grass-pasture-mowed, Hay-windrowed, Oats, Soybean-notill, Soybean-mintill, Soybean-clean, Wheat, Woods, Buildings-Grass-Trees-Drives, and Stone-Steel-Towers. Figure 2 illustrates the Indian Pines hyperspectral image.

Salinas The final experimentation is conducted on Salinas hyperspectral image. The scene is acquired from the Salinas valley in Southern California, USA and is composed of 224 bands by using AVIRIS. Each band is 512×217 pixels in the spectral range from 0.4 to 2.5 µm. This image is composed of 16 classes, namely: Broccoli-green-weeds-1, Broccoli-green-weeds-2, Fallow, Fallow-rough-plow, Fallow-smooth, Stubble, Celery, Grapes-untrained, Soil-vinyard-develop, Corn-senesced-green-weeds, Lettuce-romaine-4wk, Lettuce-romaine-5wk, Lettuce-romaine-6wk, Lettuce-romaine-7wk, Vineyard-untrained and Vineyard-vertical-trellis. Figure 3 illustrates the Salinas hyperspectral image.

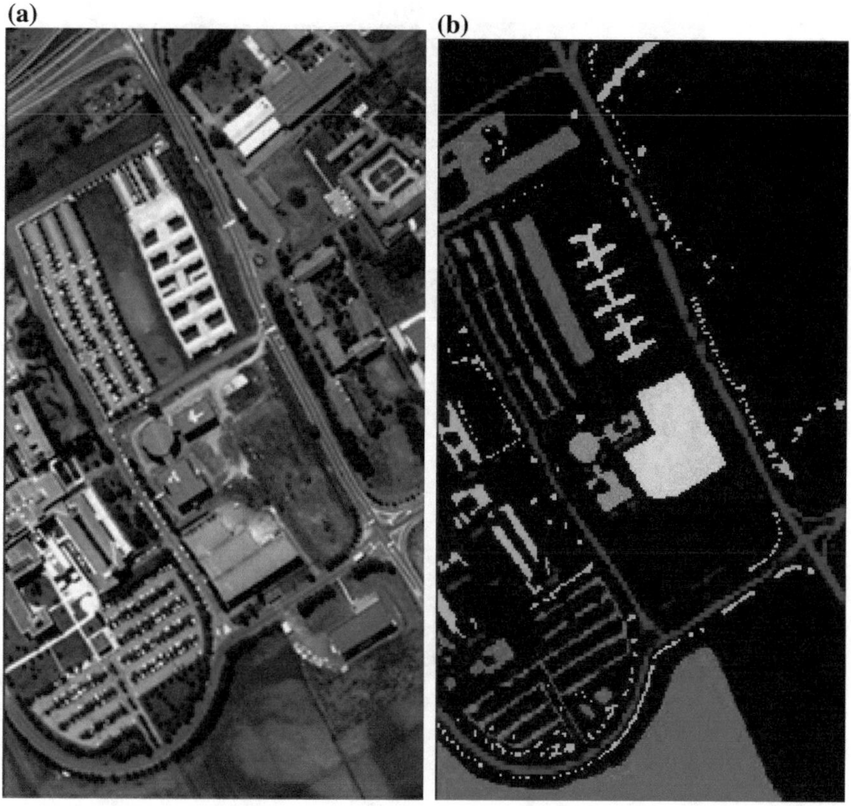

Fig. 1. Pavia University. (a) RGB image. (b) Ground truth

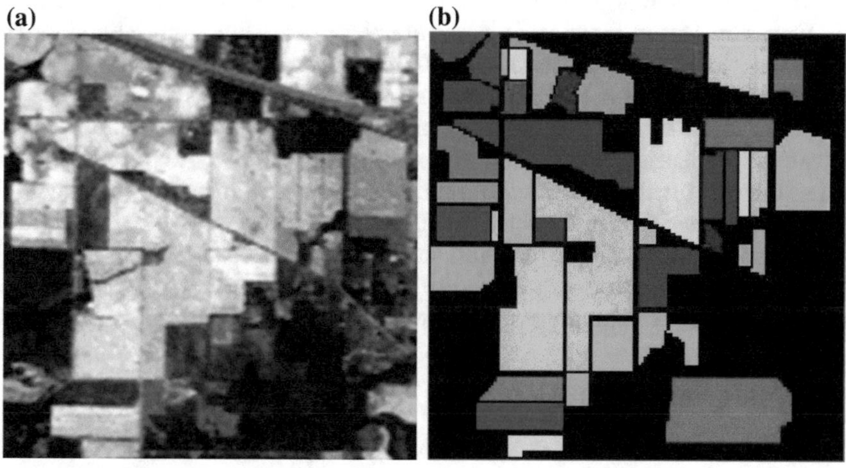

Fig. 2. Indian Pine. (a) RGB image. (b) Ground truth

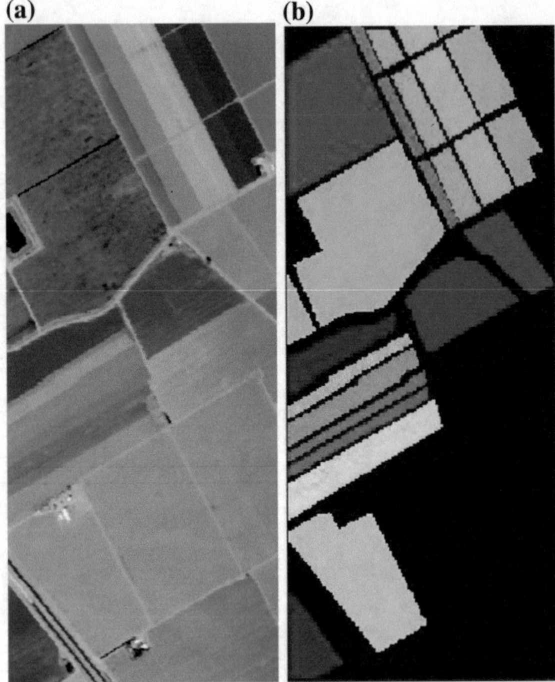

Fig. 3. Salinas. (a) RGB image. (b) Ground truth

3.2 Parameters Setting

The proposed approach for spectral band selection is set as follows: the number of gray wolf population is set to 30. We use the Support Vector Machine (C-SVM) with Gaussian kernel ($\sigma = 2$) to compute the fitness function. The algorithm stops when the number of iterations (500 iterations) is reached. For the fitness function, the values of α and β are equal to 1.

In each classifier system, the number of instances used for training, testing and validation phases must be determined. In this work, we propose to split the total number of pixels into three subsets: 10% of the instances are considered for the training phase. 30% of instances are used for test phase (used to calculate the fitness function in the algorithm) and the remaining 60% are considered for the validation. The validation step is used to measure the classification accuracy rate under the final selected bands which is considered as the optimal band subset.

3.3 Results and Discussion

The experimental and analysis of the results are presented in this section. The experimentations are conducted in term of Overall Accuracy (OA), Average Accuracy(AA) and Individual Class Accuracy (ICA). The proposed approach is a wrapper approch, to evaluate the performance, we compare it with four wrapper approaches: genetic algorithm (GA), binary practical swarm optimization (BPSO), binary gravitational search algorithm (BGSA), and a binary bat algorithm (BBA).

Tables 1, 2, and 3 present the overall accuracy, average accuracy and individual class accuracy obtained by the proposed approach and compared with other approaches reported from [2,22].

Table 1. OA, AA and ICA obtained by the proposed approach and four wrapper approaches for Pavia University

Class	Wrapper approaches				This study
	GA	BPSO	BGSA	BBA	
Asphalt	82,28	89,21	90,44	89,77	90,50
Meadows	96,91	90,98	91,64	92,62	95,79
Gravel	43,75	72,85	77,38	73,01	72,94
Trees	79,73	88,52	87,87	85,42	89,72
Painted metal sheets	99,16	99,75	94,25	96,75	98,88
Bare soil	58,25	70,45	70,70	71,00	74,06
Bitumen	75,47	80,59	84,96	83,20	86,09
Self-Bloking Bricks	83,47	81,70	80,02	80,35	84,71
Shadows	100	98,01	99,82	97,82	100,00
AA	79,89	85,78	86,12	85,77	88,08
OA	84,57	87,38	87,97	87,28	89,79

Tables 1, 2, and 3 show the classification accuracy rate obtained by the proposed approach and four band selection methods based on wrapper model and reported from [2,22].

The first column of tables represents the name of classes for each hyperspectral image. The four columns represent the wrapper approach used for comparison which are genetic algorithm, practical swarm optimization, binary gravitational search algorithm, and a binary bat algorithm. The last column represents the results obtained by the proposed approach.

The rows contain the individual class accuracy and the two last columns contain the average accuracy and the overall accuracy respectively.

From Tables 1, 2, and 3 and for each hyperspectral image, we can clearly observe that the best classification accuracy rate is obtained by the proposed approach with $88,08\%$ of average accuracy and $89,79\%$ of overall accuracy for Pavia University hyperspectral image.

For Indian pines hyperspectral image, the average accuracy reached $76,58\%$ and $78,34\%$ of overall accuracy. For the Salinas hyperspectral image, the aver-

Table 2. OA, AA and ICA obtained by the proposed approach and four wrapper approaches for Indian Pine

Class	Wrapper approaches				This study
	GA	BPSO	BGSA	BBA	
Alfalfa	0,00	60,71	50,57	60,71	82,14
Corn-on till	52,93	60,91	69,89	61,61	69,31
Corn-min till	28,16	64,65	69,67	56,42	68,07
Corn	14,74	46,15	46,85	46,15	51,05
Grass/pasture	82,17	88,62	88,96	89,31	90,34
Grass/tree	94,35	93,15	95,89	94,97	95,89
Grass/pasture-mowed	17,39	70,23	88,23	52,94	82,35
Hay-windrowed	96,87	98,60	97,90	95,47	96,17
Oats	0,00	50,00	50,00	50,00	41,67
Soybeans-no till	59,38	70,68	78,93	66,95	78,08
Soybeans-min till	74,13	77,66	82,41	75,49	79,77
Soybeans-clean till	32,00	59,26	58,98	58,42	58,71
Wheat	94,51	90,74	91,86	95,12	95,93
Woods	92,98	91,17	42,38	92,35	89,06
Bldg-grass-tree-drives	11,65	48,70	46,55	48,70	55,60
Stone-steel towers	73,33	92,85	89,28	87,50	91,07
AA	51,54	72,75	71,77	70,76	76,58
OA	64,86	74,55	74,03	73,89	78,34

age accuracy produced by the proposed approach 96, 36 and 92, 15% for overall accuracy.

The analysis of the results demonstrates the effectiveness of the proposed approach compared to other wrapper approach.

The classification map plays an important role in performance evaluation of band selection approach. Figure 4 illustrates the classification map obtained by the proposed.

Figure 4 shows the classification map obtained by the proposed approach for each hyperspectral image. The classification maps are very satisfactory and similar to the ground truth and we remark that the results are encouraged. Note that all these results are provided by using a small set of training, we have considered 10% of pixels which are used to train the model.

Table 3. OA, AA and ICA obtained by the proposed approach and four wrapper approaches for Salinas

Class	Wrapper approaches				This study
	GA	BPSO	BGSA	BBA	
Brocoli_green_weeds_1	98,26	99,17	99,75	98,67	99,50
Brocoli_green_weeds_2	99,60	99,55	99,77	99,41	99,96
Fallow	98,04	99,40	99,57	98,73	99,33
Fallow_rough_plow	99,73	99,40	99,28	99,40	99,04
Fallow_smooth	95,01	98,50	98,13	98,31	98,57
Stubble	99,94	99,91	99,87	99,95	99,96
Celery	99,37	99,72	99,67	99,58	99,63
Grapes_untrained	79,34	81,70	81,42	79,58	82,46
Soil_vinyard_develop	99,38	99,48	99,57	99,62	99,68
Corn_green_weeds	91,61	96,49	95,57	95,17	96,75
Lettuce_romaine_4wk	92,05	99,84	98,43	98,90	99,38
Lettuce_romaine_5wk	98,83	99,56	99,91	100	100,00
Lettuce_romaine_6wk	97,82	98,72	99,45	98,54	99,27
Lettuce_romaine_7wk	93,57	97,97	96,41	97,97	96,42
Vinyard_untrained	57,88	70,90	72,85	67,75	72,62
Vinyard_vertical_trellis	97,86	99,07	99,63	98,15	99,17
AA	93,64	96,21	96,20	95,61	96,36
OA	88,57	91,70	91,86	90,67	92,15

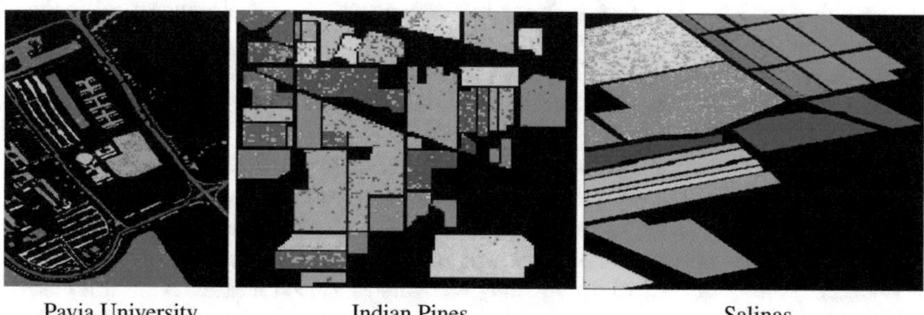

Fig. 4. The classification maps obtained by the proposed approach for each hyperspectral image

4 Conclusions and Perspectives

In this paper, a band selection approach based on a new binary version of Gray wolf optimizer BGWO is proposed. The Binarization of GWO is based on the transfer function. The spectral band selection problem is formulated as a binary optimization problem and we use the BGWO to solve this problem. The fitness function is designed based on the classification accuracy rate obtained by using support vector machine classifier and SNR measure. The proposed approach has been tested over three most widely used hyperspectral images. The proposed approach is also compared with four wrapper band selection approach. The analysis of the results obtained by our approach has proven that the BGWO is able to decrease irrelevant bands during the bands selection process and improves significantly the classification accuracy rate. We have also shown that the accuracy rate provided by the proposed approach is way higher than or exceeds those of other approaches. We have used a small number of training instances and the proposed approach has proven stable as the accuracy rate remained almost high. As futur work, is very interesting to test the proposed approach on different type of datasets such as gene selection and cancer diagnosis, also, the proposed fitness function can be improved by adding other term such as the correlation between the features.

References

1. Medjahed, S.A., Ouali, M., Saadi, T.A., Benyettou, A.: An optimization-based framework for feature selection and parameters determination of SVMs. Int. J. Inf. Technol. Comput. Sci. (IJITCS) **7**(5), 1–9 (2015)
2. Medjahed, S.A., Saadi, T.A., Benyettou, A., Ouali, M.: Gray wolf optimizer for hyperspectral band selection. Appl. Soft Comput. **40**, 178–186 (2016)
3. Medjahed, S.A., Ouali, M.: Band selection based on optimization approach for hyperspectral image classification. Egypt. J. Remote. Sens. Space Sci. (2018)
4. Liu, H., Yang, S., Gou, S., Liu, S., Jiao, L.: Terrain classification based on spatial multi-attribute graph using polarimetric SAR data. Appl. Soft Comput. **68**, 24–38 (2018)
5. Boedihardjo, A.P., Lu, C.T., Chen, F.: Fast adaptive kernel density estimator for data streams. Knowl. Inf. Syst. **42**(2) (2015)
6. Zhang, Q., Tian, Y., Yang, Y., Pan, C.: Automatic spatialspectral feature selection for hyperspectral image via discriminative sparse multimodal learning. IEEE Trans. Geosci. Remote Sens. **53**(1) (2015)
7. Wang, S., Pedrycz, W., Zhu, Q., Zhu, W.: Subspace learning for unsupervised feature selection via matrix factorization. Pattern Recognit. **48**(1) (2015)
8. Zhu, P., Zuo, W., Zhang, L., Hu, Q., Shiu, S.C.: Unsupervised feature selection by regularized self-representation. Pattern Recognit. **48**(2) (2015)
9. Wang, S., Pedrycz, W., Zhu, Q., Zhu, W.: Unsupervised feature selection via maximum projection and minimum redundancy. Knowl. Based Syst. **75**(1) (2015)
10. Lee, J., Kim, D.W.: Mutual information-based multi-label feature selection using interaction information. Expert. Syst. Appl. **42**(4) (2015)
11. Lewis, D.D.: Feature selection and feature extraction for text categorization. In: Proceedings of Speech and Natural Language Workshop, pp. 212–217. Morgan Kaufmann (1992)
12. Battiti, R.: Using mutual information for selecting features in supervised neural net learning. IEEE Trans. Neural Netw. **5**(4), 537–550 (1994)
13. Yang, H., Moody, J.: Feature selection based on joint mutual information. In: Proceedings of International ICSC Symposium on Advances in Intelligent Data Analysis, pp. 22–25 (1999)

14. Peng, H., Long, F., Ding, C.: Feature selection based on mutual information: criteria of max-dependency, max-relevance, and min-redundancy. IEEE Trans. Pattern Anal. Mach. Intell. 1226–1238 (2005)
15. Brown, G., Pocock, A., Zhao, M.J., Luj, M.: Conditional likelihood maximisation: a unifying framework for information theoretic feature selection. J. Mach. Learn. Res. **13**, 27–66 (2012)
16. Nakamura, R.Y.M., Pereira, L.A.M., Costa, K.A., Rodrigues, D., Papa, J.P.: BBA: A binary bat algorithm for feature selection. Images (SIBGRAPI) (2012)
17. Behjat, A.R., Mustapha, A., Nezamabadi–pour, H., Sulaiman, M.N., Mustapha, N.: Feature subset selection using binary gravitational search algorithm for intrusion detection system. Intell. Inf. Database Syst. **7803** (2013)
18. Lin, S.W., Lee, Z.J., Chen, S.C., Tseng, T.Y.,.: Parameter determination of support vector machine and feature selection using simulated annealing approach. Appl. Soft Comput. **8**(4) (2008)
19. Zabidi, A., Khuan, L.Y., Mansor, W., Yassin, I.M., Sahak, R.: Binary particle swarm optimization for feature selection in detection of infants with hypothyroidism. In: International Conference of the IEEE Engineering in Medecine and Biology Socienty (EMBC) (2011)
20. Mirjalili, S., Mirjalili, S.M.: Lewis, A. Grey wolf optimizer. Adv. Eng. Softw. **69**(1) (2014)
21. Crawford, B., Soto, R., Astorga, G., Garcia, J., Castro, C., Paredes, F.: Putting continuous metaheuristics to work in binary search spaces. Complexity **17**, 1–19 (2017)
22. Medjahed, S.A., Saadi, T.A., Benyettou, A., Ouali, M.: Binary Cuckoo search algorithm for band selection in hyperspectral image classification. IAENG Int. J. Comput. Sci. **42**(3), 183–191 (2015)

Metaheuristics and Optimization

New Solutions for the Density Classification Task in One Dimensional Cellular Automata

Zakaria Laboudi[1(✉)] and Salim Chikhi[2]

[1] RELA(CS)[2] Laboratory, University of Oum El-Bouaghi, Route de Constantine BP 321, 04000 Oum El-Bouaghi, DZ, Algeria
laboudi.zakaria@univ-oeb.dz, laboudizak@gmail.com

[2] MISC Laboratory, University of Constantine 2, Nouvelle Ville Ali Mendjeli BP 67A, 25100 Constantine, DZ, Algeria
salim.chikhi@univ-constantine2.dz

Abstract. The density classification task is one of the most studied benchmark problems to analyze emergent collective computations resulting from local interactions within cellular automata. Solutions for this task were produced by means of different training methods, in particular the automatic design through evolutionary algorithms. This is tied to the fact that there is still a lack of thorough understanding of computations' nature within cellular automata, which impedes writing efficient local rules. In this paper, we propose a new procedure for solving the density classification task using handwritten local rules in the case of one dimensional cellular automata of radius $r = 4$. The experimental results show that the newly designed rules outperform the currently best known solutions. This is important since it helps, on the one hand, to deepen our knowledge about selecting appropriate local rules to solve computational tasks and, to improve our general understanding of computations carried out by cellular automata, on the other hand.

Keywords: Emergent computation · Cellular automata
Density classification task · Symmetry property · Number-conserving property

1 Introduction

Cellular automata (CAs) are artificial systems that serve to study and analyze the behavior of complex systems, where decentralized computations emerge only from local communication between cells [1]. In this context, several benchmark problems were adopted such as *the firing squad synchronization problem* [2], *the parity problem* [3], *the synchronization problem* [4] and *the density classification task* (DCT for short) [4–11]. Commonly, the effectiveness of the computation carried out is granted by reaching predefined global configurations, after sufficiently long time; leading then emergent computations to appear through information transmission across time and space. This is a non-trivial task for CAs since the existence of any given global property cannot be locally detected, unlike other centralized systems.

However, in spite of their computational abilities, writing local rules for CAs to produce a specific global behavior or to perform a given computation is - in many cases

S. Chikhi et al. (Eds.): MISC 2018, LNNS 64, pp. 93–105, 2019.
https://doi.org/10.1007/978-3-030-05481-6_7

- regarded as a hard task. This comes from the fact that there is still a lack of thorough understanding of the mechanisms leading to suitable emergent computations for carrying out computational tasks. Accordingly, recent researches have been turned out to the automatic methods and tools that search for such rules, often in a blindly way.

In this paper, we deal with the DCT in one dimensional CAs following its standard definition. Solutions for this task were designed using both manual and automatic training mechanisms [5, 10–14]. So, we try to enrich the list of existing solutions by designing new more efficient ones - as compared to the currently best known rules - using a human-engineering procedure. The primary aim is to make sure that we have reached a sufficient level of knowledge about the mechanisms by which local rules are selected so they lead to a better problem solving. Indeed, if we manage to develop efficient handwritten solutions for the DCT, this may open a perspective on our general understanding of computations within CAs and eventually improve our ability to solve other computational tasks.

The rest of this paper is organized as follows. We start with some basic concepts and properties around CAs in Sect. 2. In Sect. 3, we give an overview on the DCT and address some related works, where we make a review of some well-known solutions. After that, we provide, in Sect. 4, details on how new efficient solutions can be designed using CAs of radius $r = 4$. Also, we summarize the experimental tests, give a set of numerical results and, analyze and discuss the obtained results. The paper ends with some concluding remarks and perspectives.

2 Basic Concepts and Properties

In this section, we present some basic concepts and properties around CAs.

2.1 Cellular Automata

A cellular automaton consists of a lattice of N cells, where each cell can be in one of k possible states at a given time t. Configurations are formed by assigning states to cells of the lattice, and then evolved following some local transition rules, where each cell updates its state at time step $t + 1$ depending on its current state and the states of a number n of its neighboring cells at time t. Lattices are usually designed following boundary conditions, where the cells at one end are linked to the cells on the other end in order to simulate infinite configurations.

The number of states k and the size of the neighborhood n allow enumerating all local neighborhood configurations. Consequently, any local state transition rule Φ can be represented by a lookup table that lists for each possible local neighborhood the state update of the central cell, often following a lexicographical order [7]. Table 1 shows a typical lookup table of a one dimensional CA rule of radius $r = 1$ (i.e. number of neighbors at each side of any cell of the lattice), denoted by Φ_{examp}; with $k = 2$ and $n = 3$ in this example. These lookup tables are usually coded as strings and then turned into decimal numbers, according to their output' states. For instance, the binary and the decimal representations of Rule Φ_{examp} are 01000111 and 71, respectively.

Table 1. A typical lookup table for Rule Φ_{examp}.

Neighborhood configurations	000	001	010	011	100	101	110	111
Output' states	0	1	0	0	0	1	1	1

Let $w \in \{0, 1, \ldots, k-1\}$ denotes a possible state of a cell. We refer to State $w = 0$ as the quiescent state, and any other state as an active state. We call *generation* (*annihilation*, respectively) any configuration of the neighborhood for which the central cell will be *activated* (*deactivated*, respectively) in the next time step. Also, we call *preservation* any configuration of the neighborhood for which the central cell *will not change its state* in the next time step [15].

The behavioral analysis of any given CA rule Φ is performed through space-time diagrams, where we consecutively accumulate the evolution of each initial configuration (IC) for t time steps in order to graphically display the trace. An example is illustrated in Fig. 1, where cells are plotted with white and black colors to refer to States $w = 0$ and $w = 1$, respectively.

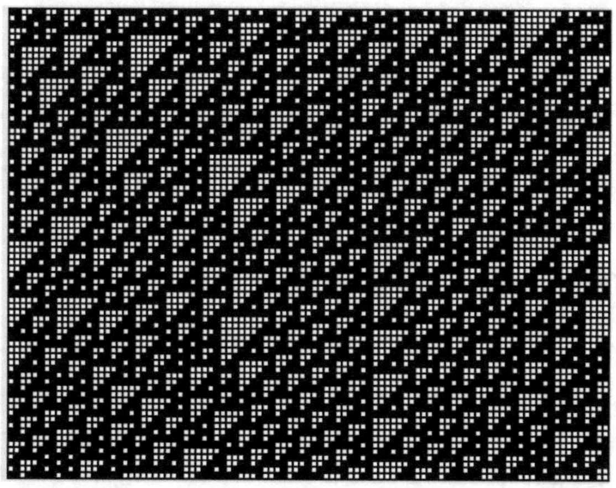

Fig. 1. A space-time diagram of a random IC evolved by the elementary rule Φ_{110}.

2.2 The Symmetry Property

Starting from the lookup table of any given CA rule Φ, it is possible to derive other rules equivalent in behavior using some transformations applied on state permutation, directions and the combination of the two. For one-dimensional binary CAs, we distinguish three possible combinations: black-white swapping, left-right swapping and the combination of these two transformations. An example is given on Fig. 2, where we consider a local state transition with its symmetric transformations using CAs of radius $r = 1$.

Fig. 2. A typical local state transition with its symmetric transformations.

- **Black-white swapping:** it consists in altering the lookup table by switching the cells' states in a way that every black cell (cell in the state 1) becomes white (cells in the state 0) and vice versa.
- **Left-right swapping:** it consists in altering the lookup table by reversing each neighborhood configuration from left to right without changing the output' states.
- **The combination of these two transformations.**

We call *symmetric rule* any one-dimensional binary CA rule whose generations and annihilations have the combined transformations; Rule Φ_{examp} is a symmetric rule.

2.3 The Number-Conserving Property

For CAs, the number-conserving is a property for which the sum of cells' states in the lattice is maintained over time [16]. For binary CAs, the evolution of the lattice leads at any time step t to a pattern for which the number of cells in each state remains invariant, which makes the CA become state-conserving [17] (i.e. each state that was in the lattice at time t is still in the lattice at time $t + 1$, potentially in another cell); Rule Φ_{examp} is a state-conserving rule.

3 The Density Classification Task

3.1 Overview

The density classification task (DCT) is one of the most studied benchmark problems for analyzing emergent computations performed by CAs. For this task, it is requested to find a CA rule that can decide for each possible IC composed of N cells with two possible states $\{0, 1\}$, if there are more black cells (cells in the state 1) than white ones (cells in the state 0). When there are more black cells, the CA should converge towards homogenous fixed-point pattern where all the cells become black, the inverse otherwise. Mitchell et al. have proposed in [7] a rigorous definition for this task by introducing a parameter denoted ρ to refer to the density of black cells in each IC. A critical value ρc is then used with $\rho c = ½$ for this task.

The fitness of any DCT solution Φ is usually measured by its performance denoted by $P_N^M(\Phi)$. It consists on calculating the portion of M ICs of size N that are correctly classified under the dynamics specified by Φ. And as it is believed that the most difficult ICs are those where the density ρ is close to 0.5, test cases are often randomly generated following uniform Bernoulli distribution (UBD) with parameter $\rho = 0.5$ [5, 15]. The aim is to cover a wide range of ICs since it is not always possible to test the entire space of all ICs, in particular when N grows.

In practical terms, the DCT serves to study complex systems in a way that we try to understand the source of emergent computations resulting from local interactions, in order to deepen our knowledge about designing artificial fully decentralized systems. Also, it may be regarded as a decisional problem for many real-world applications such as solving the majority vote in distributed systems.

3.2 Related Works on the DCT

As it was demonstrated that, following the standard definition of the DCT, there is no CA rule that can perform this task for all configurations of any given size N [18], efforts were turned to search for rules that classify a wide range of ICs. Earlier imperfect solutions for this task were designed by different training mechanisms ranging from manual design to automatic methods; we cite for instance: Rules Φ_{GKL} [19], $\Phi_{Davis95}$ [20], Φ_{Das95} [20], Φ_{DMC} [21], Φ_{COE1} [22], Φ_{COE2} [22], Φ_{GP1995} [20], Φ_{GEP1} [23], Φ_{GEP2} [23], Φ_{MM401} [15], Φ_{MM0711} [12] and Φ_{MM0802} [24]. But, since it was suggested that solutions for this task should be symmetric [5, 13] and close to number-conserving rules [10], some thousands of unknown solutions could be discovered. Rules that we refer to as Φ_{B33} [25], Φ_{WdO1} [5], Φ_{LCL} [25] and Φ_{ALTER1} [10] are very good imperfect solutions. All these rules are of radius $r = 3$.

However, as the symmetry and the number-conserving properties were recommended by analyzing some of the existing solutions such as Rule Φ_{GKL} and by using some correlations, there was still a lack of thorough understanding of the real connection between density determination and these properties. Therefore, the authors in [14] managed to decode the computational mechanisms of some good solutions of radius $r = 3$ (such as Rules Φ_{WdO1} and Φ_{ALTER1}). In other words, they gave explanations about the way by which the symmetry and the number conserving properties are used to solve that task. These mechanisms were then applied on one dimensional CAs of radius $r = 4$. Consequently, some new more efficient solutions - as compared to the best known ones in the case of CAs of radius $r = 3$ - could be designed; we cite for instance Rule Φ_{M_ALTER1} and Rules $\Phi_{R4Bi\ =\ 1..6}$. Table 2 resumes the performances of some reference solutions.

Table 2. Performances of some solutions for the DCT [14].

Rule ID	$P_{499}^{100.000}$	Rule ID	$P_{499}^{100.000}$	Rule ID	$P_{499}^{100.000}$
Φ_{B33}	82.78%	Φ_{WdO1}	85.36%	Φ_{LCL}	85.70%
Φ_{ALTER1}	85.75%	Φ_{M_ALTER1}	87.28%	Φ_{R4B1}	88.71%
Φ_{R4B2}	87.85%	Φ_{R4B3}	87.81%	Φ_{R4B4}	87.84%
Φ_{R4B5}	87.69%	Φ_{R4B6}	87.76%	–	–

4 Designing New Solutions for the DCT Using CAs of Radius $r = 4$

Next, we show how new efficient solutions for the DCT are designed using CAs of radius $r = 4$. Thus, we give and discuss some performance measurements to make comparison with the currently best known solutions. Also, we make an objective evaluation of our proposal against existing methods for solving the DCT. Finally, we address implementation details and running time measurements.

4.1 Context and Motivation

In spite of their behavioral and computational abilities (different kinds of dynamics, universality…etc.), understanding the nature of computations within CAs is, however, still an elusive problem. This comes from the fact that very little is known about what is required for any given CA rule Φ to efficiently perform predefined computations [3, 16]. For instance, most of the earlier solutions of the DCT were designed using automatic methods such as evolutionary algorithms, where the number of human-engineered solutions is limited to very few rules with relatively low performances (Rules Φ_{GKL} [19], $\Phi_{Davis95}$ [20] and Φ_{Das95} [20]).

Consequently, if we manage to manually design new efficient solutions, this helps to deepen our general understanding of computations within CAs to solve the DCT and possibly other similar computational tasks. Moreover, the ability of manually designing new solutions for the DCT certainly allows more control of the strategies leading to good problem solving.

We note here that our choice to work on CAs of radius $r = 4$ seems reasonable since most of the DCT solutions which are of radius $r = 3$ are now known; turning to CAs of radius $r = 4$ can then be a good test for designing new unknown solutions.

4.2 Computational Mechanisms of Good Solutions of Radius r = 3

As reported in [14], the mechanism by which CAs efficiently solve the DCT is based on states' exchanges between the central cells of generations and annihilations for which the pattern connecting them is symmetric with respect to the middle, except for some local updating rules that generate converging-conditions - i.e. patterns that make ICs converge towards homogenous fixed-point patterns. Table 3 shows symmetrical exchanges in Rule Φ_{ALTER1} - which is of radius $r = 3$ - depending on the number of adjacent repetitions of symmetric patterns.

Note that the generations and the annihilations processes are gathered in segments since they have some parts in common; for more details about this representation of rules, the reader should refer to Reference [14]. The remaining exchanges in Rule Φ_{ALTER1} do not preserve the symmetry of patterns, as illustrated in Table 4.

Table 3. Symmetrical cells' exchanges in Rule Φ_{ALTER1}.

Symmetric pattern	Number of repetitions	Cells' exchanges
$(1 \mid 0)$	One repetition	$\#, 0, 0, [\underline{\mathbf{1}} \mid \underline{\mathbf{0}}], 0, 1, \#$ $\#, 0, 1, [\underline{\mathbf{1}} \mid \underline{\mathbf{0}}], 1, 1, \#$ $\#, [0, 0, \underline{\mathbf{1}} \mid \underline{\mathbf{0}}, 1, 1], \#$
	Two repetitions	$\#, [1, 1, \underline{\mathbf{0}} \mid \underline{\mathbf{1}}, 0, 0], \#$
	More than two repetitions	$\#, 0, 0, [\underline{1}, 0, (1 \mid 0)^*, 1, \underline{\mathbf{0}}], 0, \#$ $\#, \#, 1, [\underline{1}, 0, (1 \mid 0)^*, 1, \underline{\mathbf{0}}], 0, \#$ $\#, 0, 0, [\underline{1}, 0, (1 \mid 0)^*, 1, \underline{\mathbf{0}}], 1, 1, \#$ $\#, \#, 1, [\underline{1}, 0, (1 \mid 0)^*, 1, \underline{\mathbf{0}}], 1, 1, \#$
$\mathbf{(1, 1 \mid 0, 0)}$	More than one repetition	$\#, [0, 1, \underline{\mathbf{1}} \mid \underline{\mathbf{0}}, 0, 1], \#$
$\mathbf{(1, 1, 1 \mid 0, 0, 0)}$	One or more repetitions	$\#, \#, [1, \underline{1}, 1 \mid 0, \underline{0}, 0], \#, \#$

The underlined states represent cells' updating rules (i.e. generations and annihilations). The symbols '[' and ']' represent delimiters that outline the symmetric patterns. The symbol # refers to a cell state that can be either 0 or 1.

Table 4. Asymmetrical cells' exchanges in Rule Φ_{ALTER1}.

Asymmetrical cells' exchanges producing converging conditions		
$\#, \#, 1, [\underline{1}, 1 \mid \underline{\mathbf{0}}, 0], 1, \#$	$\#, 0, 0, \underline{\mathbf{0}} \mid \underline{1}, 1, 0, \#$	$0, 0, 0, \underline{\mathbf{1}} \mid 0, 0, 0, \#$
$\#, 0, [1, \underline{\mathbf{1}} \mid 0, \underline{\mathbf{0}}], 0, \#, \#$	$\#, 1, 0, \underline{\mathbf{0}} \mid \underline{1}, 1, 1, \#$	$\#, 1, 1, 1 \mid \underline{0}, 1, 1, 1$

By using cells' exchanges preserving the symmetry of patterns, ICs, in particular congested ones where density $\rho \approx 0.5$, undergo a series of transformations before producing converging-conditions (i.e. Generation g = 1, 1, 1, $\underline{\mathbf{0}}$, 1, 1, 1 and Annihilation a = 0, 0, 0, $\underline{1}$, 0, 0, 0, which represent isolated cells). Only when such patterns appear that computations are carried out since the CA considers them as a criterion for density determination. The more the CA generates these patterns slowly, the more it efficiently solves the DCT on large configurations, because, it provides enough time for cells' self-organization before starting the actual computations.

4.3 Designing New Efficient Solutions

In a similar way, we design new solutions for the DCT in the case of CAs of radius $r = 4$, where we write their local rules by hand. Thus, we firstly enumerate all symmetrical cells' exchanges in neighborhood configurations - which are now more than in CAs of radius $r = 3$ - and then put them in a set denoted by S. We also construct a set of asymmetrical cell's updating local rules - as for the ones given in Table 4 - in order to converge towards homogenous fixed-point patterns; we refer to that set as A. Now, we need to select a subset $B \subseteq S$ and merge its elements with the ones of Set A to get a potential solution for the DCT. As a result, three new unknown solutions could be designed after many attempts, which we refer to as Φ_{R4BL1}, Φ_{R4BL2} and Φ_{R4BL3}; Table 5 shows their decimal representation.

Table 5. Decimal representation of Rules $\Phi_{R4BLi\ =\ 1..3}$.

Rule ID	Decimal representation
Φ_{R4BL1}	13240764716511451922804012119534133644618392230736020 23668936036310130973153641847379734856284181992302845 4408023206811309762336731194865667155623139115072
Φ_{R4BL2}	13240763118176765770247614894874500153556893905772158 25312079107522507792709985719779414724836690271038061 8836969014108639879765279920274096453109097201728
Φ_{R4BL3}	13240763118176765770247614894874500153556893905772158 25312079107522507792330558201766577015672732867361593 3472433063251116169762834974161325155680276971584

In Table 6, we give both symmetrical and asymmetrical exchanges between the central cells of the generations and the annihilations of some number-conserving CA rules $NCCA_{i\ =\ 1..20}$ in addition to converging-conditions, such that when these rules are joined, we get the processes of Rule Φ_{R4BL1}. We note that the segments where the delimiters symbols '[' and ']' are unfound represent asymmetrical exchanges between cells.

Table 6. Cells' exchanges in Rule Φ_{R4BL1}.

NCCA$_{1}$
#, [0, 0, 0, **1** | **0**, 1, 1, 1], #

NCCA$_{2}$
#, [1, 0, 0, **1** | **0**, 1, 1, 0], #

NCCA$_{3}$
#, [0, 0, 1, **1** | **0**, 0, 1, 1], #

NCCA$_{4}$
#, [1, 0, 1, **1** | **0**, 0, 1, 0], #

NCCA$_{5}$
#, 0, [0, 1, **1** | **0**, 0, 1], 0, #
#, 1, [0, 1, **1** | **0**, 0, 1], 1, #

NCCA$_{6}$
#, [0, 1, 1, **0** | **1**, 0, 0, 1], #

NCCA$_{7}$
#, 0, [1, 1, **0** | **1**, 0, 0], 0, #
#, 1, [1, 1, **0** | **1**, 0, 0], 1, #

NCCA$_{8}$
#, [0, 1, 1, **0** | **1**, 0, 1, 1], #
#, [0, 0, 1, **0** | **1**, 0, 0, 1], #

NCCA$_{9}$
#, [1, 0, 0, **0** | **1**, 1, 1, 0], #

NCCA$_{10}$
#, [0, 0, 0, **0** | **1**, 1, 1, 1], #

NCCA$_{11}$
#, 0, 0, 0, [**1** | **0**], 0, 1, 0, #
#, 1, 0, 1, [**1** | **0**], 1, 1, 1, #

NCCA$_{12}$
#, 0, 0, 0, [**1** | **0**], 0, 1, 1, #
#, 0, 0, 1, [**1** | **0**], 1, 1, 1, #

NCCA$_{13}$
#, 0, 0, 0, [**1** | **0**], 0, 0, 1, #
#, 0, 1, 1, [**1** | **0**], 1, 1, 1, #

NCCA$_{14}$
#, 1, 0, [1, **1** | **0**, 0], 0, 1, #
#, 0, 1, [1, **1** | **0**, 0], 1, 0, #

NCCA$_{15}$
#, 0, 0, 0, **0**, 0, **1**, 1, 0, #, #
1, 0, 0, 0, **0**, **1**, 1, 0, #, #
#, #, 1, 0, **0**, 1, **1**, 1, 1, 1, #
#, #, 1, 0, **0**, **1**, 1, 1, 1, 0

NCCA$_{16}$
#, 1, 0, 0, **0**, **1**, 1, 0, 0, #
#, 1, 1, 0, **0**, **1**, 1, 1, 0, #

NCCA$_{17}$
#, 1, 0, 0, **0**, **1**, 1, 0, 1, #
#, 0, 1, 0, **0**, **1**, 1, 1, 0, #

Converging-conditions
0, 0, 0, 0, **1**, 0, 0, 0, 0, #
#, 1, 1, 1, 1, **0**, 1, 1, 1, 1

NCCA$_{18}$
#, #, #, [1, **1**, 1 | 0, **0**, 0], #, #, #

NCCA$_{19}$
#, #, 0, 0, [**1**, 0, (1 | 0)*, 1, **0**], 0, #, #, #
#, #, 0, 0, [**1**, 0, (1 | 0)*, 1, **0**], 1, 1, #, #
#, #, #, 1, [**1**, 0, (1 | 0)*, 1, **0**], 0, #, #, #
#, #, #, 1, [**1**, 0, (1 | 0)*, 1, **0**], 1, 1, #, #

NCCA$_{20}$
0, 0, 0, 1, **1**, 0, **0**, 0, #, #, #
#, 1, 0, 0, [**1**, 1 | 0, **0**], 0, #, #, #
#, #, 1, 0, [**1**, 1 | 0, **0**], 0, #, #, #
#, #, #, 1, [**1**, 1, **0**, 0], 1, 1, 1
#, #, #, 1, [**1**, 1, 0, **0**], 1, 1, 0, #
#, #, #, 1, [**1**, 1, 0, **0**], 1, 0, #, #

4.4 Numerical Results and Discussion

Now, we compare the performances of Rules $\Phi_{R4BLi\,=\,1..3}$ and Rule Φ_{R4B1} [14] using ICs of different sizes ranging from $N = 149$ to $N = 699$. We mention here that we could experimentally check that Rules $\Phi_{R4BLi=1..3}$ outperform Rules $\Phi_{R4Bi\,=\,2..6}$ [14], Φ_{M_ALTER1} [14], Φ_{B33} [25], Φ_{WdO1} [5], Φ_{LCL} [25] and Φ_{ALTER1} [10]. Likewise, the experimental results presented in [14] shows that Rule Φ_{R4B1} performs better (see Table 1). Consequently, we need only to make tests on Rules $\Phi_{R4BLi\,=\,1..3}$ and Rule Φ_{R4B1}. For each test case, these rules are tested on a set of 100.000 ICs, generated following UBD; Table 7 resumes the obtained results.

Table 7. Performances of Rules Φ_{R4BL1}, Φ_{R4BL2}, Φ_{R4BL3} and Φ_{R4B1}.

ICs size	Classification type	Φ_{R4BL1}	Φ_{R4BL2}	Φ_{R4BL3}	Φ_{R4B1}	Total
$N = 149$	CC	92.220	92.148	91.949	91.251	100.000
	MC	7.776	7.805	7.953	7.175	
	NC	4	47	98	1.574	
$N = 199$	CC	91.307	90.961	90.817	90.429	100.000
	MC	8.690	8.993	9.091	8.112	
	NC	3	46	92	1.459	
$N = 249$	CC	90.380	90.240	89.881	89.547	100.000
	MC	9.610	9.702	10.004	8.863	
	NC	10	58	115	1590	
$N = 299$	CC	90.139	89.626	89.369	89.185	100.000
	MC	9.850	10.309	10.496	9.674	
	NC	11	65	135	1.141	
$N = 349$	CC	89.577	89.062	88.722	88.918	100.000
	MC	10.421	10.851	11.106	10.224	
	NC	2	87	172	858	
$N = 399$	CC	89.289	88.809	88.671	88.991	100.000
	MC	10.704	11.090	11.161	10.407	
	NC	7	101	168	602	
$N = 449$	CC	89.433	88.649	88.331	88.881	100.000
	MC	10.557	11.220	11.500	10.668	
	NC	10	131	169	451	
$N = 499$	CC	89.264	88.306	87.769	88.571	100.000
	MC	10.733	11.571	12.017	11.133	
	NC	3	123	214	296	
$N = 599$	CC	88.888	87.776	87.571	88.429	100.000
	MC	11.097	12.067	12.017	11.411	
	NC	15	157	246	160	
$N = 699$	CC	88.404	87.555	87.189	87.922	100.000
	MC	11.584	12.274	12.518	12.003	
	NC	12	171	293	75	

CC, MC and NC mean number of ICs that were correctly classified, misclassified (i.e. ICs that converged towards a wrong homogenous fixed-point pattern) and not classified (i.e. ICs that did not converge towards homogenous fixed-point patterns), respectively.

- **Discussion**

According to Table 7, we observe that Rule Φ_{R4BL1} is more stable toward ICs size variations, not only because it outperforms the other rules in all tests' instances, but also it records the lowest rate for the number of ICs that did not converge towards homogenous fixed-point patterns. For ICs of size $N < 399$, Rules $\Phi_{R4BLi\,=\,2..3}$

outperform Rule Φ_{R4B1}; otherwise, this latter records better performances. It seems that Rule Φ_{R4B1} requires more space in order to converge towards fixed-point patterns, and thus it has not the ability to classify small-sized configurations.

Indeed, it is noteworthy to observe the variations of Number NC according to ICs size N. That kind of ICs leads the CA to fall in a dead-end where homogenous fixed-point patterns are never reached. For the case of Rules $\Phi_{R4BLi\ =\ 2..3}$, the experimental results show us that Number NC increases in accordance with the growth of ICs size N. This is explained by the existence of some patterns such that when they appear in evolved configurations, the resulting configurations are partially converging towards homogenous fixed-point patterns. Consequently, the more ICs size N grows the more the frequency of appearance of such patterns increases, which directly influences the performances. In contrast, Rule Φ_{R4B1} requires more space to converge towards homogenous fixed-point patterns since Number NC decreases when N grows. For small-sized ICs, the behavior of the CA would be close to a number-conserving rule rather than the one whose dynamics lead to homogenous fixed-point patterns.

4.5 Evaluation of the Proposed Procedure

Now, we need to analytically assess the potential advantages and limitations of our proposal against other approaches and techniques.

4.5.1 The Advantages

The main advantages of our procedure to design efficient DCT solutions can be resumed as follows:

- **The efficiency:** Instead of exploring a whole space of rules (see for instance [5, 7, 8, 10, 14]), our procedure is limited to a reduced search space where we deal with few tens of rules. This makes the design process simpler and more effective since it allows locating the region from rules' space with high precision.
- **The scalability:** In fact, there is a strong correlation between the ability of solving the DCT for ICs having sufficiently large size N and the used radius r. Clearly, by expanding the radius r, our proposal is the best since the number of potential rules is much smaller than in existing methods that lead to exponential growth of the number of rules and therefore a risk of combinatory explosion.
- **Rules' analysis:** Once designed, it is possible to make analysis on solutions and the computations carried out by CAs (for instance, detection of attractors) through cells' exchanges in generations and annihilations processes. Indeed, most of the existing methods make such analysis through space-time diagrams and performance measurements, and therefore they do not give explanations about the source of emergent computations.

4.5.2 The Limitations

- As the elements of sets S and A are generated by hand, there is a risk that some elements would be missed, especially those supporting non-trivial symmetric exchanges (i.e. rules using regular expressions) given that the number of symmetric CAs that are of radius $r = 4$ is very important.

– The selection of a subset $B \subseteq S$ is not clear since it is difficult to predict the resulting global behavior of joined rules. Nevertheless, it is possible to overcome this limitation by using automatic search methods to explore the possible combinations in Set S by means of exact methods or heuristics.

4.6 Implementation Details and Running Time Measurements

Tests were implemented using the Java language and run on HP Z640 station, doted by 64 Go of memory and Xeon processor composed of 20 cores. And as these tests are CPU time consuming, we should make use of tools that efficiently exploit the available processors. For this purpose, we have used the AtejiPX Java library, which is freely downloadable as a plug-in for the Eclipse IDE, due to its advantages comparing to other libraries (simplicity of use, high level of abstraction for parallel and distributed computation primitives…etc.).

For each test, the set of ICs - which is of size 100 K ICs - is equally divided into m subsets assigned to cores that partially evaluate each CA rule in parallel; m is the number of used cores. Once computations in all cores completed, the partial results are gathered in order to compute the performance of each rule. Running time variations according to ICs size N are plotted in Fig. 3.

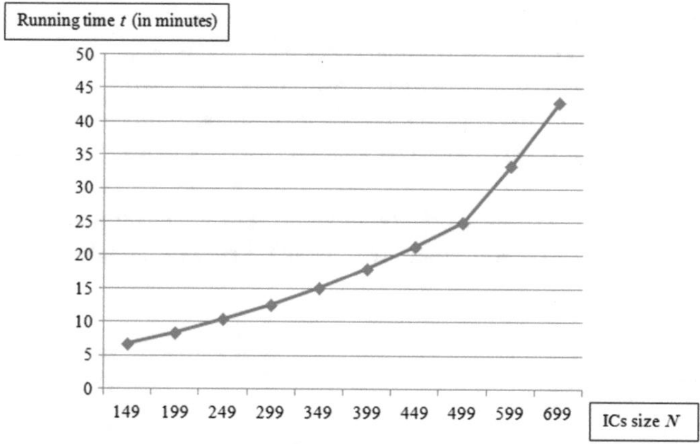

Fig. 3. Running time variations according to ICs size N.

- **Comments**

Figure 3 shows that the running time increases in accordance with the growth of ICs size N. This is expected since the amount of processed data becomes bigger and thus ICs require more time to converge towards homogenous fixed-point patterns. Consequently, without using parallel architectures, running such programs on simple machines would be difficult due to highly CPU time consuming. Nevertheless, there are some factors that may impede parallel processing of information; we cite for instance:

- The nature of CA rules, where each of them has a different speed of convergence towards homogenous fixed-point patterns. This may increase the unemployment rate in some cores, which directly influences the acceleration.
- Accessing shared data by the processes installed in cores and more generally the influence of memory access mode in the used hardware architecture since the cores read and write data through a common memory space.
- The system is not totally isolated where there are some other tasks that are simultaneously running with the current tests.

5 Conclusion

Designing CAs rules to solve computational tasks is often regarded as a difficult procedure due to our insufficient knowledge about their computations' nature. Accordingly, recent researches have been turned out to the automatic tools that blindly search for suitable rules. In order to overcome that dead-end, we have proposed an effective human-engineering procedure that allows designing new efficient solutions for the DCT: Rules $\Phi_{R4BLi\ =\ 1..3}$ that could outperform Rule Φ_{R4B1}, in particular Rule Φ_{R4BL1}. This is very important since it helps to deepen and even to open new perspectives on our general understanding of computations within CAs.

In practical terms and to improve the general performances of implemented tests, we relied on some hardware and software parallel tools in order to overcome running time and memory space requirements. Nevertheless, it seems unavoidable to use advanced high performance computing (HPC) techniques to process larger instances of tests (solving the DCT in 2D-CAs for example) such as Clusters, Grids and Clouds infrastructures.

In future, we plan to study DCT solutions in 2D-CAs since their general computational characteristics are still unknown, which impedes developing strategies to design efficient solutions.

References

1. Hoekstra, A.G., Kroc, J., Sloot, P.M. (eds.): Simulating Complex Systems by Cellular Automata. Understanding Complex Systems. Springer, Berlin (2010)
2. Stephen, W.: A New Kind of Science, p. 1035. Wolfram Media, Champaign, IL (2002)
3. Betel, H., de Oliveira, Pedro P.B., Flocchini, Paola: Solving the parity problem in one-dimensional cellular automata. Nat. Comput. **12**(3), 323–337 (2013)
4. Oliveira, G.M., Martins, L.G., de Carvalho, L.B., Fynn, E.: Some investigations about synchronization and density classification tasks in one-dimensional and two-dimensional cellular automata rule spaces. Electron. Notes Theor. Comput. Sci. **252**, 121–142 (2009)
5. Wolz, D., De Oliveira, P.P.B.: Very effective evolutionary techniques for searching cellular automata rule spaces. J. Cell. Autom. **3**(4), 289–312 (2008)

6. Packard, N.H.: Adaptation towards the edge of chaos. In: Kelso, J.A.S., Mandell, A.J., Schlesinger, M.F. (eds.) Dynamic Patterns in Complex Systems. World Scientific, Singapore, pp. 293–301 (1988)
7. Mitchell, M., Hraber, P., Crutchfield, J.P.: Revisiting the edge of chaos: evolving cellular automata to perform computations. Complex Syst. **7**, 89–130 (1993)
8. Mitchell, M., Crutchfield, J.P., Hraber, P.T.: Evolving cellular automata to perform computations: mechanisms and impediments. Physica D **75**(1), 361–391 (1994)
9. De Oliveira, P.P.B.: Conceptual connections around density determination in cellular automata. In: Cellular Automata and Discrete Complex Systems. Lecture Notes in Computer Science, vol. 8155, pp. 1–14 (2013)
10. Kari, J., Le Gloannec, B.: Modified traffic cellular automaton for the density classification task. Fundam. Inf. **116**(1–4), 141–156 (2012)
11. Le Gloannec, B.: Around Kari's traffic cellular automaton for the density classification. Project Report, ENS Lyon, France (2009)
12. Marques-Pita, M., Mitchell, M., Rocha, L.M.: The role of conceptual structure in designing cellular automata to perform collective computation. In: Unconventional Computation. Lecture Notes in Computer Science, vol. 5204, pp. 146–163 (2008)
13. De Oliveira, P.P.B., Bortot, J.C., Oliveira, G.M.: The best currently known class of dynamically equivalent cellular automata rules for density classification. Neurocomputing **70** (1), 35–43 (2006)
14. Laboudi, Z., Chikhi, S.: Computational mechanisms for solving the density classification task by cellular automata. J. Cell. Autom. **14**, 1–2 (2019)
15. Márques, M., Manurung, R., Pain, H.: Conceptual representations: what do they have to say about the density classification task by cellular automata? Comput. Mech. 1–15 (2006)
16. De Oliveira, P.P.B.: On density determination with cellular automata: results, constructions and directions. J. Cell. Autom. **9**(5–6), 357–385 (2014)
17. Etienne, M.: State-Conserving Cellular Automata. Project Report, ENS Lyon, France (2011)
18. Land, M., Belew, R.: No two-state CA for density classification exists. Phys. Rev. Lett. **74** (25), 5148 (1995)
19. Gacs, P., Kurdyumov, G., Levin, L.: One-dimensional homogenuous media dissolving finite islands. Probl. Inf. Transm. **14**(3), 92–96 (1978)
20. Andre, D., Bennett III, F.H., Koza, J.R.: Discovery by genetic programming of a cellular automata rule that is better than any known rule for the majority classification problem. In: Proceedings of the 1st Annual Conference on Genetic Programming, pp. 3–11. MIT Press (1996)
21. Das, R., Mitchell, M., Crutchfield, J.P.: A genetic algorithm discovers particle-based computation in cellular automata. In: International Conference on Parallel Problem Solving from Nature, pp. 344–353. Springer, Berlin, Heidelberg (1994)
22. Juille, H., Pollack, J.B.: Coevolving the ideal trainer: application to the discovery of cellular automata rules. In: Proceedings of the Third Annual Conference, San Francisco. Morgan Kauffmann, University of Wisconsin (1998)
23. Ferreira, C.: Gene expression programming: a new adaptive algorithm for solving problems. Complex Syst. **13**(2), 87–129 (2001)
24. Marques-Pita, M., Rocha, L.M.: Conceptual structure in cellular automata-the density classification task. In: ALIFE, pp. 390–397 (2008)
25. Laboudi, Z., Chikhi, S.: Scalability property in solving the density classification task. J. Inf. Technol. Res. **10**(2), 59–75 (2017)

A Chaotic Binary Salp Swarm Algorithm for Solving the Graph Coloring Problem

Yassine Meraihi[1(✉)], Amar Ramdane-Cherif[2], Mohammed Mahseur[3], and Dalila Achelia[1]

[1] Automation Department, University of M'Hamed Bougara Boumerdes, Avenue of Independence, 35000 Boumerdes, Algeria
yassine.meraihi@yahoo.fr, dacheli2000@yahoo.fr

[2] LISV Laboratory, University of Versailles St-Quentin-en-Yvelines, 10-12 Avenue of Europe, 78140 Velizy, France
rca@lisv.uvsq.fr

[3] Faculty of Electronics and Informatics, University of Sciences and Technology Houari Boumediene, El Alia Bab Ezzouar, 16025 Algiers, Algeria
mahseur.mohammed@gmail.com

Abstract. This paper proposes a new Chaotic Binary Salp Swarm Algorithm (CBSSA) to solve the graph coloring problem. First, the Binary Salp Swarm Algorithm (BSSA) is obtained from the original Salp Swarm Algorithm (SSA) using the S-Shaped transfer function (Sigmoid function) and the binarization method. Second, the most popular chaotic map, namely logistic map, is used to replace the random variables used in the mathematical model of SSA. The aim of using chaotic map is to avoid the stagnation to local optima and enhance the exploration and exploitation capabilities. We use the well-known DIMACS benchmark to evaluate the performance of our proposed algorithm. The simulation results show that our proposed algorithm outperforms other well-known algorithms in the literature.

Keywords: Graph coloring problem · Salp Swarm algorithm
Binary Salp swarm algorithm · Chaotic maps
Combinatorial optimization problem

1 Introduction

The graph coloring problem (GCP) is one of the most interesting NP-Hard combinatorial optimization problem in mathematics, operations research, and computer science. It has been widely used to model and solve many significant real-world problems like time tabling [1], scheduling [2,3], computer register allocation [4,5], radio frequency assignment [6,7], communication networks [8], and many others. It consists in coloring each node of a given graph with the restriction that no two adjacent nodes are colored with the same color and the number

S. Chikhi et al. (Eds.): MISC 2018, LNNS 64, pp. 106–118, 2019.
https://doi.org/10.1007/978-3-030-05481-6_8

of different colors used is minimized. The minimum number of colors by which a graph can be colored is called its chromatic number and it is denoted by $X(G)$. The graph coloring problem is known to be an NP-Hard problem [9], so several methods have been developed to solve this problem including greedy constructive approaches, local search heuristics, metaheuristics, and hybrid approaches.

The two most constructive algorithms used to solve the GCP are the recursive largest first algorithm (RLF) proposed by Leighton [10] and the largest saturation degree algorithm (DSATUR) developed by Brlaz [11]. These constructive methods are based on greedy approach which color the nodes of the graph one at time using a predefined greedy function. They are recently used to generate initial solutions for advanced metaheuristics.

Tabu Search Algorithm proposed by Hertz and de Werra [12] was the first local search algorithm applied to solve the graph coloring problem. It is called TABUCOL and has been enhanced by several researchers and used as a subcomponent of more graph coloring algorithms.

Moreover, many efficient metaheuristics such as Genetic Algorithm (GA) [13], Cuckoo Search algorithm (CS) [14,15], Artificial Bee Colony (ABC) [16], Bat Algorithm (BA) [17], Memetic Algorithm (MA) [18], and combining algorithms [19–21] have been employed for solving the Graph coloring problem.

Abbasian and Mouhoub [13] proposed a Hierarchical approach based on Parallel Genetic Algorithms (HPGAs) to solve the graph coloring problem. In this approach, a novel estimator is implemented to predict an upper-bound for the graphs chromatic number. Furthermore, an extension of the genetic algorithm, namely the genetic modification (GM) and the parental success crossover operator are proposed. Simulation results showed that the proposed approach is very accurate and faster for solving the graph coloring problem.

Djelloul et al. [14] proposed a discrete binary version of cuckoo search algorithm for solving the graph coloring problem. In this approach, a binary representation of the search space is adopted and sigmoid function and probability model are employed in order to generate binary values. Computational results demonstrated the feasibility and the efficiency of the proposed algorithm. Mahmoudi and Lotfi [15] proposed a Modified Cuckoo Search Algorithm (MCOA) to solve the graph coloring problem. CAO is discretized by redefining, over discrete space, the standard arithmetic operators such as addition, subtraction, and multiplication that exist in COA migration operator based on the distances theory. Experimental results revealed the high performance of the proposed algorithm compared with other well-known heuristic search methods.

Faraji and Javadi [16] proposed a new approach based on Bee Behavior in nature (BEECOL). Computational results revealed better performances of BEECOL compared to ACO algorithm. The proposed algorithm has the capability of establishing a proper connection between accuracy and speed of coloring the graph.

Djelloul et al. [17] proposed a discrete binary bat algorithm to solve the graph coloring problem. The bat algorithm is discretized using sigmoid function and binarization method. Experimental results showed the feasibility and the effectiveness of the proposed algorithm.

Lü and Hao [18] proposed a Memetic Algorithm denoted by MACOL to solve the graph coloring problem. The proposed algorithm integrated several distinguished features such as an adaptative multi-parent crossover operator and a distance-and-quality based replacement criteria for pool updating. Computational results demonstrated that the proposed algorithm achieves highly competitive results compared with 11 state-of-the-art algorithms.

Mabrouk et al. [19] proposed a parallel approach based on the combination of a genetic algorithm with the tabu search algorithm. Experimental results demonstrated that the proposed approach permitted to obtain very satisfactory solutions for the graph coloring problem. Douiri and Elbarnoussi [20] proposed a new hybrid approach called DBG. DBG is based on the combination of genetic algorithm with a local search algorithm. The proposed algorithm is based on the reduction of variables through a multiplier w of the surrogate constraint. Numerical results showed that DBG achieves highly competitive results compared to other best existing algorithms in the literature. Fidanova and Pop [21] proposed an improved hybrid Ant Colony Optimization (ACO) algorithm for solving the partition graph coloring problem. The hybrid approach is obtained by executing a local search procedure after every ACO iteration. Computational results showed that the proposed algorithm achieves solid results in very short-times compared to state-of-the-art algorithms.

In this paper, we propose a new chaotic binary salp swarm algorithm (CBSSA) to solve the graph coloring problem. First, the binary salp swarm algorithm is obtained using the sigmoid transfer function and the binarization method [22]. Second, chaotic map is employed to choose the right values of the random parameters c_1 and c_2. To the best of our knowledge, this is the first work introducing salp swarm algorithm to solve the graph coloring problem.

The remainder of this paper is organized as follows. In Sect. 2, we describe the graph coloring problem formulation. In Sect. 3, we present the salp swarm algorithm and the binary salp swarm algorithm. In Sect. 4, we propose the new chaotic binary salp swarm algorithm for the graph coloring problem. The experimental results and analysis of different algorithms in graph coloring problem are given in Sect. 5. Finally, Conclusion and future works are given in Sect. 6.

2 Graph Coloring Problem Formulation

Let $G = (V, E)$ be an undirected graph where V denotes the set of nodes and E denotes the set of the edges representing links between nodes. $|V| = n$ is the number of nodes and $|E| = l$ is the number of edges, with $e = (i, j) \in E$ represents the edge that connects node i with node j. Formally, a legal K-coloring of nodes in the graph G is given by a vector $S = [C(v_1), C(v_2), \dots, C(v_i), C(v_j), \dots, C(v_n)]$ such as $C(v_i) \neq C(v_j)$ for all $e = (i, j) \in E$ where $C(v_i)$ represents the color assigned to the node i[17]. If $C(v_i) = (v_j)$, then node i and node j are in conflict and are called conflicting nodes. The graph coloring problem can be represented as a minimization problem whose main objective is to minimize the total number of colors by which the graph can be legally colored.

3 Salp Swarm Algorithm and Binary Salp Swarm Algorithm

3.1 Salp Swarm Algorithm

Salp Swarm Algorithm (SSA), introduced by Mirjalili et al. in 2017 [23], is a newly nature-inspired metaheuristic algorithm used to solve various optimization problems that possess uni-modal and multi-modal functions [22,24–28]. This optimization algorithm is inspired by the swarming behavior of Salps in the sea [23]. Salps belong to the family of Salpidae and move very similar to jelly fish. Generally, salps live in group and form a swarm called salp chain. This chain is divided into two group: leader and followers. The leader takes the position at the beginning of the chain, while the remainder of salps are the followers. In the mathematical model of the SSA, the position of the leader is updated using the following equation:

$$x_i^1 = \begin{cases} F_i + c_1((ub_i - lb_i)c_2 + lb_i), & c_3 \geq 0.5 \\ F_i - c_1((ub_i - lb_i)c_2 + lb_i), & c_3 < 0.5 \end{cases} \tag{1}$$

Where x_i^1 denotes the position of leader in the *i-th* dimension, F_i is the position of food source in the *i-th* dimension, lb_i, ub_i are the lower and upper boundaries of the *i-th* dimension, c_1, c_2, and c_3 are random numbers. c_1 is an important parameter that balances exploration and exploitation. It is mathematically defined as follows:

$$c_1 = 2e^{-(\frac{4t}{T})^2} \tag{2}$$

Where t denotes the current iteration and T is the maximum number of iterations. The random numbers c_2 and c_3 are uniformly generated in the range [0, 1].

The position of the follower salps are updated using the following equation:

$$x_i^j = \frac{1}{2}(x_i^j + x_i^{j-1}) \tag{3}$$

Where $j \geq 2$ and x_i^j represent the position of the *j-th* salp in the *i-th* dimension The pseudo code of the original Salp Swarm Algorithm is illustrated in Algorithm 1 [23].

Algorithm 1. The pseudo-code of the Salp Swarm Algorithm

Generate initial population of Salps $x_i (i = 1, 2, \ldots, n)$ randomly considering ub and lb
while ($t < Maximum\ number\ of\ iterations$) **do**
  Calculate the fitness function of each salp $f(x_i)$
  Set the best salp to x_{best}
  Update the value of c_1 according to equation (2)
  for each salp (x_i) **do**
    if x_i is the leader **then**
      Update the position of the leading salp using equation (1)
    else
      Update the position of the follower salp using equation (3)
    end if
  end for
  Update the positions of the salps based on the upper and lower bounds
end while
Return the best salp position x_{best}

3.2 Binary Salp Swarm Algorithm

The Salp Swarm Algorithm was designed to solve optimization problems with continuous search space while the graph coloring problem has a discrete binary search space. So, a binary version of SSA, namely BSSA, is needed for solving the graph coloring problem. In discrete binary space, position updating means a switching between "0" and "1" values [29,30]. In order to obtain values "0" or "1", a transfer function and a binarization method are used. The most frequently transfer function, named sigmoid function, is adopted in this paper. It is defined as follows:

$$S(x_i^j(t+1)) = \frac{1}{1 + exp(-x_i^j(t+1))} \tag{4}$$

The result of this operation is a real number between "0" and "1". A binarization method is required to update position to obtain a value "0" or "1". The binarization method is defined as follows:

$$x_i^j(t+1) = \begin{cases} 1, & if \ \ rand() < S(x_i^j(t+1)) \\ 0, & if \ \ rand() \geq S(x_i^j(t+1)) \end{cases} \tag{5}$$

Where $x_i^j(t+1)$ denotes the position of the *j-th* salp at iteration $t+1$ in the *i-th* dimension

The pseudo code of the Binary Salp Swarm Algorithm is illustrated in Algorithm 2.

Algorithm 2. The pseudo-code of the Binary Salp Swarm Algorithm

Generate initial population of Salps $x_i (i = 1, 2, \ldots, n)$ randomly considering ub and lb
while ($t < Maximum\ number\ of\ iterations$) **do**
    Calculate the fitness function of each salp $f(x_i)$
    Set the best salp to x_{best}
    Update the value of c_1 according to equation (2)
    for each salp (x_i) **do**
        if x_i is the leader **then**
            Update the position of the leading salp using equation (1)
        else
            Update the position of the follower salp using equation (3)
        end if
        Calculate S-Shaped transfer (Sigmoid) function value using equation (4)
        Update the position of the salp using equation (5)
    end for
    Update the positions of the salps based on the upper and lower bounds
end while
Return the best salp position x_{best}

4 Chaotic Binary Salp Swarm Algorithm for the Graph Coloring Problem

4.1 Chaotic Binary Salp Swarm Algorithm

In this section, a new Chaotic Binary Salp Swarm algorithm (CBSSA) is proposed in order to avoid the stagnation in local optima, provide a good balance between the exploitation and exploration, and enhance the efficiency of the BSSA. We adopt logistic map which is the most representative chaotic map with simple operations and well dynamic randomness [31,32] to generate the random numbers c_1 and c_2 in the mathematical model of SSA. The random number generated by logistic map is given by:

$$L_{t+1} = aL_t(1 - L_t), 0, 1, 2, \ldots \tag{6}$$

Where L_t is the chaotic map value at the *i-th* iteration, it is in [0, 1]. Here a is a control parameter and $0 < a < 4$. The chaotic system is very sensitive to the initial values, to make the logistic map in a complete chaos, we adopt the parameter $a = 4$ and L_t does not belong to $0, 0.25, 0.5, 0.75, 1$ otherwise the logistic equation does not show chaotic behavior [33]. In our experiments, we take $a = 4$ and $L_0 = 0.7$ as used in [31,32,34] The updated position of the salp using chaotic map is given by the following equation:

$$x_i^1 = \begin{cases} F_i + c_1((ub_i - lb_i)L_{c2} + lb_i), & L_{c3} \geq 0.5 \\ F_i - c_1((ub_i - lb_i)L_{c2} + lb_i), & L_{c3} < 0.5 \end{cases} \tag{7}$$

In the proposed algorithm CBSSA, the population of salps are initialized using constructive modified RLF algorithm to get diverse solutions. At each iteration, CBSSA calculates and evaluates the fitness function of each salp to obtain the best salp. The values of the random parameters c_2 and c_3 are updated using logistic map. Subsequently, CBSSA updates the position of the salp and adopts the sigmoid transfer function and the binarization method to obtain binary solutions of the graph. After this step, the positions of the salps are updated based on the upper and lower bounds and the best salp solution is kept for the next iteration.

The process of searching the best solution is iterated until the maximum number of iterations MIt is reached or the best feasible solution has not changed. The pseudo code of the Chaotic Binary Salp Swarm Algorithm is illustrated in Algorithm 3.

Algorithm 3. The pseudo-code of the Chaotic Binary Salp Swarm Algorithm

Initialize the population using constructive modified RLF algorithm
while ($t < Maximum\ number\ of\ iterations$) **do**
    Calculate the fitness function of each salp $f(x_i)$
    Set the best salp to x_{best}
    Update the value of c_1 according to equation (2)
    Update the values of c_2 and c_3 using logistic map according to equation (6)
    for each salp (x_i) **do**
        if x_i is the leader **then**
            Update the position of the leading salp using equation (7)
        else
            Update the position of the follower salp using equation (3)
        end if
        Calculate S-Shaped transfer (Sigmoid) function value using equation (4)
        Update the position of the salp using equation (5)
    end for
    Update the positions of the salps based on the upper and lower bounds
end while
Return the best salp position x_{best}

4.2 Representation of the Solution

The graph coloring solution is represented by $N * C$ binary Matrix X, where N represents the total number of nodes in the graph and C the number of colors. The $N * C$ binary matrix is defined as follows:

$$x_{ij} = \begin{cases} 1; \text{ if the node } i \text{ is colored with the color } j. \\ 0; \quad \text{otherwise.} \end{cases} \tag{8}$$

4.3 Objective Function

For a K-coloring solution S such as $S = [C(v_1), C(v_2), C(v_i), \dots, C(v_n)]$, the objective function $f(S)$ is the number of the conflicting nodes produced by S. It is given as follows:

$$f(s) = \sum_{i=1}^{n} \sum_{j=1}^{n} conflict_{ij} \tag{9}$$

Where the conflicting matrix $conflict_{ij}$ is given by:

$$conflict_{ij} = \begin{cases} 1; & \text{if the } C(v_i) = C(v_j) \text{ and } e = (i, j) \in E. \\ 0; & \text{otherwise.} \end{cases} \tag{10}$$

A K-coloring solution S with $f(S) = 0$ represents a valid solution. A solution S is found if the number of conflicts $f(S) = 0$.

4.4 Initialization of the Population

The constructive modified RLF method proposed by Mabrouk et al. [19] is used to generate the initial population. This approach builds the classes of colors sequentially, it selects nodes from one color class at time. Assume that Ncl is the next color class to be assigned, Y is the set of uncolored nodes which can be included in the independent set Ncl, m is the number of colors, and B the set of colors that cannot be included in the independent set Ncl.

1. If $(Ncl > m)$ then color the remaining nodes with colors randomly chosen in the Set $1, 2, \dots, m$.
2. Else, Choose randomly $v \in Y$ and color v with the color Ncl.

4.5 Complexity Analysis of CBSSA

In this subsection, the complexity of the proposed algorithm CBSSA is carried out. CBSSA uses $O(N)$ to generate N solutions. The process of choosing the best solution is iterated until the maximum number of iterations MIt is reached or the best valid solution has not changed. So the worst case complexity of our proposed algorithm is $O(N \times MIt \times n)$, where N represents the number of population of solutions, MIt denotes the maximum number of iterations, and n the number of nodes.

5 Experimental Results

Our proposed algorithm was implemented with Visual C++ language and the experiments were executed on a PC with an Intel Core i5 2.5 GHz-CPU and 8 GB RAM based platform running Windows7. In order to evaluate the performance and the effectiveness of our proposed algorithm CBSSA, we have conducted a

Table 1. The computational results of the proposed algorithm CBSSA

Graph No	Graph Name	$\|V\|$	$\|E\|$	K^*	K_{Ours}	T_{avg}	*succ*
1	myciel3	11	20	4	4	0.38	20/20
2	myciel4	23	71	5	5	0.74	20/20
3	myciel5	47	236	6	6	1.73	20/20
4	myciel6	95	755	7	7	1.68	20/20
5	queen5_5	25	160	5	5	2.08	20/20
6	queen6_6	36	290	7	7	2.17	20/20
7	queen7_7	49	476	7	7	5.67	20/20
8	queen8_8	64	728	9	10	9.43	18/20
9	queen9_9	81	2112	10	10	12.18	17/20
10	1-Insertions_4	67	232	4	5	2.46	20/20
11	2-Insertions_3	37	72	4	4	1.83	20/20
12	3-Insertions_3	56	110	4	4	3.25	20/20
13	4-Insertions_3	79	156	3	4	3.37	20/20
14	mug88_1	88	146	4	4	2.23	20/20
15	mug88_25	88	146	4	4	1.39	20/20
16	huck	74	301	11	11	4.32	20/20
17	jean	80	254	10	10	3.98	20/20
18	david	87	406	11	11	2.36	20/20
19	anna	138	493	11	11	2.48	20/20
20	games120	120	638	9	9	3.12	20/20
21	myciel7	191	2360	8	8	3.96	20/20
22	2-Insertions_4	149	541	4	5	5.06	20/20
23	mug100_1	100	166	4	4	2.72	20/20
24	mug100_25	100	166	4	4	2.32	20/20
25	miles250	128	387	8	8	4.38	20/20
26	miles500	128	1170	20	20	5.18	20/20
27	miles750	128	2113	31	31	5.37	20/20
28	miles1000	128	3216	42	42	5.52	20/20
29	DSJC125.1	125	736	5	5	5.37	20/20
30	DSJC125.5	125	3891	17	17	9.24	17/20

series of experiments on a benchmark of 30 graphs from the well-known DIMACS graph coloring benchmarks. The results of CBSSA were compared with four other relevant coloring algorithms MCOACOL [15], BBCOL [17], DBG [20] and HPGAs [13]. The best value K_{Ours} and the average execution time T_{avg} are obtained by running the simulation 20 times for each instance. The population size is set to 50 and the maximum number of iterations is set to 2000.

Table 1 reports the computational results of our proposed algorithm CBSSA. In this table, $|V|$ denotes the number of nodes, $|E|$ the number of edges, K^* the best-known chromatic number results reported in the literature, K_{Ours} the chromatic number result of our proposed algorithm, T_{avg} the average execution time in second, and *succ* the success rate.

According to Table 1, we observe that the results obtained by our algorithm CBSSA are similar to the best-known results except for three instances (queen8_8, 1-Insertions_4, 4-Insertions_3), the average execution time does not exceed 13 seconds for all instances. It is also seen that the success rate in almost cases is 100%, except for three instances (queen8_8, queen9_9, DSJC125.5).

The computational results of CBSSA were compared with four relevant algorithms given in the literature. The results are given in Table 2.

Table 2. The computational results of CBSSA with four relevant algorithms

Graph No	Graph name	$\|V\|$	$\|E\|$	K^*	CBSSA	MCOACOL [15]	BBCOL [17]	DBG [20]	HPGAs [13]
1	myciel3	11	20	4	4	4	4	4	4
2	myciel4	23	71	5	5	5	5	5	5
3	myciel5	47	236	6	6	6	6	6	6
4	myciel6	95	755	7	7	7	*	7	7
5	queen5_5	25	160	5	5	5	5	5	5
6	queen6_6	36	290	7	7	8	8	7	8
7	queen7_7	49	476	7	7	7	*	7	8
8	queen8_8	64	728	9	10	10	*	9	10
9	queen9_9	81	2112	10	10	11	*	10	*
10	1-Insertions_4	67	232	4	5	5	*	*	5
11	2-Insertions_3	37	72	4	4	4	*	4	4
12	3-Insertions_3	56	110	4	4	4	*	4	4
13	4-Insertions_3	79	156	3	4	4	*	*	*
14	mug88_1	88	146	4	4	4	*	4	4
15	mug88_25	88	146	4	4	4	*	4	4
16	huck	74	301	11	11	11	11	11	11
17	jean	80	254	10	10	10	10	10	10
18	david	87	406	11	11	11	11	11	11
19	anna	138	493	11	11	11	11	11	11
20	games120	120	638	9	9	9	9	9	9
21	myciel7	191	2360	8	8	8	*	8	8
22	2-Insertions_4	149	541	4	5	5	*	*	5
23	mug100_1	100	166	4	4	4	*	4	4
24	mug100_25	100	166	4	4	4	*	4	4
25	miles250	128	387	8	8	8	8	8	8
26	miles500	128	1170	20	20	20	20	20	20
27	miles750	128	2113	31	31	31	*	*	31
28	miles1000	128	3216	42	42	42	*	*	43
29	DSJC125.1	125	736	5	5	6	*	6	*
30	DSJC125.5	125	3891	17	17	19	*	17	*

Our proposed algorithm gives similar results compared to those obtained by MCOACOL algorithm on 26 instances and better results on 4 instances (queen6_6, queen9_9, DSJC125.1, DSJC125.5)

When comparing with the results obtained by BBCOL algorithm on 12 instances, our algorithm gives better results for one instance (queen6_6) and similar results for 11 instances.

When Compared with the results obtained by DBG algorithm on 25 instances, CBSSA is better in 1 instance (DSJC125.1), worse in 1 instance (queen8_8), and similar in 23 instances.

Compared to HPGAs algorithm on 26 instances, our proposed algorithm CBSSA provides better results for 3 instances (queen6_6, queen7_7, miles1000) and similar results for 23 instances.

6 Conclusion

In this paper, we have proposed a Chaotic Binary Salp Swarm Algorithm to solve the graph coloring problem. In this regard, we introduced chaotic map in order to enhance the exploitation and exploration capabilities and ensure the diversity of the solutions. We adopt the most representative chaotic map, namely logistic map, to choose the right values of c_1 and c_2 in the mathematical model of SSA. The well-known DIMACS benchmark instances are used to assert the performance of the proposed algorithm in comparison with four existing algorithms MCOACOL, BBCOL, DBG, and HPGAs. The experimental results prove the performance and effectiveness of our proposed algorithm CBSSA compared to MCOACOL, BBCOL, DBG, and HPGAs. Future works will be to apply CBSSA to solve other optimization problems such as Knapsack problem and detection intrusion problem.

References

1. de Werra, D.: An introduction to timetabling. Eur. J. Oper. Res. **19**(2), 151–162 (1985)
2. Lotfi, V., Sarin, S.: A graph coloring algorithm for large scale scheduling problems. Comput. Oper. Res. **13**(1), 27–32 (1986)
3. Dowsland, K.A., Thompson, J.M.: Ant colony optimization for the examination scheduling problem. J. Oper. Res. Soc. **56**(4), 426–438 (2005)
4. Chaitin, G.J., Auslander, M.A., Chandra, A.K., Cocke, J., Hopkins, M.E., Markstein, P.W.: Register allocation via coloring. Comput. Lang. **6**(1), 47–57 (1981)
5. de Werra, D., Eisenbeis, C., Lelait, S., Marmol, B.: On a graph-theoretical model for cyclic register allocation. Discret. Appl. Math. **93**(2–3), 191–203 (1999)
6. Gamst, A.: Some lower bounds for a class of frequency assignment problems. IEEE Trans. Veh. Technol. **35**(1), 8–14 (1986)
7. Smith, D.H., Hurley, S., Thiel, S.U.: Improving heuristics for the frequency assignment problem. Eur. J. Oper. Res. **107**(1), 76–86 (1998)
8. Woo, T.K., Su, S.Y., Newman-Wolfe, R.: Resource allocation in a dynamically partitionable bus network using a graph coloring algorithm. IEEE Trans. Commun. **39**(12), 1794–1801 (1991)

9. Garey, M.R., Johnson, D.S.: Computers and intractability: a guide to the theory of npcompleteness (series of books in the mathematical sciences). Comput. Intractability **340** (1979)
10. Leighton, F.T.: A graph coloring algorithm for large scheduling problems. J. Res. Natl. Bur. Stand. **84**(6), 489–506 (1979)
11. Brlaz, D.: New methods to color the vertices of a graph. Commun. ACM **22**(4), 251–256 (1979)
12. Hertz, A., de Werra, D.: Using tabu search techniques for graph coloring. Computing **39**(4), 345–351 (1987)
13. Abbasian, R., Mouhoub, M.: A hierarchical parallel genetic approach for the graph coloring problem. Appl. Intell. **39**(3), 510–528 (2013)
14. Djelloul, H., Layeb, A., Chikhi, S.: A binary cuckoo search algorithm for graph coloring problem. Int. J. Appl. Evol. Comput. (IJAEC) **5**(3), 42–56 (2014)
15. Mahmoudi, S., Lotfi, S.: Modified cuckoo optimization algorithm (MCOA) to solve graph coloring problem. Appl. Soft Comput. **33**, 48–64 (2015)
16. Faraji, M., Javadi, H.H.S.: Proposing a new algorithm based on bees behavior for solving graph coloring. Int. J. Contemp. Math. Sci. **6**(1), 41–49 (2011)
17. Djelloul, H., Sabba, S., Chikhi, S.: Binary bat algorithm for graph coloring problem. In: 2014 Second World Conference on Complex Systems (WCCS), pp. 481–486. IEEE (2014)
18. Lü Z., Hao, J.K.: A memetic algorithm for graph coloring. Eur. J. Oper. Res. **203**(1), 241–250 (2010)
19. Mabrouk, B.B., Hasni, H., Mahjoub, Z.: On a parallel genetictabu search based algorithm for solving the graph colouring problem. Eur. J. Oper. Res. **197**(3), 1192–1201 (2009)
20. Douiri, S.M., Elbernoussi, S.: Solving the graph coloring problem via hybrid genetic algorithms. J. King Saud Univ. Eng. Sci. **27**(1), 114–118 (2015)
21. Fidanova, S., Pop, P.: An improved hybrid ant-local search algorithm for the partition graph coloring problem. J. Comput. Appl. Math. **293**, 55–61 (2016)
22. Faris, H., Mafarja, M.M., Heidari, A.A., Aljarah, I., AlaM, A.Z., Mirjalili, S., Fujita, H.: An efficient binary Salp Swarm algorithm with crossover scheme for feature selection problems. Knowl. Based Syst. **154**, 43–67 (2018)
23. Mirjalili, S., Gandomi, A.H., Mirjalili, S.Z., Saremi, S., Faris, H., Mirjalili, S.M.: Salp Swarm algorithm: a bio-inspired optimizer for engineering design problems. Adv. Eng. Softw. **114**, 163–191 (2017)
24. Sayed, G.I., Khoriba, G., Haggag, M.H.: A novel chaotic Salp Swarm algorithm for global optimization and feature selection. Appl. Intell. 1–20 (2018)
25. El-Fergany, A.A.: Extracting optimal parameters of PEM fuel cells using Salp Swarm optimizer. Renew. Energy **119**, 641–648 (2018)
26. Abusnaina, A.A., Ahmad, S., Jarrar, R., Mafarja, M.: Training neural networks using Salp Swarm algorithm for pattern classification, p. 17. ACM (2018)
27. Rizk-Allah, R.M., Hassanien, A.E., Elhoseny, M., Gunasekaran, M.: A new binary Salp Swarm algorithm: development and application for optimization tasks. Neural Comput. Appl. 1–23 (2018)
28. Ibrahim, A., Ahmed, A., Hussein, S., Hassanien, A.E.: Fish image segmentation using Salp Swarm algorithm, pp. 42–51. Springer, Cham (2018)
29. Mirjalili, S., Mirjalili, S.M., Yang, X.S.: Binary bat algorithm. Neural Comput. Appl. **25**(3–4), 663–681 (2014)
30. Mirjalili, S., Lewis, A.: S-shaped versus V-shaped transfer functions for binary particle swarm optimization. Swarm Evol. Comput. **9**, 1–14 (2013)

31. Lei, X., Du, M., Xu, J., Tan, Y.: Chaotic fruit fly optimization algorithm. In: International Conference in Swarm Intelligence, pp. 74–85. Springer, Cham (2014)
32. Kanso, A., Smaoui, N.: Logistic chaotic maps for binary numbers generations. Chaos Solitons Fractals **40**(5), 2557–2568 (2009)
33. Tamiru, A.L., Hashim, F.M.: Application of bat algorithm and fuzzy systems to model exergy changes in a gas turbine. In: Artificial Intelligence Evolutionary Computing and Metaheuristics, pp. 685–719. Springer, Heidelberg (2013)
34. Heidari, A.A., Abbaspour, R.A., Jordehi, A.R.: An efficient chaotic water cycle algorithm for optimization tasks. Neural Comput. Appl. **28**(1), 57–85 (2017)

Gene Selection for Microarray Data Classification Using Hybrid Meta-Heuristics

Nassima Dif(✉), Mohamed walid Attaoui, and Zakaria Elberrichi

EEDIS Laboratory, Djillali Liabes University, Sidi Bel Abbes, Algeria
difnassima05@gmail.com, atwm19@gmail.com, elberrichi@gmail.com

Abstract. The hybridization of metaheuristics got a lot of interest lately. The crucial step lies in the choice of the hybrid methods. The major purpose is to make a tradeoff between exploitation and exploration concepts to create a more robust method. Hybrid metaheuristics are used as a solution to many optimization problems such as feature selection. In this paper we propose an hybridization between metaheuristics (PeSOA, FA, DE, AIS, BAT) for a best gene selection in microarray datasets. The main objective is to prove the efficiency of the proposed hybridization compared to the hybrid methods. The experimentations showed that PeSOA-C and HFA were competitive to their hybrid methods, on the other side, AIS-BAT was less promising compared to AIS. As results, we obtained a perfect 100% in case of Leukemia, Ovarian Cancer, Lymphoma and MLL-Leukemia datasets by the HFA hybridization with only 2%–4% of selected genes.

Keywords: Gene selection · Feature selection · Metaheuristics Hybridization · Microarray dataset · Classification · Data mining PeSOA · FA · DE · AIS · BAT

1 Introduction

DNA microarrays provide a format to measure the expression level of thousands of genes [1]. Gene expression analysis is an important task, since the cell physiology changes affect the gene expression pattern [2]. In oncology, microarrays can identify different types of tumours and assist in treatment selection strategies [3].

The main challenge in DNA microarray analysis is the massive output of data. Machine learning (ML) algorithms were widely used for DNA Microarrays analysis [4], for their capacity in knowledge extraction.

Microarrays datasets are characterized by a large number of genes associated with a few number of samples. In such type of datasets often there is only a few number of relevant genes for the classification task [5,6]. However, we know that the inclusion of non discriminate genes for cancer prediction can lead to false decisions [7].

S. Chikhi et al. (Eds.): MISC 2018, LNNS 64, pp. 119–132, 2019.
https://doi.org/10.1007/978-3-030-05481-6_9

Some classical machine learning algorithms have no embedded feature selection process which can reduce their classification performances [8]. In this case, a feature selection preprocess is required.

Gene selection is the process of selecting a subset of informative genes from the original set. The selection process improves the performances and ease the learning task.

There are four feature selection methods: filter [9], wrapper [10], hybrid [11] and embedded [12]. In a Filter approach, the search does not depend on the learning algorithm, while wrapper methods use the classifier performances to guide the search. These characteristics make filter methods less accurate but faster compared to wrapper. Hybrid methods create a co-operation between filters and wrappers to make a tradeoff between performances and time complexity. In embedded methods, the feature selection is performed in synchronization with the learning process.

Feature selection belongs to the class of NP-hard combinatorial optimization problems [13]. For a set of n genes there are 2^n subsets. So, for high-dimensional datasets, exact methods become computationally difficult.

To reduce the time complexity, different alternative stochastic methods have been proposed in literature: probabilistic [14], heuristic [15], metaheuristic [16] and hybridization of metaheuristics [17].

Glover introduced the term meta-heuristic in 1986 to differentiate between taboo search and other heuristics. Unlike exact method, metaheuristics do not guarantees the best global solution. The use of metaheuristic for a best gene selection got a lot of interest: Genetic Algorithm (GA) [8], Particle Swam Optimization (PSO) [18], Ant Colony Optimization [19], Multi-Verse Optimizer (MVO) [20] for examples.

Metaheuristics are based on two concepts: exploitation and exploration of the search space. In general, swarm optimization algorithms are characterized by better exploitation and the evolutionary algorithms focus better on the exploration of the search space. However, the exploration can slow the convergence in case of high multidimensional datasets. In the other side, the exploitation can cause a premature convergence. Thus, a good method is able to make a tradeoff between the two concepts. For that purpose, hybrid methods are proposed.

A hybrid method consists of at least two distinct methods. In such methods, the crucial part lies in the choice of its components according to their characteristics.

This paper tackles two main problems: the first is the use of metaheuristics for gene selection problem (bat algorithm: BAT, artificial immune system: AIS, firefly algorithm: FA, differential evolution: DE), the second is to create co-operation between those metaheuristics by different hybridizations (PeSOA-C, FA- DE, BAT-AIS and TS-GA).

This paper is organized as follows, the next session presents some related works to gene selection. The Sect. 2 details the principle of the used metaheuristics. Section 3 explains briefly the principal of the used methods. Section 4: the proposed hybridizations. Section 5: results and discussion. Section 6: conclusion.

2 Related Works

The proposed works in gene selection domain can be categorized into 3 principle categories: filters [21], wrappers [7,16,18,22–28] and filter/wrapper hybridization [11,28–30]. The majority of the works have used the SVM classifier for model generation.

Yang et al. [21] proposed a filter approach based on the Bayesian error (Filter Based Bayes Error: BBF) to select the relevant genes in four microarray datasets: Colon, DLBCL, Leukemia, Prostate, Lymphoma. In the wrapper category, the search strategy can be performed by a metaheuristic [25,27], a modified version of a metaheuristic [16,18,24] or hybridization between metaheuristics [7,22,23,26].

Deepthi et al. [25] proposed a method based on FA metaheuristic and K-MEANS algorithm for gene selection, the obtained results have proved its efficiency compared to Random Projection and PCA. Chen et al. [27] used the PSO metaheuristic to select informative genes in 3 microarrays datasets in ALL-AML, Lung Cancer and SRBCT.

The limitations of some metaheuristics especially in the case of microarrays datasets encouraged the researchers community to enhance and to propose more robust method. The main contribution was to propose other versions of metaheuristics according to the application filed. Prasad et al. [18] proposed a refined PSO wrapper (RPSW), the refinement helps to reduce the search space in case of high-dimensional datasets and to select in a real time a few number of relevant genes. The proposed approach was more promising compared to other wrapper methods: PSO-GA and GSA. Zhu et al. [24] proposed a modified version of GA metaheuristic (MBGA). MBGA uses Markov Blanket mimic operators to add or remove genes from the generated solutions by GA. the obtained results on 11 microarrays datasets have shown the efficiency of MBGA. Li et al. [22] proposed a modified version of ACO metaheuristic (ACO-S). The proposed approach was tested on 5 microarray datasets: Leukemia, Colon, Breast cancer, Lung carcinoma and Brain cancer. The comparative study between ACO-S and other metaheuristics: GA, ACO, PSO, SA has shown the efficiency of this method in terms of performances and dimensionality reduction.

To create more robust methods, researchers were interested by the hybridization between metaheuristics to create co-operation between the hybrid method and mainly to make a tradeoff between exploration and exploitation characteristics.

Hybrid metaheuristic based on GA algorithm got a lot of interest [22,23,26]. Das et al. [22] proposed an hybrid method based on Harmony Search (HS) and GA metaheuristics, the purpose of this hybridization is to perform a preselection step by HS and to provide GA with the good selected solution by HS. The proposed approach was tested on 5 microarray datasets (Colon, SRBT, Leukeumia, Prostate Tumor, Lung Cancer). The comparative study between the hybrid method and its component: GA and HS prove the efficiency of co-operation in such application. Alshamlan et al. [23] proposed an hybridization (GBC) between Artificial Bee Colony (ABC) and GA metaheuristics. The main purpose of this hybridization is to make a tradeoff between the exploration of

GA and the exploitation of GBC and to improve the local search of ABC. The proposed method was tested on 6 microarray datasets (Colon, Leukemia, Lung, SRBCT, Lymphoma, Leukemia2). The comparative study between GBC and other feature selection methods (MRMR-ABC, ABC) proves its efficiency. Talbi et al. [26] proposed an hybridization (GPSO) between GA and PSO metaheuristics. The objective of the proposed approach is to improve the exploration of PSO by introducing crossover and mutation functions in the local search process of PSO. GPSO was performed on 4 microarray datasets: Colon Tumor, Breast Cancer, Ovarian Cancer and Lung Cancer. The obtained results by GPSO and other feature selection methods (GA, EA, PSO) proves its efficiency.

Sharbaf et al. [7] proposed an hybridization (CLACOFS) between cellular learning algorithm (CLA) and ant colony optimization (ANT). The proposed method was tested on 4 microarray datasets: ALL-AML leukemia, prostate tumor, MLL-leukemia and ALL-AML-4. The obtained results by CLACOFS proves its efficiency in terms of dimensionality reduction.

The hybridization of multiples of metaheuristics in a wrapper approach may be expensive in terms of run time complexity, especially for a high-dimensional dataset. To reduce time complexity and to ensure the efficiency of the method, other researchers have been interested by filter/wrapper hybridizations. The purpose of such methods is to reduce first the search space by a filter method, then perform an exhaustive search by a more robust wrapper approach.

Dashtban et al. [28] proposed an hybridization between a filter method based on Laplacian and fisher score and a wrapper approach based on GA. The obtained results on 5 microarray datasets (Breast, DLBCL, Prostate, Leukemia, SRBCT) proves the efficiency of the proposed approach. Other framework of wrappers/filter hybridizations (ECPM) was proposed by Boucheham et al. [30]. ECPM is based on three metaheuristics: PSO, GA and ACO.

3 Materials and Methods

3.1 Differential Evolution

DE [31] is a population-based algorithm, defined as the only algorithm designed to deal with continuous optimization problems [32]. As GA, DE is an evolutionary algorithm characterized by three main operations: selection, crossover and mutation. DE uses two main parameters: crossover rate (CR) and mutation factor (F), these two parameters control the exploration and exploitation of DE. This metaheuristic got a lot of interest to solve different feature selection problems [33–35].

3.2 Artificial Immune System

AIS is a bio inspired computation method that mimics human immune system. AIS is designed to solve optimization problems. This metaheuristic uses the clonal selection, this operation is similar to genetic algorithm process without crossover. Some properties affinity proportional mutation and proliferation make the difference between AIS and GA. AIS was used in many optimization problems such as parameters optimization and feature selection [36].

3.3 Firefly Algorithm

FA is a population-based algorithm proposed by Yang [37]. FA mimics the behavior of fireflies in their communication process via flash lights. Artificial fireflies behavior is based on the following three rules:

- All fireflies are unisex and can be attracted to each other.
- The attractiveness of a firefly is proportional to its brightness. The attractiveness decreases when the distance increases between two fireflies.
- The value of the objective function design the firefly attractiveness value.

3.4 BAT Algorithm

BAT [38] is a bio-inspired metaheuristic, it mimics the behavior of bats in their search for prey. The search process is based on communication and co-operation. However, the common purpose of bats is to find the best hunting strategy. This algorithm was previously used in feature selection problems [39].

3.5 Penguin Search Optimization Algorithm

PESOA [40] mimics the collaboration behavior of penguins in their hunting strategies. This metaheuristics is characterized by a set of solutions (penguins). Each penguin has a position, a reserve of oxygen, a group, and a probability to leave its current group. After the search process and expiration of the oxygen reserve, the penguin return to his group to share information about interesting areas. PESOA was used to solve many optimization problems such as: travelling salesman problem [41], the construction of association rules in machine learning [42] and feature selection [43].

4 Hybridizations

4.1 Modified Version of PeSOA (PeSOA-C)

As the majority of swarm intelligence algorithms, PeSOA is characterized by the exploitation of the search space. However, PeSOA risks to get trapped quickly into a local optimum. To prevent rapid convergence, an improved variant of PeSOA algorithm was proposed (PeSOA-C). This variant incorporates the two-point crossover function of GA into PeSOA. The purpose of this modification is to improve PeSOA exploration and to make a tradeoff between exploration and exploitation. The algorithm 1 illustrates PeSOA-C method.

Algorithm 1. PeSOA-C

```
Inputs: dataset, number of iterations Nmax = 50, number of solutions M = 50,
  objective function f.
Output: the best solution.
Begin
Initialize population pop of M penguin regrouped in G groups.
Evaluate each penguin with SVM classifier and 10-cross-validation.
Initialize the Best solution: Best.
Initialize the XLocalBest solution for each group.
while (n < Nmax) do
    Perform PESOA algorithm.
    for each (group Gr ∈ pop) do
        for each (penguin p ∈ Gr do) do
            Pick randomly a penguin p2 from Gr.
            p3 ← 2-Point-Crossover(p, p2).
            p ← 2-Point-Crossover(p3, Best).
            Evaluate p with SVM classifier and 10-cross-validation.
        Update Best solution.
End.
```

4.2 FA and DE Hybridization (HFA)

FA is characterized by the exploration of the search space [44]. Despite FA advantages, it can be trapped quickly into several local optimum [45]. To solve FA limitations, Zhang et al. [44] proposed an hybridization between FA and DE metaheuristics. The purpose of this hybridization is to take advantage from FA exploration and the local search of DE. Unlike the previous sequential hybridization, HFA is a parallel hybridization. The Fig. 1 explains the principle of this hybridization. First the initial population is divided into two groups: G1, and G2, G1 contains the best individuals and G2 contains the worst ones. Then, G1 is treated by FA and G2 by DE. In a third Step, the outputs are combined. This process is repeated until reaching the maximum number of iterations. Our ambition is to take advantages from this hybridization and to exploit HFA for gene selection in microarrays datasets.

4.3 BAT and AIS Hybridization (AIS-BAT)

BAT has three major problems: the lack of diversification, the problem of getting trapped local optimum and the fast convergence in case of high-dimensional spaces [46]. The proposed hybridization introduce the two operators of AIS: cloning and mutation. We propose to apply the mutation and cloning functions to generate new bats. The Algorithm 2 details the proposed hybridization.

5 Results and Discussion

In this section, the metaheuristics (PeSOA, FA, DE, BAT, AIS) and the proposed hybridizations (PeSOA-C, HFA, BAT-AIS) were performed on 11 microarray

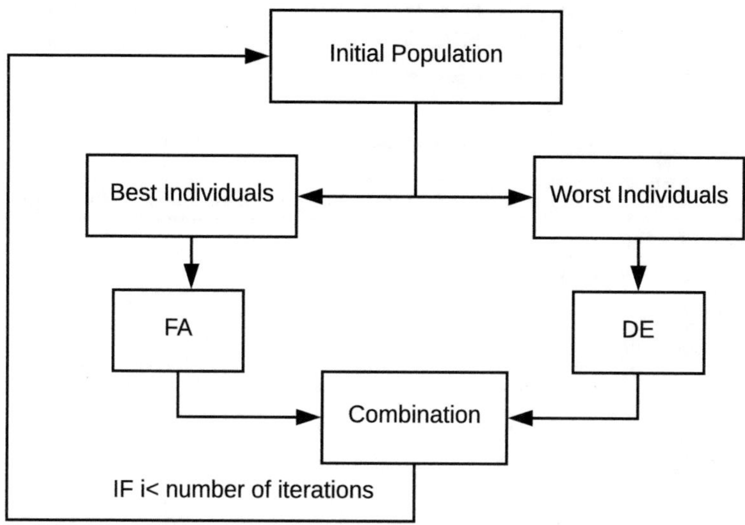

Fig. 1. HFA hybridization.

Algorithm 2. BAT-DE

Inputs: dataset, number of iterations Nmax = 50, number of solutions M = 50, objective function f.
Output: the best solution.
Begin
Initialize parameters: velocity $v_i = 0$, frequency $f_i = 0.1$, loudness A_i = Random(1,2) and pulse rate r_i = Random(0,1).
Initialize population pop of M bats.
Evaluate each bat with SVM classifier and 10-cross-validation.
Initialize the Best solution: Best.
while $(n < Nmax)$ **do**
  for each (bat ab $\in$ pop) **do**
    Generate new bats through BAT algorithm equations.
    Generate a random number rand $\in$ (0,1).
    if $(rand > r_i)$ **then**
      Clone each bat and mutate the clones with different rate.
      Evaluate each clone.
      if $(f_{clone} > f_i)$ **then**
        Replace initial subset with its clone.
    if $(f(ab) > f(best)$ and $rand < A_i)$ **then**
      Best $\leftarrow$ best solution
      Increase r_i and reduceA_i
End.

---

datasets. These datasets are described in Table 1. We used the support vector machine (SVM), this algorithm is conceded as the most powerful classifier in the microarray field [29]. For evaluation, we used the 10-cross-validation method and F-measure.

Table 1. Microarrays datasets description.

Datasets	Number of genes	Number of samples	Number of classes
Colon tumor [47]	2000	60	2
Central nervous system	7129	60	2
Leukemia [48]	7129	72	2
Bresat cancer [49]	24484	97	2
Lung cancer [50]	12533	181	2
Ovarian cancer [51]	15154	253	2
Leukemia_3c [48]	7129	72	3
Leukemia_4c [48]	7129	72	4
Lymphoma [52]	4026	62	3
MLL-Leukemia [53]	12582	72	3
SRBCT [54]	2308	83	4

The Table 2 summarizes the obtained results: f-measure value associated with the number of selected genes. The comparative study between results before and after genes selection shows that all performances were improved compared to metaheuristics and their hybridizations. We observe a remarkable improvement in case of Central Nervous System (from 0.677 to 0.847) by PeSOA-C method with only 7% of selected genes.

The comparative study between the results of PeSOA algorithms and its modified version (PeSOA-C) shows a 3% of improvement in f-measure value for the Central Nervous System dataset and 2% in case of Breast Cancer dataset. The same performance in the remaining cases with a reduction in dimensionality for Ovarian Cancer, Leukemia-3c and MLL-leukemia datasets.

The comparative study between the results of FA, DE and their hybridization (HFA) shows that HFA was more promising in the majority of cases, expected for: Colon Tumor, Central Nervous System and SRBCT datasets. We observe that HFA proved its efficiency especially in dimensionality reduction. However, HFA has the same F-measure values as FA or DE in the majority of cases.

The obtained results by the metaheuristics BAT and AIS and their hybridization shows that AIS has performed better than its sequential hybridization with BAT.

The Table 3 summarize the best obtained performances, and the percentage of selected genes for the best subset in case of each dataset, the results show that the best performances were obtained by the hybridization of metaheuristics. we observe a remarkable reduction in dimensionality, however, the

Table 2. Obtained results.

Dataset	Before	PeSOA	PeSOA-C	FA	DE	HFA	BAT	AIS	BAT-AIS
1	0.854 (2000)	0.918 (120)	0.919 (422)	0.919 (127)	0.903 (155)	0.919 (550)	0.887 (1315)	0.904 (989)	0.872 (1004)
2	0.677 (7129)	0.815 (400)	0.847 (502)	0.831 (884)	0.844 (374)	0.828 (829)	0.740 (4588)	0.740 (3539)	0.725 (3540)
3	0.958 (7129)	1.0 (483)	1.0 (700)	1.0 (880)	0.986 (1139)	1.0 (301)	0.986 (3614)	0.986 (3508)	0.986 (3546)
4	0.680 (12600)	0.762 (1910)	0.793 (2925)	0.803 (3052)	0.793 (1770)	0.813 (2036)	0.732 (16031)	0.721 (12315)	0.721 (12295)
5	0.954 (12600)	0.975 (613)	0.975 (1196)	0.970 (1691)	0.975 (1643)	0.975 (1090)	0.959 (8770)	0.959 (6324)	0.959 (6397)
6	1.0 (15154)	1.0 (4786)	1.0 (4586)	1.0 (581)	1.0 (368)	1.0 (355)	1.0 (10024)	1.0 (7497)	1.0 (7501)
7	0.972 (7129)	0.985 (1141)	0.986 (315)	0.985 (892)	0.985 (808)	0.985 (577)	0.985 (5094)	0.985 (3460)	0.985 (3512)
8	0.927 (7129)	0.971 (503)	0.971 (565)	0.971 (1562)	0.957 (942)	0.971 (844)	0.943 (4795)	0.957 (3489)	0.986 (6097)
9	1.0 (4028)	1.0 (1686)	1.0 (1686)	1.0 (105)	1.0 (111)	1.0 (99)	1.0 (2941)	1.0 (1968)	1.0 (2024)
10	0.972 (12582)	1.0 (849)	1.0 (550)	1.0 (555)	1.0 (1192)	1.0 (553)	0.972 (9124)	0.986 (6370)	0.986 (5846)
11	0.988 (2308)	1.0 (315)	1.0 (315)	1.0 (85)	1.0 (52)	1.0 (120)	1.0 (1681)	1.0 (1580)	1.0 (1032)

Table 3. Best obtained results.

Datasets	Initial performance	Final performance	Percentage of number of selected genes	Metaheuristic or Hybridization
Colon tumor	0.854	0.919	6%	FA
Central nervous system	0.677	0.847	7%	PeSOA-C
Leukemia	0.958	1.0	4%	HFA
Breast cancer	0.680	0.813	8%	HFA
Lung cancer	0.954	0.975	4%	PeSOA
Ovarian cancer	1.0	1.0	2%	HFA
Leukemia_3c	0.972	0.986	4%	PeSOA-C
Leukemia_4c	0.927	0.986	85%	BAT-AIS
Lymphoma	1.0	1.0	2%	HFA
MLL-Leukemia	0.972	1.0	4%	PeSOA-C
SRBCT	0.988	1.0	2%	DE

Table 4. Comparison with the state of the art

Datasets	MBEGA [21] %	MRMR-GBC [23] %	HS-GA [22]	CLA-ACO [7] %	ICA-FBFE [55] %	LFS-GA [28] %	MRMR-MOEDA [29]	ACO-S [16] %	FA [25]	GPSO [26] %	Ours
Colon tumor	85.66	98.38 (20)	0.986 (197)	–	90.1 (30)	–	0.87 (19)	81.42 (69)	–	100 (2)	0.919 (127)
Central nervous system	72.1	–	–	–	–	–	–	–	–	–	0.847 (502)
Leukemia	95.89	100 (5)	1.0 (708)	95.95 (2)	94.2 (35)	88.2% (15)	1.0 (6)	91.68 (84)	0.8 (11)	–	1.0 (301)
Breast cancer	80.74	–	–	–	–	91.0 (6)	–	89.25 (85)	–	86.35 (4)	0.813 (2036)
Lung cancer	98.96	100 (8)	–	–	91.23 (80)	–	–	–	–	99 (4)	0.975 (613)
Ovarian cancer	99.71	–	–	–	–	–	–	–	–	99.44 (4)	1.0 (355)
Leukemia_3c	96.64	100 (8)	–	–	–	–	1.0 (4)	–	–	–	0.986 (6097)
Leukemia_4c	91.93	–	–	–	–	–	–	–	–	–	0.986 (6097)
Lymphoma	97.68	–	–	–	–	–	–	–	0.73 (13)	–	1.0 (99)
MLL-leukemia	94.33	–	–	–	–	–	–	–	0.77 (22)	–	1.0 (550)
SRBCT	99.23	100 (6)	0.9882 (266)			85% (18)	0.96 (4)		0.75 (12)		1.0 (52)

percentage of selected genes is less than 8% excepted for Leukemia_3c dataset. We obtained a perfect 100% in case of Leukemia, Ovarian Cancer, Lymphoma and MLL-Leukemia datasets. In a second step we compared the best obtained results with other state of the art results. The results are illustrated in Table 4. The different methods are categorized into wrappers, filters and the hybridization filters/wrappers. We observe that the obtained results were more promising compared to: MBEGA, CLA-ACOA, ICA-FBFE, FA methods, and competitive to HS-GA, LFS-GA, ACO-S, MRMR-MOEDA methods, while GPSO, MRMR-GBC outperformed the proposed hybridizations.

6 Conclusion

Gene selection in microarray datasets is a challenging task. To solve this problem we proposed sequential and parallel hybrid metaheuristics: PeSOA-C, HFA, AIS-BAT. The experiments showed that the selection of informative genes improves the performances. Our best results were obtained in most cases by the hybridizations PeSOA-C and HFA. We also found that some metaheuristics perform better than their hybridization (AIS-BAT). Which is very understandable, since the stochastic nature of metaheuristics.

Perspectives to this work are to test the proposed hybridization on other application domains and to develop more robust hybridizations.

References

1. Harrington, C.A., Rosenow, C., Retief, J.: Monitoring gene expression using DNA microarrays. Curr. Opin. Microbiol. **3**(3), 285–291 (2000)
2. van Hal, N.L., Vorst, O., van Houwelingen, A.M., Kok, E.J., Peijnenburg, A., Aharoni, A.,., Keijer, J., : The application of DNA microarrays in gene expression analysis. J. Biotechnol. **78**(3), 271–280 (2000)
3. Michiels, S., Koscielny, S., Hill, C.: Prediction of cancer outcome with microarrays: a multiple random validation strategy. Lancet **365**(9458), 488–492 (2005)
4. Cho, S.B., Won, H.H.: Machine learning in DNA microarray analysis for cancer classification. In: Proceedings of the First Asia-Pacific Bioinformatics Conference on Bioinformatics, vol. 19, pp. 189–198. Australian Computer Society, Inc. (2003)
5. Mao, Z., Cai, W., Shao, X.: Selecting significant genes by randomization test for cancer classification using gene expression data. J. Biomed. Inform. **46**(4), 594–601 (2013)
6. Zhang, H., Li, L., Luo, C., Sun, C., Chen, Y., Dai, Z., Yuan, Z.: Informative gene selection and direct classification of tumor based on chi-square test of pairwise gene interactions. BioMed Res. Int. (2014)
7. Sharbaf, F.V., Mosafer, S., Moattar, M.H.: A hybrid gene selection approach for microarray data classification using cellular learning automata and ant colony optimization. Genomics **107**(6), 231–238 (2016)
8. Alba, E., Garcia-Nieto, J., Jourdan, L., Talbi, E.G.: Gene selection in cancer classification using PSO/SVM and GA/SVM hybrid algorithms. In: IEEE Congress on 2007 Evolutionary Computation, CEC 2007, pp. 284–290. IEEE (2007)

9. Peng, H., Long, F., Ding, C.: Feature selection based on mutual information criteria of max-dependency, max-relevance, and min-redundancy. IEEE Trans. Pattern Anal. Mach. Intell. **27**(8), 1226–1238 (2005)
10. Maldonado, S., Weber, R.: A wrapper method for feature selection using support vector machines. Inf. Sci. **179**(13), 2208–2217 (2009)
11. Apolloni, J., Leguizamn, G., Alba, E.: Two hybrid wrapper-filter feature selection algorithms applied to high-dimensional microarray experiments. Appl. Soft Comput. **38**, 922–932 (2016)
12. Anaissi, A., Kennedy, P.J., Goyal, M.: Feature selection of imbalanced gene expression microarray data. In: 2011 12th ACIS International Conference on Software Engineering, Artificial Intelligence, Networking and Parallel/distributed Computing (snpd), pp. 73–78. IEEE (2011)
13. Siedlecki, W., Sklansky, J.: On automatic feature selection. In: Handbook of Pattern Recognition and Computer Vision, pp. 63–87 (1993)
14. Liu, H., Setiono, R.: A probabilistic approach to feature selection-a filter solution. In: ICML, vol. 96, pp. 319–327 (1996)
15. Zhong, N., Dong, J., Ohsuga, S.: Using rough sets with heuristics for feature selection. J. Intell. Inf. Syst. **16**(3), 199–214 (2001)
16. Li, Y., Wang, G., Chen, H., Shi, L., Qin, L.: An ant colony optimization based dimension reduction method for high-dimensional datasets. J. Bionic Eng. **10**(2), 231–241 (2013)
17. Jona, J.B., Nagaveni, N.: Ant-cuckoo colony optimization for feature selection in digital mammogram. Pak. J. Biol. Sci. **17**(2), 266 (2014)
18. Prasad, Y., Biswas, K.K (eds.): In: AAAI, pp. 4288–4289 (2015)
19. Glover, F.: Future paths for integer programming and links to artificial intelligence. Computers operations research **13**(5), 533–549 (1986)
20. Dif, N., Elberrichi, Z.: Microarray data feature selection and classification using an enhanced multi-verse optimizer and support vector machine (2017)
21. Yang, K., Cai, Z., Li, J., Lin, G.: A stable gene selection in microarray data analysis. BMC Bioinform. **7**(1), 228 (2006)
22. Das, K., Mishra, D., Shaw, K.: A metaheuristic optimization framework for informative gene selection. Inform. Med. Unlocked **4**, 10–20 (2016)
23. Alshamlan, H.M., Badr, G.H., Alohali, Y.A.: Genetic Bee Colony (GBC) algorithm: a new gene selection method for microarray cancer classification. Comput. Biol. Chem. **56**, 49–60 (2015)
24. Zhu, Z., Ong, Y.S., Dash, M.: Markov blanket-embedded genetic algorithm for gene selection. Pattern Recognit. **40**(11), 3236–3248 (2007)
25. Deepthi, P.S., Thampi, S.M.: A metaheuristic approach for simultaneous gene selection and clustering of microarray data. Intell. Syst. Technol. Appl., pp. 449–461. Springer, Cham (2016)
26. Talbi, E.G., Jourdan, L., Garcia-Nieto, J., Alba, E.: Comparison of population based metaheuristics for feature selection: application to microarray data classification. In: 2008 IEEE/ACS International Conference on Computer Systems and Applications, AICCSA 2008, pp. 45–52. IEEE (2008)
27. Chen, K.H., Wang, K.J., Tsai, M.L., Wang, K.M., Adrian, A.M., Cheng, W.C., Chang, K.S.: Gene selection for cancer identification: a decision tree model empowered by particle swarm optimization algorithm. BMC Bioinform. **15**(1), 49 (2014)
28. Dashtban, M., Balafar, M.: Gene selection for microarray cancer classification using a new evolutionary method employing artificial intelligence concepts. Genomics **109**(2), 91–107 (2017)

29. Lv, J., Peng, Q., Chen, X., Sun, Z.: A multi-objective heuristic algorithm for gene expression microarray data classification. Expert. Syst. Appl. **59**, 13–19 (2016)
30. Boucheham, A., Batouche, M., Meshoul, S.: An ensemble of cooperative parallel metaheuristics for gene selection in cancer classification. In: International Conference on Bioinformatics and Biomedical Engineering, pp. 301–312. Springer, Cham (2015)
31. Storn, R., Price, K.: Differential evolutiona simple and efficient heuristic for global optimization over continuous spaces. J. Glob. Optim. **11**(4), 341–359 (1997)
32. Do, D.T., Lee, S., Lee, J.: A modified differential evolution algorithm for tensegrity structures. Compos. Struct. **158**, 11–19 (2016)
33. Khushaba, R.N., Al-Ani, A., Al-Jumaily, A.: Differential evolution based feature subset selection. In: 2008 19th International Conference on Pattern Recognition. ICPR 2008, pp. 1–4. IEEE (2008)
34. Sikdar, U.K., Ekbal, A., Saha, S.: Differential evolution based feature selection and classifier ensemble for named entity recognition. Proc. COLING **2012**, 2475–2490 (2012)
35. Chattopadhyay, S., Mishra, S., Goswami, S.: Feature selection using differential evolution with binary mutation scheme, pp. 1–6. IEEE (2016)
36. Shojaie, S., Moradi, M.H.: An evolutionary artificial immune system for feature selection and parameters optimization of support vector machines for ERP assessment in a P300-based GKT. In: 2008 Biomedical Engineering Conference, CIBEC 2008, Cairo International, pp. 1–5. IEEE (2008)
37. Yang, X.S.: Nature-Inspired Metaheuristic Algorithms. Luniver Press (2010)
38. Yang, X.S.: A new metaheuristic bat-inspired algorithm. In: Nature Inspired Cooperative Strategies for Optimization (NICSO 2010), pp. 65–74. Springer, Heidelberg (2010)
39. Mishra, S., Shaw, K., Mishra, D.: A new meta-heuristic bat inspired classification approach for microarray data. Procedia Technol. **4**, 802–806 (2012)
40. Gheraibia, Y., Moussaoui, A.: Penguins search optimization algorithm (PeSOA). In: International Conference on Industrial, Engineering and Other Applications of Applied Intelligent Systems, pp. 222–231. Springer, Heidelberg (2013)
41. Mzili, I., Riffi, M.E.: Discrete penguins search optimization algorithm to solve the traveling salesman problem. J. Theor. Appl. Inf. Technol. **72**(3) (2015)
42. Gheraibia, Y., Moussaoui, A., Djenouri, Y., Kabir, S., Yin, P.Y.: Penguins search optimisation algorithm for association rules mining. J. Comput. Inf. Technol. **24**(2), 165–179 (2016)
43. Bidi, N., Elberrichi, Z.: Using Penguins search optimization algorithm for best features selection for biomedical data classification. Int. J. Organ. Collect. Intell. (IJOCI) **7**(4), 51–62 (2017)
44. Zhang, L., Liu, L., Yang, X.S., Dai, Y.: A novel hybrid firefly algorithm for global optimization. PloS One **11**(9), e0163230 (2016)
45. Mazen, F., Abul Seoud, R.A., Gody, A.M.: Genetic algorithm and firefly algorithm in a hybrid approach for breast cancer diagnosis. Int. J. Comput. Trends Technol. (IJCTT) **32**, (2) (2016)
46. Fister Jr, I., Fister, D., Yang, X.S.: A hybrid bat algorithm (2013). arXiv preprint arXiv:1303.6310
47. Alon, U., Barkai, N., Notterman, D.A., Gish, K., Ybarra, S., Mack, D., Levine, A.J.: Broad patterns of gene expression revealed by clustering analysis of tumor and normal colon tissues probed by oligonucleotide arrays. Proc. Natl. Acad. Sci. **96**(12), 6745–6750 (1999)

48. Golub, T.R., Slonim, D.K., Tamayo, P., Huard, C., Gaasenbeek, M., Mesirov, J.P., Bloomfield, C.D.: Molecular classification of cancer: class discovery and class prediction by gene expression monitoring. Science **286**(5439), 531–537 (1999)
49. Van't Veer, L.J., et al.: Gene expression profiling predicts clinical outcome of breast cancer. Nature **415**(6871), 530 (2002)
50. Bhattacharjee, A., Richards, W.G., Staunton, J., Li, C., Monti, S., Vasa, P., Loda, M.: Classification of human lung carcinomas by mRNA expression profiling reveals distinct adenocarcinoma subclasses. Proc. Natl. Acad. Sci. **98**(24), 13790–13795 (2001)
51. Petricoin III, E.F., Ardekani, A.M., Hitt, B.A., Levine, P.J., Fusaro, V.A., Steinberg, S.M., Liotta, L.A.: Use of proteomic patterns in serum to identify ovarian cancer. Lancet **359**(9306), 572–577 (2002)
52. Alizadeh, A.A., Eisen, M.B., Davis, R.E., Ma, C., Lossos, I.S., Rosenwald, A., Powell, J. I.: Distinct types of diffuse large B-cell lymphoma identified by gene expression profiling. Nature **403**(6769), 503 (2000)
53. Armstrong, S.A., Staunton, J.E., Silverman, L.B., Pieters, R., den Boer, M.L., Minden, M.D., Korsmeyer, S. J.: MLL translocations specify a distinct gene expression profile that distinguishes a unique leukemia. Nat. Genet. **30**(1), 41 (2002)
54. Khan, J., Wei, J.S., Ringner, M., Saal, L.H., Ladanyi, M., Westermann, F., Meltzer, P. S.: Classification and diagnostic prediction of cancers using gene expression profiling and artificial neural networks. Nat. Med. **7**(6), 673 (2001)
55. Aziz, R., Verma, C.K., Srivastava, N.: A fuzzy based feature selection from independent component subspace for machine learning classification of microarray data. Genomics Data **8**, 4–15 (2016)

Optimization of PID Sliding Surface Using Ant Lion Optimizer

Diab Mokeddem(✉) and Hakim Draidi

Department of Electrical Engineering, Faculty of Technology, University of Ferhat Abbas Setif-1, Setif, Algeria
mokeddem_d@yahoo.fr, hakimdraidi19@gmail.com

Abstract. In this paper, a sliding mode control SMC system with a proportional integral derivative PID sliding surface is presented. The main contribution in this work is to determine the optimal values of the PID sliding surface parameters using biologically-inspired algorithm, namely Ant lion optimization (ALO). This technique guarantee a robust sliding mode controller insensitive to uncertainty conditions, nonlinear dynamics, external disturbances and allowing the system to reach maximum switching and minimum chattering. The proposed system stability during reaching phase and sliding phase is mathematically confirmed by Lyapunov theorem. Simulation results of ALO tuning of PID sliding surface proved better tracking performance of the desired trajectory compared to conventional SMC.

Keywords: Optimization · Ant lion optimizer ALO · Sliding mode controller SMC · PID sliding surface · Genetic algorithm · PSO

1 Introduction

Technology progress made industrial processes requirement more and more complex with multivariable, nonlinear dynamics, parameter uncertainties and external disturbance. Thus, modeling such systems mathematically with adequate control law is a serious challenge. Various control strategies have been proposed to solve these problems such as adaptive control, feedback linearization control, sliding mode control and artificial intelligence techniques.

Sliding Mode Control (SMC) is successfully used in various engineering fields since the late 1970s (Bandyopadhyay et al. [1]; Utkin [2]; Young et al. [3]), due to its insensitivity to parameter variations and disturbances affecting the plant. Perruquetti and Barbot [4], Utkin [2] defines the SMC as an efficient tool for designing robust controllers for linear or nonlinear control systems operating under uncertainty conditions.

In the early 1950s, S.V. Emel'yanov introduced the basic steps of sliding mode control theory as Variable Structure Control (VSC) (Zinober [5]). Firstly, a stable sliding surface is selected according to control objectives and desired properties of the closed loop system. Then, a discontinuous control law is synthesized to restrict the state trajectory of the system in the sliding surface with high frequency switching of a nonlinear control input [6]. High frequency switching over the control signal causes the

S. Chikhi et al. (Eds.): MISC 2018, LNNS 64, pp. 133–145, 2019.
https://doi.org/10.1007/978-3-030-05480-9_10

system chattering. Hence, the control law parameters should be optimized in respect to the reaching rate and the chattering.

In this context, Eker [7] adopted a SMC system with a PID sliding surface to control the speed of an electromechanical system. Also Aliakbari et al. [8] designed an adaptive second-order SMC with a PID sliding surface to overcome the faults in the heat recovery steam generator boilers. Another adaptive fuzzy SMC approach with a PID sliding surface is proposed by Amer et al. [9], for the control of robot manipulators. All the aforementioned works motivate us to enhance the performance of SMC controller by optimizing the PID sliding surface parameters.

Many optimization methods attracted researchers interest especially nature inspired ones such as Genetic Algorithm (GA) [10], Particle Swarm Optimization (PSO) [11], Firefly Algorithm (FA) [12], Ant Colony Optimization (ACO) [13], artificial bee colony (ABC) [14] and Ant Lion Optimizer (ALO) [15]. ALO algorithm is recently introduced By Mirjalili in 2015 based on the hunting mechanism of ant lions in nature. It is mainly characterized by fast speed convergence.

In the recent work, the intelligent technique ALO is successfully applied, for the time, to design the optimal values of a PID sliding surface parameters to ensure maximum reaching rate and minimum chattering. The paper remainder is organised as follows: Sect. 2 presents the Ant Lion algorithm. Then, Sect. 3 demonstrates the design of a PID sliding surface parameters and the stability analysis of the SMC controller. Results discussion is reported in Sect. 4. Section 5 summarizes the conclusion.

2 Ant Lion Algorithm

In recent time, algorithms are becoming much more inspired from nature. In 2015 Mirjalili [15] proposed Ant Lion Optimizer (ALO) which mimics the hunting behavior of ant lions in nature. An ant lion larva digs a cone shaped pit in sand by moving along a circular path and throwing out sands with its massive jaw. The size of the trap depends on the hungry level of the ant lion and moon size. After digging the trap, the larva hides underneath the bottom of the cone and waits for the prey to be trapped in the pit. If the prey comes into the cone surface it will easily fall down into it. Once the ant lion realizes that a prey is in the trap, it tries to catch it (Fig. 1). There are mainly five operations in ALO algorithm namely random movement of ants, construction of trap, trapping of ants in traps, catching preys and re-construction of traps. These operations are formulated as follows:

The random positions of the ants are saved in the matrix M_{ant}

$$M_{ant} = \begin{bmatrix} A_{1,1} & A_{1,2} & \cdots & \cdots & A_{1,d} \\ A_{2,1} & A_{2,2} & \cdots & \cdots & AL_{2,d} \\ \vdots & \vdots & \vdots & \vdots & \vdots \\ \vdots & \vdots & \vdots & \vdots & \vdots \\ A_{n,1} & AL_{n,2} & \cdots & \cdots & A_{n,d} \end{bmatrix} \quad (1)$$

$A_{m,\ n}$ is the value of m-th variable (dimension) of n-th ant, n = number of ants (population size).

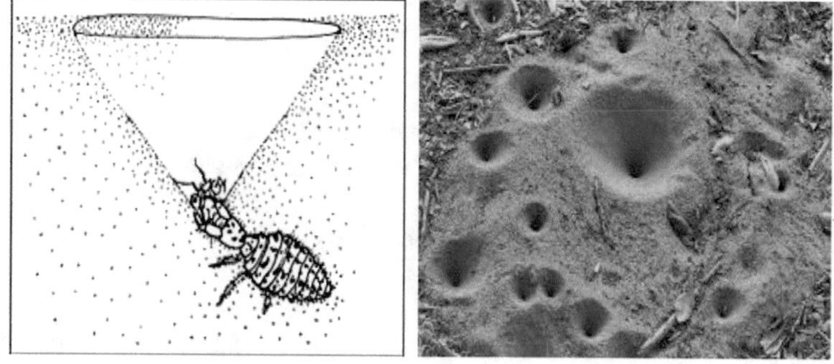

Fig. 1. Hunting behavior of ant lions.

Fitness of all ants will be stored in the matrix M_{OA} in terms of objective function *f*:

$$M_{OA} = \begin{bmatrix} f([A_{1,1} & A_{1,2} & \cdots & \cdots & A_{1,d}]) \\ f([A_{2,1} & A_{2,2} & \cdots & \cdots & A_{2,d}]) \\ \vdots & \vdots & \vdots & \vdots & \vdots \\ \vdots & \vdots & \vdots & \vdots & \vdots \\ f([A_{n,1} & A_{n,2} & \cdots & \cdots & A_{n,d}]) \end{bmatrix} \tag{2}$$

The position and fitness of ant lion are represented by the matrices $M_{antlion}$, M_{OAL} respectively.

$$M_{antlion} = \begin{bmatrix} AL_{1,1} & AL_{1,2} & \cdots & \cdots & AL_{1,d} \\ AL_{2,1} & AL_{2,2} & \cdots & \cdots & AL_{2,d} \\ \vdots & \vdots & \vdots & \vdots & \vdots \\ \vdots & \vdots & \vdots & \vdots & \vdots \\ AL_{n,1} & AL_{n,2} & \cdots & \cdots & AL_{n,d} \end{bmatrix} \tag{3}$$

$$M_{OAL} = \begin{bmatrix} f([AL_{1,1} & AL_{1,2} & \cdots & \cdots & AL_{1,d}]) \\ f([AL_{2,1} & AL_{2,2} & \cdots & \cdots & AL_{2,d}]) \\ \vdots & \vdots & \vdots & \vdots & \vdots \\ \vdots & \vdots & \vdots & \vdots & \vdots \\ f([AL_{n,1} & AL_{n,2} & \cdots & \cdots & AL_{n,d}]) \end{bmatrix} \tag{4}$$

2.1 Random Movement of Ants

Ants randomly move to search for food. Random movement of ants is given by Eq. (5)

$$X(t) = [0, \text{cumsum}(2r(t_1) - 1), \text{cumsum}(2r(t_2) - 1), \ldots, \text{cumsum}(2r(t_n) - 1)] \tag{5}$$

where *cumsum* represents cumulative sum, *n* = maximum number of ants, *t* = step of random walk (iteration) and

$$r(t) = \begin{cases} 1, rand > 0.5 \\ 0, rand \leq 0.5 \end{cases} \tag{6}$$

To restrict the random movement within the search space, normalized form is used, which is based on min–max normalization. Position of ants can be updated by Eq. (7).

$$X_i^t = \frac{(X_i^t - a_i) \times (d_i - c_i^t)}{(b_i - a_i)} + c_i \tag{7}$$

where a_i, b_i are minimum and maximum of random walk of ants c_i^t, d_i represents minimum and maximum i-th variable at t-th iteration.

2.2 Trapping of Ants

The mathematical expression of the trapping of the ants to the ant lion's pits is given by Eqs. (8) and (9)

$$c_i^t = Antlion_j^t + c^t \tag{8}$$

$$d_i^t = Antlion_j^t + d^t \tag{9}$$

where

c^t, d^t are the minimum and maximum of all variables at t-th iteration, and $Antlion_j^t$ shows the position of the selected j-th ant lion at t-th iteration.

2.3 Construction of Trap

The fittest ant lion is selected using the roulette wheel method Ant's sliding towards ant lion is given by Eqs. (10) and (11)

$$c^t = \frac{c^t}{I} \tag{10}$$

$$d^t = \frac{d^t}{I} \tag{11}$$

where

$I = 10^w \frac{t}{maxiter}$, t is current iteration, $maxiter$ is the maximum number of iterations and w is a constant whose value is given by:

$$w = \begin{cases} 2 \text{ if } t > 0.1\, maxiter \\ 3 \text{ if } t > 0.5\, \text{maxiter} \\ 4 \text{ if } t > 0.75\, maxiter \\ 5 \text{ if } t > 0.9\, maxiter \\ 6 \text{ if } t > 0.95\, maxiter \end{cases} \tag{12}$$

2.4 Catching the Prey and Reconstruction of the Pit

Ant lion catches the ant when it reaches the pit bottom and consumes it. After this, ant lion has to update the position in order to catch a new prey. This process is represented by Eq. (13).

$$Antlion_j^t = Ant_i^t \text{ if } f(Ant_i^t) > f(Antlion_j^t) \tag{13}$$

2.5 Elitism

Elitism is an important characteristic of evolutionary algorithms that allows them to maintain the best solution(s) obtained at any stage of optimization process. Best ant lion obtained is treated as elite, which is the fittest ant lion. Elite should affect the ant lion in every stage (random movement). For this, every ant is assumed to associate with an ant lion and elite by Roulette wheel is given by Eq. (14).

$$Ant_i^t = \frac{R_A^t + R_E^t}{2} \tag{14}$$

R_A, R_E represents the random walk around the selected ant lion and elite at t-th iteration simultaneously and Ant_i^t indicate the position of i-th ant at t-th iteration.

3 Sliding Mode Controller Design Using PID Sliding Surface

Sliding mode control (SMC) is an efficient technique for designing robust controllers for linear or nonlinear systems operating under uncertainty conditions and external disturbances. SMC is a Variable Structure System (VSS) consisting of two main steps [6].

The first step is designing a sliding surface so that the plant restricted to the sliding surface has a desired system response. This means the state variables of the plant dynamics are constrained to satisfy another set of equations which define the switching surface.

The second step is constructing a switched feedback gains necessary to drive the plant's state trajectory to the sliding surface. These constructions are built on the generalized Lyapunov stability theory.

Consider a plant as:

$$\ddot{\theta} = -a\dot{\theta} - b\theta + cu(t) \tag{15}$$

where $\theta(t)$ is position signal and $u(t)$ is control input

Defining $x_1 = \theta$, $x_2 = \dot{\theta}$ can be written as:

$$\begin{cases} \dot{x}_1 = x_2 \\ \dot{x}_2 = -ax_2 - bx_1 + cu(t) \\ y = x_1 \end{cases} \tag{16}$$

The trajectories are enforced to lie on the sliding surfaces. Let the tracking error in a closed-loop control system be *e*(*t*), then a sliding proportional integral derivative PID surface in the space of error can be defined as:

$$S(t) = K_p e(t) + K_i \int_0^t e(\tau)d\tau + K_d \frac{de(t)}{dt} \tag{17}$$

Where K_p is the proportional gain, K_i is the integration gain, and K_d is the derivative gain [16–19]. If the system trajectory has reached the sliding surface $S(t) = 0$, it remains on it while sliding into the origin $e(t) = 0, = \dot{e}(t) = 0$.

The error *e*(*t*) can be defined in terms of physical plant parameters as:

$$e(t) = \theta_r(t) - \theta(t) \tag{18}$$

where $\theta_r(t)$ is desired state.

The derivative of sliding surface defined by Eq. (17) can be given as:

$$\dot{S}(t) = K_p \dot{e}(t) + K_i e(t) + K_d \ddot{e}(t) \tag{19}$$

A necessary condition for output trajectory to remain on the sliding surface *s*(*t*) is $\dot{s}(t)$:

$$K_p \dot{e}(t) + K_i e(t) + K_d \ddot{e}(t) = 0 \tag{20}$$

Suitable parameter values of a proportional integral derivative (PID) sliding surface K_p, K_i, and K_d are optimized with Ant lion optimizer (ALO) to ensure the asymptotical

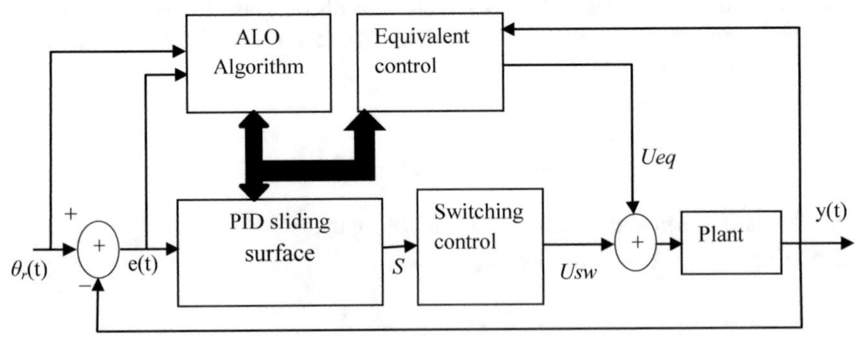

Fig. 2. Block diagram of the SMC with PID sliding surface

stability of the system. The block diagram of conventional SMC with the PID sliding surface is shown in Fig. 2.

It is necessary to search the global optimal solution in the problem while satisfying the performance criteria [20, 21] which is measured in terms of Integral Absolute Error (IAE) and given by:

$$IAE = \int_0^T |e(t)|dt \tag{21}$$

where T is the time at which the response reaches steady state.

The second derivative of the error $e(t)$ is:

$$\ddot{e}(t) = \ddot{\theta}_r(t) - \ddot{\theta}(t) \tag{22}$$

We obtain

$$K_p\dot{e}(t) + K_ie(t) + K_d\left(\ddot{\theta}_r(t) + a\dot{\theta}(t) + b\theta(t) - cu(t)\right) = 0 \tag{23}$$

Then the equivalent control $U_{eq}(t)$ is:

$$U_{eq}(t) = (cK_d)^{-1}\left(K_p\dot{e}(t) + K_ie(t) + K_d\ddot{\theta}_r(t) + K_da\dot{\theta}(t) + K_db\theta(t)\right) \tag{24}$$

The switching control can be chosen as signum function of sliding surface [6], hence the following switching control law is proposed:

$$U_{sw} = -kSsign(S) \tag{25}$$

Where k are positive coefficients and

$$sign(S) = \left\{ \begin{array}{c} 1\, if\, S > 0 \\ 0\, if\, S = 0 \\ -1\, if\, S < 0 \end{array} \right\} \tag{26}$$

The design sliding mode controller is expressed as follows:

$$\begin{aligned} U &= U_{eq}(t) + U_{sw} \\ &= (cK_d)^{-1}\left(K_p\dot{e}(t) + K_ie(t) + K_d\ddot{\theta}_r(t) + K_da\dot{\theta}(t) + K_db\theta(t)\right) - kSsign(S) \end{aligned} \tag{27}$$

To verify the stable convergence behavior of nonlinear controller, Lyapunov functions is introduced:

$$V = \frac{1}{2} S^2 \tag{28}$$

With V(0) = 0 and v(t) > 0 for $S \neq 0$

To guarantee the trajectory movement from reaching phase to sliding phase and ensure the stability, it is necessary to follow the reaching condition:

$$\dot{V}(t) < 0 \text{ for } S \neq 0 \tag{29}$$

Then we get

$$\begin{aligned}
\dot{V}(t) &= S\dot{S} = S\left(K_p\dot{e}(t) + K_i e(t) + K_d \ddot{e}(t)\right) \\
&= S\left\{K_p e(t) + K_i e(t) + K_d\left(\ddot{\theta}_r(t) + a\dot{\theta}(t) + b\theta(t) - c\left(U_{eq}(t) + U_{sw}\right)\right)\right\} \\
&= S\left\{K_p e(t) + K_i e(t) + K_d\left(\ddot{\theta}_r(t) + a\dot{\theta}(t) + b\theta(t) - c\left((cK_d)^{-1}\left(K_p\dot{e}(t) + K_i e(t) + K_d\ddot{\theta}_r(t) + K_d a\dot{\theta}(t) + K_d b\theta(t)\right) + U_{sw}\right)\right)\right\}
\end{aligned} \tag{30}$$

Then

$$\dot{V} = S(-aSsign(S)) \tag{31}$$

We have $\dot{V} < -k(|S|)^2$ for $S \neq 0$

To ensure $\dot{V}$ is a negative definite function, the following condition is needed: $k > 0$

4 Results and Discussion

In order to evaluate and validate the performance of the ALO algorithm to determine the optimal values of PID sliding surface parameters, two reference trajectories are presented.

Consider the following second order system:

$$\begin{cases} \dot{x}_1 = x_2 \\ \dot{x}_2 = -2.58x_2 - 1.6x_1 + 1.6u(t) \\ y = x_1 \end{cases} \tag{32}$$

4.1 Case 1: Unit Step Reference Trajectory

The position tracking performance of the proposed technique is achieved by feeding the plant with a step reference input signal. Figure 3 shows the controller tracking ability.

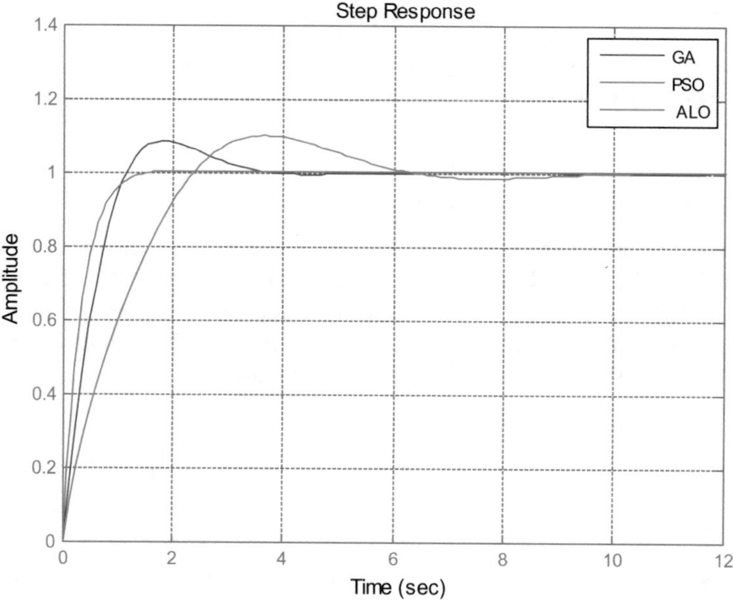

Fig. 3. Step response of SMC controller with different algorithm tuning.

Table 1. PID sliding surface parameters

Algorithm	K_p	K_i	K_d
GA	3.1639	2.9716	0.9000
PSO	1.1650	1.4503	0.6659
ALO	4.8651	3.1326	1.8262

Table 2. Step response performance for PID sliding surface.

Algorithm	Rise time	Settling time	Overshoot (%)
GA	0.84	3.11	8.55
PSO	1.80	5.72	10
ALO	0.73	1.18	0.34

The tuning of SMC with ALO algorithm is compared with some well-known algorithms including GA and PSO.

The computational procedure is implemented in MATLAB with 40 search agents and 200 iterations.

PID sliding surface parameters obtained by the methods of ALO, GA, and PSO are given in Table 1.

Figure 3 shows clearly that SMC with ALO controller has the best performance in the overshoot, settling time and rise time compared with GA and PSO results (Table 2).

4.2 Case 2: Sinusoidal Reference Trajectory

In the following section the tracking performance of the proposed technique is tested on selected trajectory defined as

$$\theta_r(t) = \sin(t) \tag{33}$$

The initial states are $x_1(t) = 0.2$, $x_2(t) = 0$

Figures 4, 5 and 6 show the position and speed tracking performance of SMC controller tuned by ALO, GA and PSO respectively with IAE optimum criterion. It's obviously clear from the figure that the tracking error achieved from the tuning of PID sliding surface with ALO is minimum compared to GA and PSO tuning, which confirm the best tracking performance of the proposed technique.

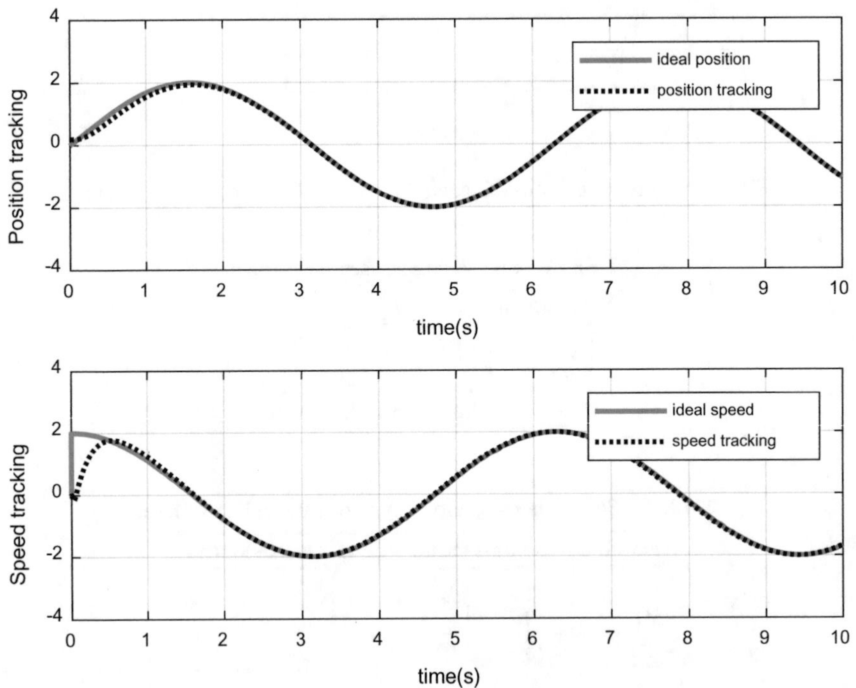

Fig. 4. Tracking positions and speed using SMC with ALO algorithm

Figure 7 show the control input using SMC with PID sliding surface tuned by Ant lion optimizer. It can be seen that the control input has negligible chattering effect.

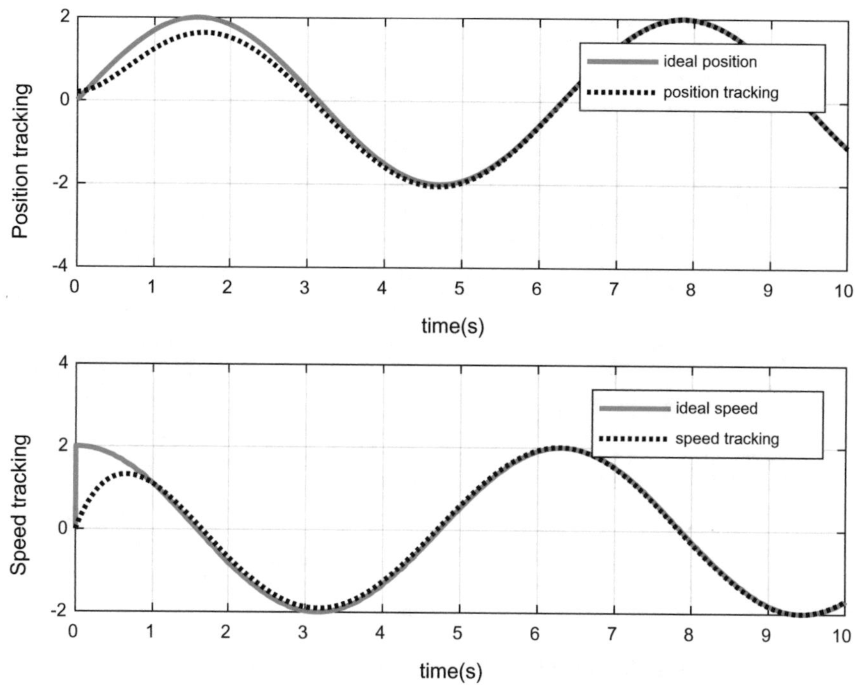

Fig. 5. Tracking positions and speed using SMC with GA algorithm

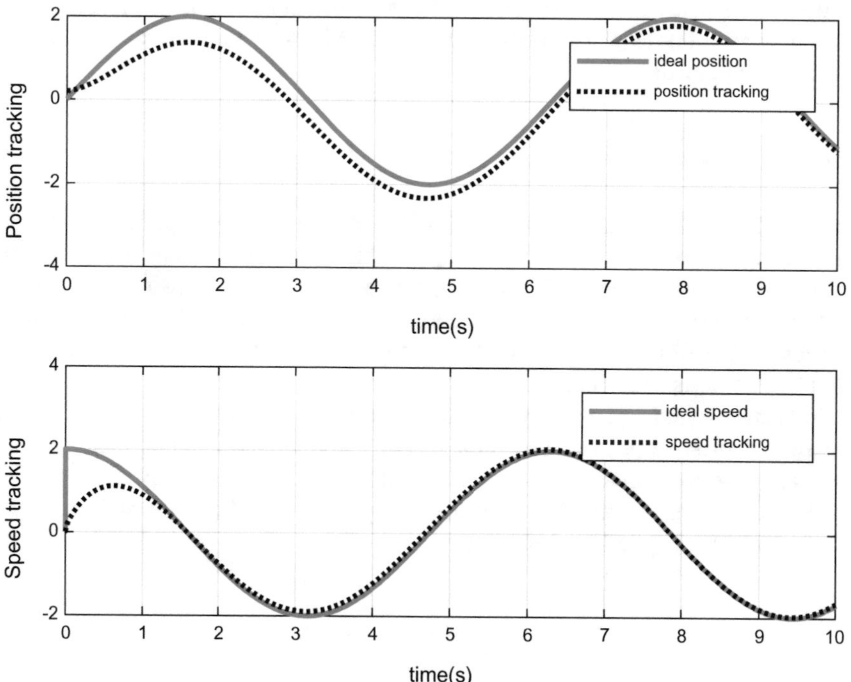

Fig. 6. Tracking positions and speed using SMC with PSO algorithm

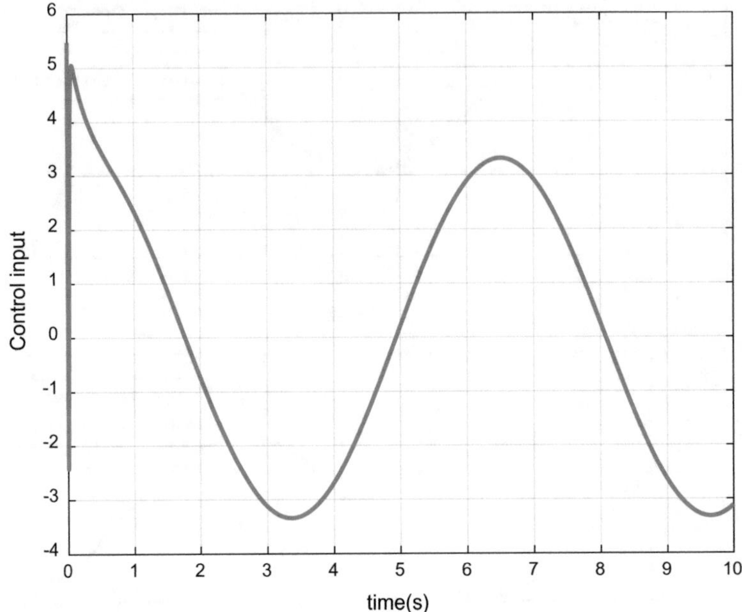

Fig. 7. Control input using SMC with ALO algorithm

It is clear from the simulation analyses that the proposed nonlinear controller approach has a better tracking to the desired trajectory in both unit step and sinusoidal function, confirming efficiency and feasibility of SMC with PID sliding surface tuned by Ant lion optimizer.

5 Conclusion

In this study, a sliding mode control with PID sliding surface is presented. Suitable parameter values of the proportional integral derivative (PID) sliding surface Kp, Ki, and Kd are optimized using Ant lion optimizer (ALO) to ensure the asymptotical stability of the system and good control performance. The proposed controller provides good tracking performance in different trajectories. Stability of the control system is guaranteed by Lyapunov theorem.

Comparative analysis was carried out to test the effectiveness of the proposed technique in terms of position and time tracking. It is proved that the optimization of the sliding surface parameters using ALO outperforms significantly GA and PSO algorithms. The results are very satisfactory within reasonable time and number of iteration and with negligible chattering.

References

1. Bandyopadhyay, B., Deepak, F., Kim, K.: Sliding mode control using novel sliding surfaces. Springer, Heidelberg (2009)
2. Utkin, V.: Sliding Modes in Control and Optimization. Springer, Heidelberg (1992)
3. Young, K., Utkin, V., Ozguner, U.: A control enginees guide to sliding mode control. IEEE Trans. Control Syst. Technol. **7**, 328–342 (1999)
4. Perruquetti, W., Barbot, J.: Sliding Mode Control in Engineering. Marcel Dekker, New York (2002)
5. Zinober, A.S.I.: Variable Structure and Lyapunov Control. Springer, London (1993)
6. Derbel, N., Q Zhu, J.G.: Applications of Sliding Mode Control. Springer, Heidelberg (2017)
7. Eker, I.: Sliding mode control with PID sliding surface and experimental application to an electromechanical plant. ISA Trans. **45**(1), 109–118 (2006)
8. Aliakbari, S., Ayati, M., Osman, J.H.S., Sam, Y.M.: Second order sliding mode fault-tolerant control of heat recovery steam generator boiler in combined cycle power plants. Appl. Therm. Eng. **50**(1), 1326–1338 (2013)
9. Amer, A.F., Sallam, E.A., Elawady, W.M.: Adaptive fuzzy sliding mode control using supervisory fuzzy control for 3 DOF planar robot manipulators. Appl. Soft Comput. **11**(8), 4943–4953 (2011)
10. Whitley, D.: A genetic algorithm tutorial. Stat. Comput. **4**(2), 65–85 (1994)
11. Kennedy, J., Eberhart, R.: Particle swarm optimization. Proc. IEEE Int. Conf. Neural Netw. **4**, 1942–1948 (1995)
12. Yang, X.S.: Firefly algorithm, stochastic test functions and design optimisation. Int. J Bio-Inspired Comput. **2**, 78–84 (2010)
13. Colorni, A., Dorigo, M., Maniezzo, V.: Distributed optimization by ant colonies. In: Proceedings of the First European Conference on Artificial Life, pp. 134–42 (1991)
14. Karaboga, D., Basturk, B.: A powerful and efficient algorithm for numerical function optimization: artificial bee colony (ABC) algorithm. J. Global Optim. **39**(3), 459–471 (2007)
15. Mirjalili, S.: The ant lion optimizer. Adv. Eng. Softw. **83**, 80–98 (2015)
16. Xiaofei, Z., Hongbin, M., Nannan, L.: Application of sliding mode controller with proportional integral switching gain in robot. In: 2017 IEEE International Conference on Unmanned Systems (ICUS), pp. 478–482 (2017)
17. Cao, Y., Chen, X.B.: An output-tracking-based discrete PID-Sliding mode control for MIMO systems. IEEE/ASME Trans. Mechatron. **19**(4), 1183–1194 (2014)
18. Li, Y., Xu, Q.: Adaptive sliding mode control with perturbation estimation and PID sliding surface for motion tracking of a piezo-driven micromanipulator. IEEE Trans. Control Syst. Technol. **18**(4), 798–810 (2010)
19. LY, I., Wang, Z., Zhu, L.: Adaptive neural network PID sliding mode dynamic control of non holonomic mobile robot. In: Proceedings of 2010 IEEE International Conference on Information and Automation, pp. 753–757 (2010)
20. Kumanan, D., Nagaraj, B.: Tuning of proportional integral derivative controller based on firefly algorithm. Syst. Sci. Control Eng. **1**, 5 (2013)
21. Solihin, M.I., Lee, F.T., Moey, K.L.: Tuning of PID controller using particle swarm optimization (PSO). Int. J. Adv. Sci. Eng. Inf. Technol. **1**(4), 458–461 (2011)

Computational Intelligence

Deep Neural Network for Supervised Inference of Gene Regulatory Network

Meroua Daoudi[1(✉)] and Souham Meshoul[2]

[1] Computer Science Dept, Constantine 2 University, Ali Mendjeli, Algeria
meroua.daoudi@univ-constantine2.dz
[2] MISC Laboratory, Constantine 2 University, Ali Mendjeli, Algeria
souham.mesoul@univ-constantine2.dz

Abstract. Inferring gene regulatory network from gene expression data is a challenging task in system biology. Elucidating the structure of these networks is a machine-learning problem. Several approaches have been proposed to address this challenge using unsupervised semi-supervised and supervised methods. Semi-supervised and supervised methods use primordially SVM. Most supervised approaches infer local model where each local model is associated with one TF. In this work, we propose a global model to infer gene regulatory networks from experimental data using deep neural network architecture. We evaluate our method on DREAM4 multifactorial datasets. The obtained results show that prediction accuracy using deep neural network outperform SVM in all tested data.

Keywords: Deep neural network · Gene regulatory network
Machine learning · Supervised learning · SVM

1 Introduction

Inferring gene regulatory network from gene expression data is an active field of research. The current interest of the molecular biology is to deepen the knowledge of the genomes. A way to understand the organism is to know the function of each gene as well as the interaction between them. Now the characterization of gene regulatory networks (GRN) is strongly supported by the scaling up of experimental methods in molecular biology. In particular, microarray technology measure changes in the expression of thousands of genes simultaneously. Inferring (GRN) from expression data make it possible to understand the relations ships between transcription factor (TF) and target genes [1].

Learning the structure of gene regulatory networks from expression data is a machine-learning problem [2]. Several approaches including supervised, unsupervised and semi-supervised techniques. There four main categories in unsupervised models can be distinguished such as Boolean model, Bayesian model, differential equation model and information theory models. The Boolean network is one of the simplest, edges are represented by Boolean function to model the interaction between genes and the state of a gene activity is represented by a binary variable. Active if the expression level of the gene is above a certain threshold, inactive otherwise. A REVerse

S. Chikhi et al. (Eds.): MISC 2018, LNNS 64, pp. 149–157, 2019.
https://doi.org/10.1007/978-3-030-05481-6_11

Engineering Algorithm REVAL [3] use the Boolean model to infer gene regulatory networks from expression data. Bayesian models are among the most effective to infer GRN, they make use of Bayes rules and consider gene expression as a random variable. For example, Friedman et al. [4] have introduced a framework for discovering the interaction between genes using Bayesian networks. Differential equation the most widely used class of dynamical models. It takes into consideration the change of concentration of metabolites over time [5]. In information theoretic models, the most proposed approach uses mutual information to capture complex regulatory relation including Relevance Network [6], Context likelihood of relatedness CLR [7] and ARACNE an algorithm for the reconstruction of gene regulatory networks [8]. The advantage of unsupervised methods is that they do not need any information about the system, but they are less efficient.

Recently due to the identification of a large number of interactions between the transcription factor and target gene. Several supervised and semi-supervised approaches have been proposed to infer gene regulatory network. There are two approaches in semi-supervised learning. The first approach learns from only positive Data and the second approach learns from both positive and unlabeled data. Using only positive data, a semi-supervised approach is proposed by Cerulo et al. [9]. This method works under the assumption that all the positive examples are randomly sampled from a uniform distribution. By another side in [1, 10, 11] the authors propose a semi-supervised approach to learn from positive and unlabeled data. Nihir Patel in [1] propose an iterative approach using random Forest (FR) and SVM to predict regulation of each TF with self-training. In [10, 11] authors use clustering techniques to extract reliable negative example from unlabeled data.

Proposed supervised approaches use primordially SVM such as sirene [12] and compareSVM [13]. sirene decomposes the problem of gene regulatory network inference into a large number of binary classification problems, each sub problem is associated to on TF, and SVM is used to predict GRN. Where compareSVM is a tool that compares four SVM kernel functions: linear, Gaussian, sigmoid and polynomial kernels, including three steps: optimization, comparison and prediction. Supervised methods require gene expression data and known regulation ship between genes, but they are more accurate than unsupervised and semi-supervised methods. They can be trained efficiently even when only a portion of interaction is known, as shown by Maetschke et al. in [14]. Figure 1 shows the classification of the methods proposed in the literature according to the learning approach used.

SVM classifier is widely used due to the efficiency of obtained results in gene regulatory networks [12, 13]. In the other side, the deep neural network is a powerful model inspired by the neural network of the brain that has been improved by the high performance in a wide range of applications [15] deep learning has been applied successfully to solve several prediction problems in bioinformatics [16]. Most supervised approaches infer local model where each local model is associated with one TF. In this work, we propose a global model to infer gene regulatory networks. In this context, we propose a deep neural network architecture to infer gene regulatory networks from DREAM4 multifactorial subchallenge, which are designed for the structure of a large scale GRN. The results show that the obtained results outperform SVM in all tested data.

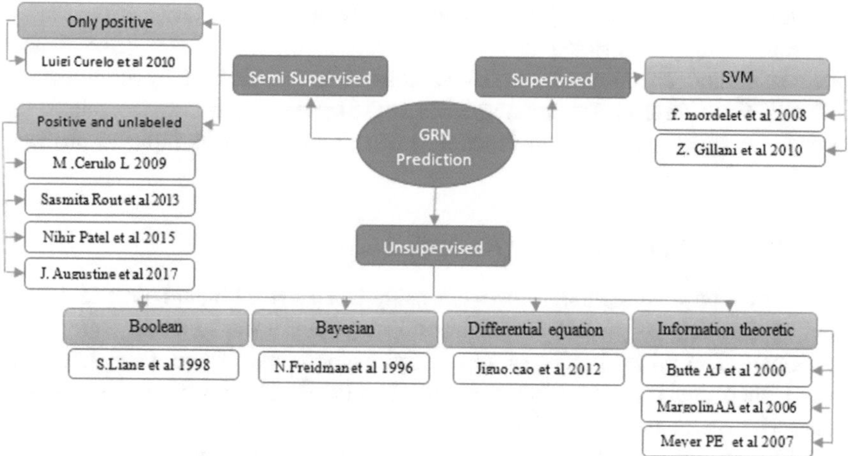

Fig. 1. Classification of proposed methods in literature according to the learning approach used

2 System and Method

2.1 Data Classification

The main purpose of the classification is to define rules for classifying objects in classes based on qualitative or quantitative variables characterizing these objects [17]. Initially, we have learning samples whose classification is known. These samples are used for learning classification rules. However, before applying these classification rules, they must be evaluated and for this task, a second independent sample, called validation or test is often used. And finally, these rules will be used to classify new or unknown objects. In another word, classification is part of the predictive techniques whose classes are predetermined. The process consists in analyzing the characteristics of a newly presented element in order to assign it to a class of a predefined set. The classification has been applied to numerous fields and applications such as medical diagnosis, image processing, agriculture, chemistry, geology, and automatic document processing. Indeed, there are several methods and algorithms used for classification such as linear and logistic regression, naïve Bayes, decision tree, k-nearest neighbour algorithm, support vector machines, and artificial neural network.

2.2 Deep Neural Network

Deep learning is a form of artificial intelligence which derived from machine learning. Thus, deep learning architectures are based on Artificial Neural Networks that are inspired by the neurons of the human brain. They are composed of several artificial neurons connected to each other [18]. So, the higher the number of neurons, the deeper the network. The power of today's computers and the explosion of data accessible have increased the effectiveness of deep learning. Over the last decade, deep learning has made progress advanced in many areas such as image recognition, speech processing and natural language processing. Furthermore, Deep Neural Network consists of a succession of fully connected layers whose consist on input layer, more than one hidden layer and an output layer. Its parameters are optimized by minimizing the misclassification error on the training datasets [15]. Then, each layer contains a set of neurons that apply an activation function, often nonlinear, at its input to produce an output.

2.3 DNN for Gene Regulatory Network Inference

The inference of gene regulatory network using a supervised model aims to develop a model M that help to predict the relationship between a transcription factor and the target genes. The principle idea of supervised methods that if two genes have a relationship, then any other two genes having similar expression profile also likely to interact with each other [12] under this principle, we concatenate the features vectors of TF and target genes to construct new feature vectors that contained two twice the number of feature of genes. Most supervised approaches infer local model where each local model is associated with one TF. In this work, we propose a global model to infer gene regulatory networks. The algorithm provided below can outline the proposed approach.

Algorithm: DNN for GRN inference

Input: Exp, Regul
For each gene Gi, Gj in R :
  Extraire expression profile Exp(Gi)
  Extraire expression profile Exp(Gi)
  Exp (Gi,Gj)= concatenate (Exp(Gi),Exp(Gj))
   If R(Gi, Gj) is positive then
    Exp (Gi,Gj) is labled as 1
   Else
    Exp (Gi,Gj) is labled as 0
auc = TrainDNN(Exp, labels)
output : auc

Where the algorithm receives as input the expression file, Exp and regulation file Regul. For each pair $G_{i,}$ Gj belongs to regul, the expression profile of both $G_{i,}$ Gj is extracted from the expression file Exp, and a new feature vector is then created. Let for example Exp1 be the expression value of gene G1 and Exp2 the expression value of gene G2, the concatenated vector of regulation G1-G2 is shown in Fig. 2.

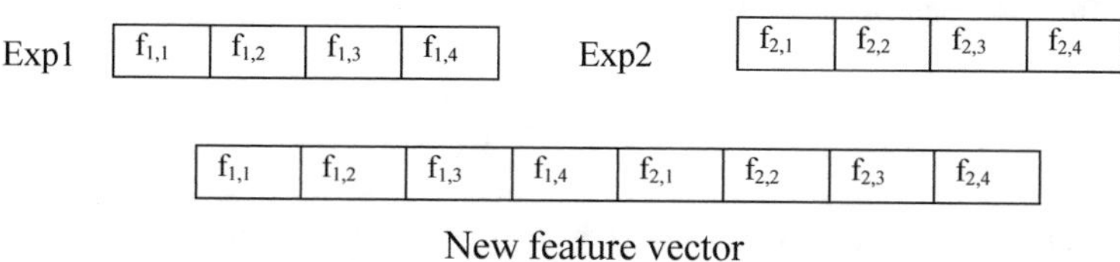

Fig. 2. Resulting feature vector of regulation G1-G2

Note that the direction of the regulation is important, which means that the resulting feature means that gene G1 regulate gene G2 and not gene G2 regulate gene G1. Next, if the regulation is labelled as positive, then the resulting vector is labelled by 1 otherwise 0. Then a deep neural network is performed to differentiate between positives and negatives regulations. The proposed model is shown in Fig. 3.

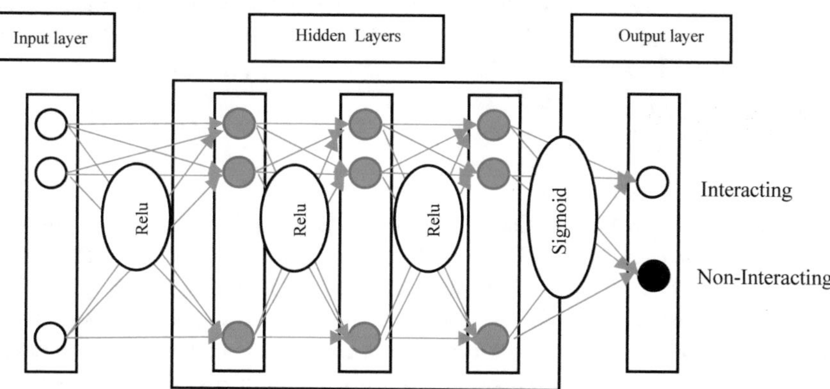

Fig. 3. The architecture of the proposed deep neural network for gene regulatory network inference

As shown in Fig. 3, we propose a fully connected neural network, which encompasses three hidden layers. Each neural is connected with all neurons in the previous layer, the input layer contains as a number of neurons the number of features in the concatenated vector which mean two twice the number of the features of genes. In this case, each concatenated vector contains 200 feature where each gene has 100 conditions. The hidden layer contains also the same number of neurons, which equal to 200. We have used as activation function the RELU function described as:

$$RELU(x) = max(x, 0)$$

Relu function is the most used function in neural networks. As the classification here is binary, the choice then is to use the sigmoid function because we predict a probability of an output. Where the probability range between zero and one. The sigmoid function can be described as:

$$Sigmoid(x) = \frac{1}{1 + e^{-x}}$$

Once the global model is trained, new regulations G_i-G_i can be predicted as positives if there is a relationship between the gene G_i and the gene G_i, and as negative if there is no regulation.

3 Experimental Study

3.1 Datasets

We evaluated the proposed approach using the DREAM4 multifactorial datasets. This challenge aims to infer five networks from Multifactorial perturbation data where each of them contained 100 genes and 100 samples. In the multifactorial experiment, a small number of genes are perturbed simultaneously. The data was simulated by GeneNetWeaver [19]. The topology of these networks was derived from the transcriptional regulatory system of E.coli and S.cerevisiae

3.2 Results

To evaluate the effectiveness of the developed inference method, tests are performed on five different benchmark of DREAM4 multifactorial subchallenge, which are designed for the structure of a large scale GRN. We measured the prediction accuracy by the area under the Receiver Operator characteristic curve (AUC). We adopt a cross validation procedure to make sure that the performance of the model is measured on prediction. We compare our proposed model with SVM using the same procedure. We use 5 cross validation and the prediction of all folds are averaged. For SVM we use the Gaussian kernel that founded the best option for prediction of GRN from microarray data and it has high accuracy and less standard derivation as shown by Gillani et al. [13]. Table 1 shows the obtained results for the proposed DNN model and SVM.

In addition, to assess the effectiveness of the proposed method, the roc curve obtained by the proposed DNN and SVM are shown in Figs. 4, 5, 6, 7 and 8. Figures 4, 5, 6, 7 and 8

Table 1. AUROC for DREAM4 multifactorial challenge

Network	Proposed DNN	SVM
Network 1	0.82	0.75
Network 2	0.87	0.81
Network 3	0.70	0.67
Network 4	0.76	0.73
Network 5	0.78	0.77

show the plot of ROC curves (AUC) of averaged AUC for all folds and AUC results obtained in each fold of cross validation procedure. The obtained results show that the prediction accuracy in term of averaged AUC of the proposed DNN is better than the averaged AUC obtained using SVM. Which means that the prediction of regulation between genes is more accurate using the proposed method comparing with SVM.

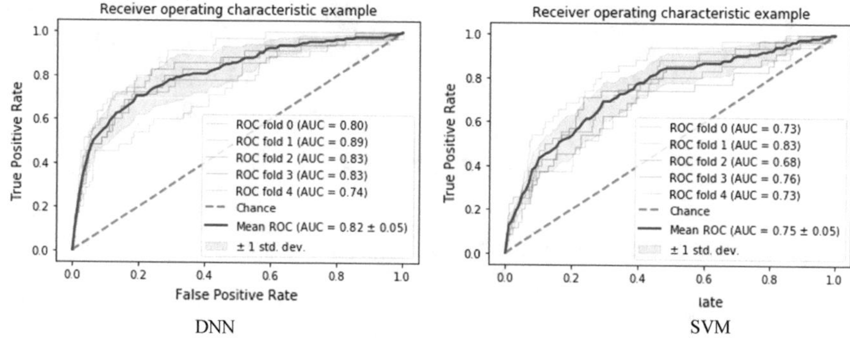

Fig. 4. Network 1

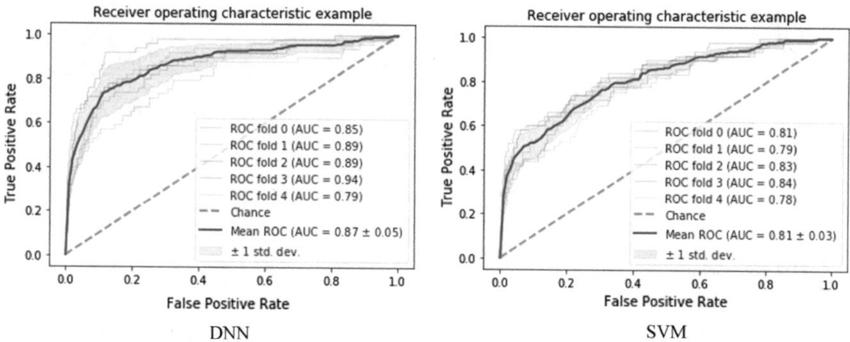

Fig. 5. Network 2

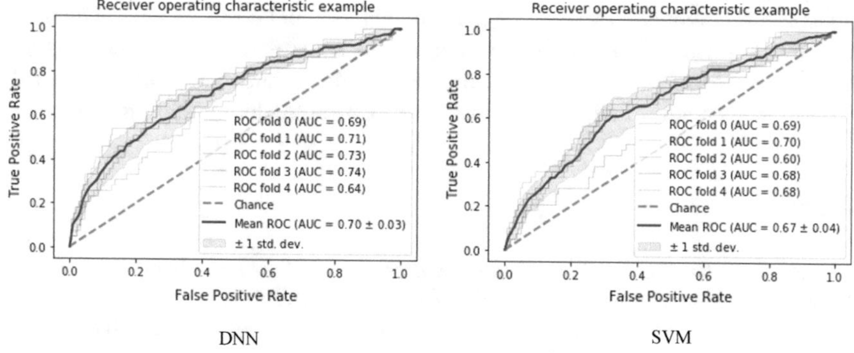

Fig. 6. Network 3

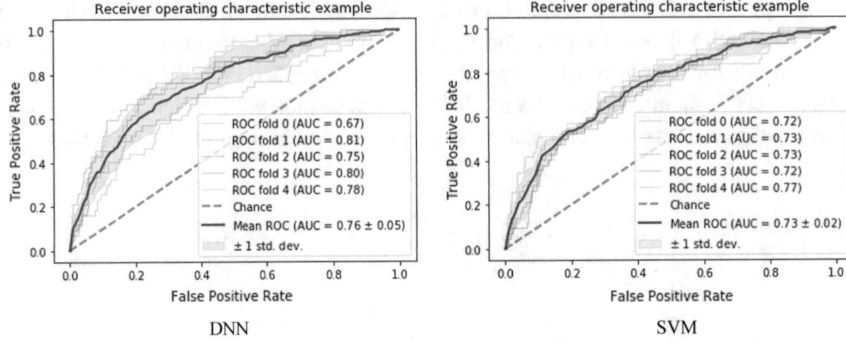

Fig. 7. Network 4

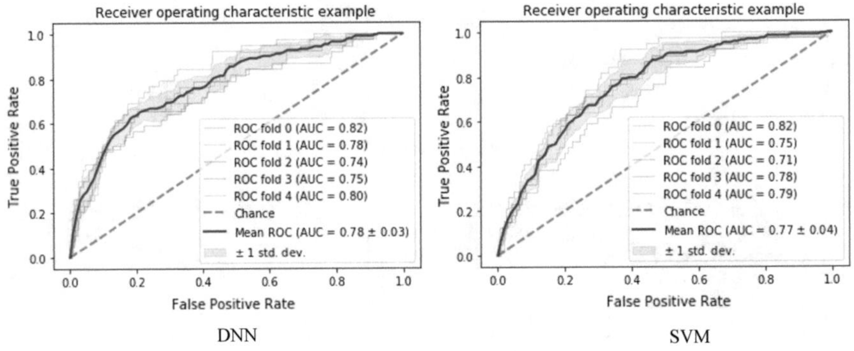

Fig. 8. Network 5

4 Conclusion

Several approaches have been proposed to infer gene regulatory network including unsupervised, semi-supervised and supervised methods. The most of the proposed supervised models are based on SVM. In this work, we propose a DNN model to infer the GRNS. With the aim to predict new regulations that containing similar expressions profiles and differentiate between negatives regulations and positives ones. The method was compared with SVM and the obtained results show that DNN outperforms SVM even in small datasets. As ongoing work, we intend to integrate others type of data presented in the same challenge and test the proposed model on large datasets.

References

1. Patel, N., Jason, T.L.Wang: Semi-supervised prediction of gene regulatory networks using machine learning algorithms. J. Biosci. **40**(4), 731–740 (2015)
2. Ristevski, B.: A survey of models for inference of gene regulatory networks. Nonlinear Anal. Model. Control. **18**(4), 444–465 (2013)

3. Liang, S., Fuhrman, S., Somogyi, R.: REVAL a general reverse engineering algorithm for inference of genetic network architectures. In: Pacific Symposium on Biocomputing, vol. 3, pp. 18–19 (1998)
4. Freidman, N., et al.: Using Bayesian network to analyze expression data [J]. Comput. Biologie **7**, 601–620 (1996)
5. Jiguo, C., et al.: Modeling gene regulation network using ordinary differential equation. In: Next Generation Microarray. Bioinformatics, pp. 185–197 (2012)
6. Butte, A.J., Kohane, I.S.: Mutual information relevance networks, functional genomic clustering using pairwise entropy measurements. In: Pacific Symposium on Biocomputing, pp. 418–429 (2000)
7. Meyer, PE., Kontos, K., Lafitte, F., Bontempi, G.: Information-theoretic inference of large transcriptional regulatory networks. EURASIP J. Bioinform. Syst. Biology (2007)
8. Margolin, A.A., et al.: ARACNE: an algorithm for the reconstruction of gene regulatory networks in a mammalian cellular context. BMC Bioinform. **7**(1), S7 (2006)
9. Curelo, L., et al.: Learning gene regulatory networks from only positive and unlabeled data. BMC Bioinform (2010)
10. Jisha, A., Jereech, A.S.: Gene regulatory network: a semi supervised approach. In: International Conference on Electronics Communication and Aerospace Technology ICECA (2017)
11. Sasmita, R., et al.: Handling unlabeled data in gene regulatory network. In: Proceeding of International Conference on Frontiers of Intelligence Computing AISC 199, pp. 113–120. Springer, Heidelberg (2013)
12. Mordelet, F., Vert, J.P.: SIRENE: supervised inference of regulatory network. Bioinformatics **24**, i76–i82 (2008)
13. Gillani, Z., et al.: Compare SVM: supervised support vector machine (SVM) inference of gene regularity network. BMC Bioinform. (2014)
14. Maetschke, S.R., et al.: Supervised, semi supervised and unsupervised inference of gene regulatory networks. Brief. Bioinform. **15**(2), 195–211 (2013)
15. Buduma, N.: Fundamentals of Deep Learning: Designing Next-Generation Machine Intelligence Algorithms. O'Reilly Media, Boston (2017)
16. Min, S., Lee, B., Yoon, S.: Deep learning in bioinformatics. Brief. Bioinform. **18**(5), 851–869 (2017)
17. Bramer, M.: Principle of Data Mining. Springer, London (2016)
18. Schmidhuber, J.: Deep learning in neural networks: An overview. Neural Netw. **61**, 85–117 (2015)
19. Shaffer, T., et al.: GeneNetWeaver: in silico benchmark generation and performance profiling of network inference methods. Bioinformatics 2263–2270 (2011)

Sentiment Analysis of Arabic Tweets: Opinion Target Extraction

Behdenna Salima(✉), Barigou Fatiha, and Belalem Ghalem

Computer Science Department, Faculty of Sciences, University of Oran 1, Ahmed Ben Bella, PB 1524 El M'Naouer, Oran, Algeria
Behdennasalima@gmail.com, fatbarigou@gmail.com, ghalemldz@gmail.com

Abstract. Due to the increased volume of Arabic opinionated posts on different social media, Arabic sentiment analysis is viewed as an important research field. Identifying the target on which opinion has been expressed is the aim of this work. Opinion target extraction is a problem that was generally very little treated in Arabic text. In this paper, an opinion target extraction method from Arabic tweets is proposed. First, as a preprocessing phase, several feature forms from tweets are extracted to be examined. The aim of these forms is to evaluate their impacts on accuracy. Then, two classifiers, SVM and Naïve Bayes are trained. The experiment results show that, with 500 tweets collected and manually tagged, SVM gives the highest precision and recall (86%).

Keywords: Opinion mining · Arabic sentiment analysis · Opinion target Machine learning · Arabic tweet

1 Introduction

Due to the emergence of Web2.0, users can share their opinions and sentiments on a variety of topics in new interactive forms where users are not only passive information receivers.

Sentiment analysis or opinion mining is the computational study of people's opinions, appraisals, attitudes, and emotions toward entities, individuals, issues, events, topics and their attributes [1]. The aim of sentiment analysis is to automatically extract users' opinions [2]. The main tasks of sentiment analysis (SA) are [3]:

- Subjectivity extraction
- Opinion polarity identification.
- Opinion element's extraction task.
- Development of resources like sentiment lexicon and annotated corpora required for previous tasks.

With the increase in the volume of Arabic opinionated posts on different social media, Arabic Sentiment Analysis (ASA) is viewed as an important research field. Most of researches in Arabic sentiment Analysis attempt to determine the overall opinion polarity. This work focuses on opinion target extraction subtask, which is a subject that has been little studied to date for ASA. This task aims to extract topics

S. Chikhi et al. (Eds.): MISC 2018, LNNS 64, pp. 158–167, 2019.
https://doi.org/10.1007/978-3-030-05481-6_12

expressed within opinions. For example, when saying in an Arabic opinion; "4 سوني أفضل جهاز" ("Sony 4 the best device"), the opinion target is "4 سوني" ("Sony 4").

In this paper, we propose a method for extracting opinion targets from Arabic tweets by modeling the problem of Opinion Target Extraction as a machine learning classification task and combining a number of the available resources for Arabic language together with tweets features.

The rest of the paper is organized as follows. Section 2 discusses related work; Sect. 3 describes the features and the method for extracting the opinion target. Section 4 focuses on the experimental results and discussion. Section 5 concludes the paper.

2 Related Work

Few works have focused on the task of opinion target extraction from documents in Arabic. In [4], the authors propose a feature-based opinion mining framework for Arabic reviews. This framework uses the semantic of ontology and lexicons in the identification of opinion features and their polarity. This approach is composed of five components: Ontology and lexicon Development, Semantic Feature Identification, Polarity Identification, Feature Polarity Identification and Opinion Mining. Experiments showed that this approach achieved a good level of performance. The accuracy improved to be much better than the result achieved by baseline approach.

In [5] proposed a generic approach that extracts the entity aspects and their attitudes for reviews written in modern standard Arabic. This approach does not exploit predefined sets of features, nor domain ontology. Rather it relies on the idea that the entity aspects and their opinion-bearing words are usually correlative. These words are used to orient the process of extracting the entity aspects. For this, the authors add sentiment tags on the roots and patterns of an Arabic lexicon. These tags are used to extract the opinion bearing words and their polarities. The proposed approach system is evaluated on the entity-level using two datasets of 500 movie reviews with accuracy 96% and 1000 restaurant reviews with accuracy 86.7%. Then the system is evaluated on the aspect-level using 500 Arabic reviews in different domains (Novels, Products, Movies, Football game events and Hotels). The proposed system achieves a recall 80%, precision 77% and F-measure 79%.

A two-stage method for annotating targets of opinions was developed in [6], by using the crowd sourcing tool Amazon Mechanical Turk. The first stage consists of identifying candidate targets "entities" in a given text. The second stage consists of identifying the opinion polarity (positive, negative, or neutral) expressed about a specific entity.

The work presented in [7] focuses on evaluating Arabic tweets for restaurant services. The authors constructed a prototype for sentiment analysis of Arabic tweets with corresponding Arabic Opinion Lexicon (AOL) and Arabic Tweet Sentiment Analyzer (ATSA). The prototype has been designed and tested using customer's opinions obtained from tweeter; it accepts tweets as input, generates polarity, and determines the tweet target as outputs. Additionally, the prototype can ascertain the type of tweet as subjective, objective, positive, or negative and give a summarization of tweet polarity.

3 Proposed Method

This work focuses on the opinion target extraction as part of the sentiment analysis task. We model the problem as a machine learning classification task. The process which follows the data mining process is composed of four steps.

3.1 Corpus Building

We need an annotated Arabic corpus for opinion target. Unfortunately, Arabic corpus required for this task is not available, therefore we decide to build our own Arabic corpus, and manually annotate it for opinion target.

We used the Twitter Archive Google Spreadsheet (TAGS)[1] to collect tweets related to opinions expressed in Arabic from the topic: "mobile phone brand". After filtering out retweets and performing some pre-processing steps to clean up unwanted content like URL, we ended up with 500 tweets. We then manually annotated the opinion target in these tweets. Table 1 shows some examples of manually annotated tweets.

Table 1. Examples of annotated tweets

	Tweet	Target
1	بطارية # أبل ذات جودة خيالية	بطارية
2	سوني # أفضل جهاز	سوني
3	سامسونج يصرح عن وجود عيب في # السماعة	سماعة
4	شحن أسرع مع شاحن # أبل	شاحن

3.2 Preprocessing

In this phase, every tweet is tokenized into words and several forms are generated to be examined. The aim of these forms is to evaluate their impacts on accuracy. The following Fig. 1 shows several forms of pre-processed tweets.

- Form (a): After removing the stop words and special characters, every tweet is transformed into a feature vector which includes all the remaining words.
- Form (b): After removing the special characters like (RT, URL, @, #), words are stemmed using the Khoja stemmer [8] and combined with stop words removal before transforming into a feature vector.
- Form (c): consists of form (b) followed by filtering of words according to their grammatical categories. We have retained the nouns and adjectives. The words are tagged using Stanford Arabic part of speech tagger[2].

---

[1] https://tags.hawksey.info/.

[2] http://nlp.stanford.edu/software/tagger.shtml.

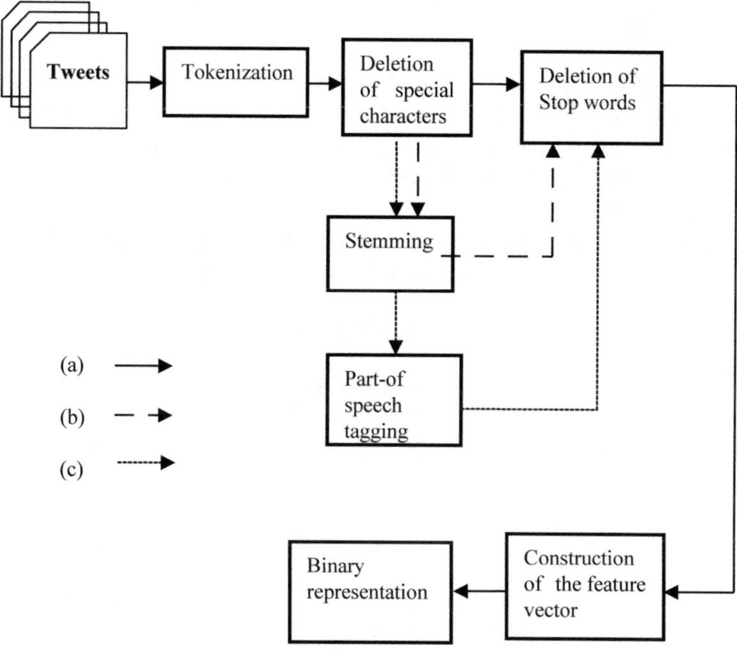

Fig. 1. Tweets preprocessing

At the end of this preprocessing step, tweets data are converted from text format into ARFF format required by the WEKA tool[3]. The tool we have used for classification step.

The size of the feature vector depends on the number of tweets and the different forms used in preprocessing as shown in Table 2.

Table 2. Size of the feature vector

Number of tweets	100	200	300	400	500
Words (form a)	455	714	1066	1214	1256
Stem (form b)	426	654	942	1065	1097
Stem and POS (form c)	317	494	719	815	839

3.3 Classification

In the literature, several machine learning techniques are used, but two of them appear to provide the best results. They are the SVM and NB classifiers [9]. The data mining tool we have used is Weka 3.7 open-source data mining software.

[3] http://www.cs.waikato.ac.nz/ml/weka.

4 Experiments and Discussions

During all the experiments that we have carried, we used 10-fold cross-validation to train the classifiers (SVM and NB). Each dataset is divided into 10 parts; one is used for testing and 9 for training in the first run. This process is repeated 10 times, using a different testing fold in each case. Experiments are carried out with different forms of preprocessed tweets.

The performance of classification model is measured by evaluating the precision, recall and F-measure. They are measured with Eqs. (1), (2) and (3).

$$\text{Precision} = \text{TP}/(\text{TP} + \text{FP}) \tag{1}$$

$$\text{Recall} = \text{TP}/(\text{TP} + \text{FN}) \tag{2}$$

$$\text{F} - \text{measure} = 2 \times (\text{Precision} \times \text{Recall})/(\text{Precision} + \text{Recall}) \tag{3}$$

4.1 Experiment 1: Impact of Simple Words by Varying the Corpus Size

The first experiment is carried out to evaluate the effect of using simple words on the performance of SVM and NB classifiers. Table 3 shows the performance obtained from each classifier.

Table 3. Classifier performance using simple words and according to the size of the dataset

	Precision		Recall		F-measure	
	NB	SVM	NB	SVM	NB	SVM
100 tweets	0.550	0.534	0.630	0.640	0.573	0.545
200 tweets	0.557	0.541	0.650	0.675	0.590	0.572
300 tweets	0.648	0.663	0.680	0.705	0.641	0.661
400 tweets	0.658	0.673	0.697	0.720	0.663	0.682
500 tweets	0.672	0.680	0.701	0.741	0.676	0.691

As shown in Figs. 2 and 3, the passing from 100 tweets to 500 tweets improve significantly the performance. In the case of NB: Precision increased 12%, Recall increased 7.1%, and F-measure increased 10,3%. In the case of SVM: Precision increased 14%, Recall increased 10%, and F-measure increased 14,6%.

4.2 Experiment 2: Impact of Stemming

This experiment is carried out to evaluate the effectiveness of stemming in the classification process.

As illustrated in Table 4, Figs. 4 and 5:

- Both the NB and SVM classifiers performed best compared to experiment 1. This means that stemming enables to generate a representative feature vector of tweets.

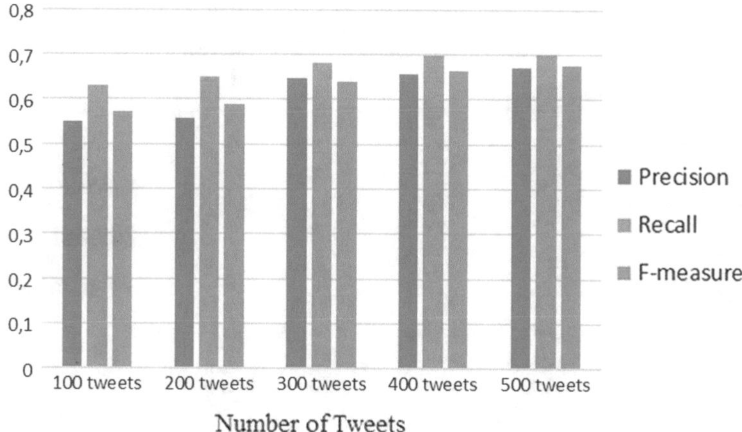

Fig. 2. Experiment 1 with NB.

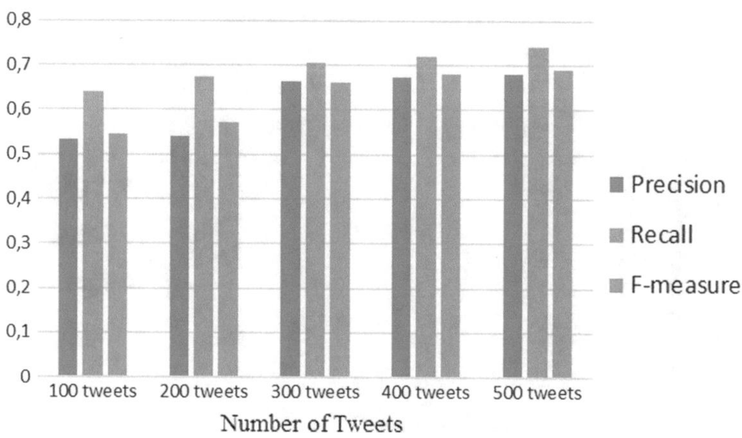

Fig. 3. Experiment 1 with SVM.

Table 4. Classifier performance using stemming by varying dataset size

	Precision		Recall		F-measure	
	NB	SVM	NB	SVM	NB	SVM
100 tweets	0,653	0,552	0,66	0,64	0,631	0,568
200 tweets	0,742	0,756	0,72	0,76	0,707	0,736
300 tweets	0,727	0,756	0,727	0,777	0,711	0,76
400 tweets	0,775	0,81	0,773	0,813	0,768	0,805
500 tweets	0,821	0,86	0,814	0,86	0,816	0,857

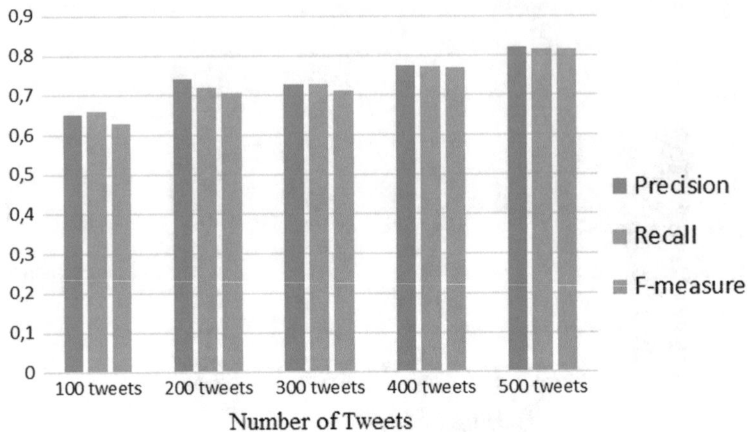

Fig. 4. Experiment 2 with NB.

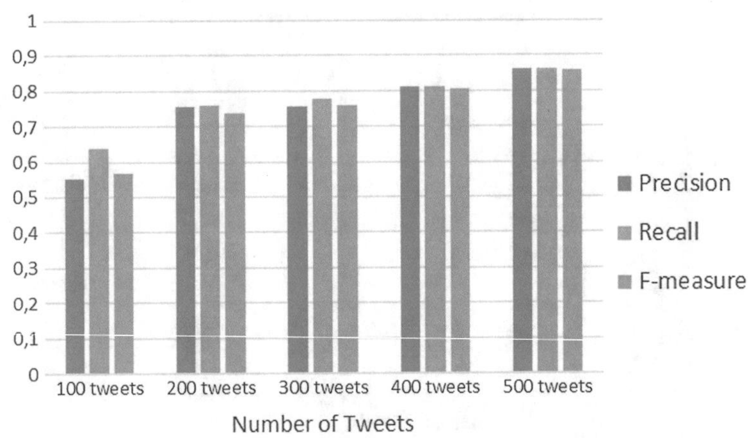

Fig. 5. Experiment 2 with SVM

For SVM, we observe a performance improvement in F-measure from 69,1% (using simple words) to 85.67% for 500 tweets. And an improvement in F-measure from 67,6% to 81,6% for NB.

- The performance results of the two classifiers are close but SVM provides the best results.

4.3 Experiment 3: Impact of Stemming and Part of Speech Tagging

This experiment was carried out to assess the effect of stemming followed by filtering of words according to their grammatical categories. We have retained only nouns and adjectives, we have noticed that noun phrases are regarded as opinion target candidates.

As illustrated in Table 5, Figs. 6 and 7:

Table 5. Classifier performance using stemming and POS by varying dataset size

	Precision		Recall		F-measure	
	NB	SVM	NB	SVM	NB	SVM
100 tweets	0,659	0,605	0,66	0,63	0,638	0,603
200 tweets	0,659	0,713	0,66	0,735	0,637	0,702
300 tweets	0,655	0,662	0,66	0,693	0,648	0,662
400 tweets	0,719	0,673	0,725	0,68	0,709	0,67
500 tweets	0,803	0,706	0,8	0,708	0,797	0,701

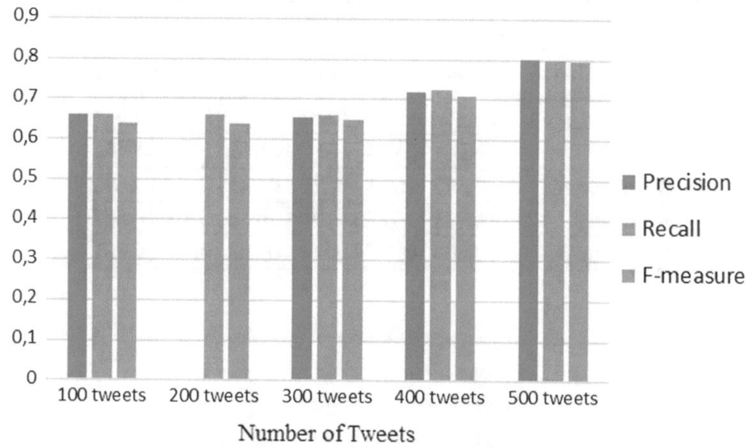

Fig. 6. Experiment 3 with NB

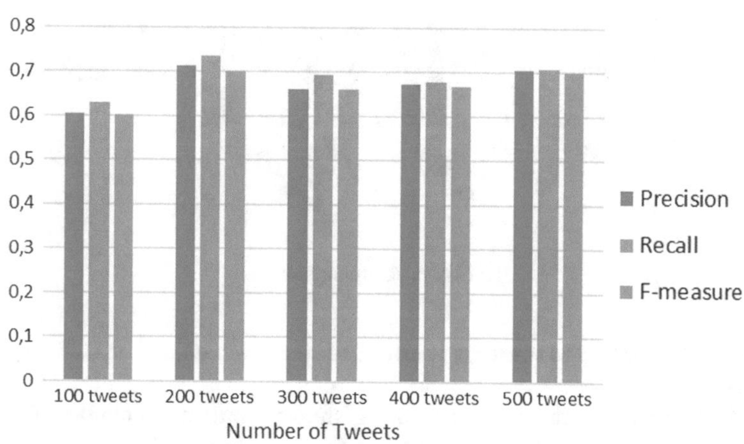

Fig. 7. Experiment 3 with SVM

* Unlike previous experiments, we observe a performance drop in precision from 86% to 70.6% for SVM.
* The best results are acquired using 500 tweets.
* The classifier NB performs better than SVM.

4.4 Discussions

In the three experiments, results show that the best results are acquired when the corpus of 500 tweets is used. Table 6 shows the precision, recall and F-measure obtained from each classifier for 500 tweets.

As illustrated in Table 6 and Fig. 8, we can state the following findings:

Table 6. Classifier comparison using 500 tweets

	Precision		Recall		F-measure	
	NB	SVM	NB	SVM	NB	SVM
Simple word	0,672	0,68	0,701	0,741	0,676	0,691
Stemming	0,821	0,86	0,814	0,86	0,816	0,857
Stemming & POS	0,803	0,706	0,8	0,708	0,797	0,701

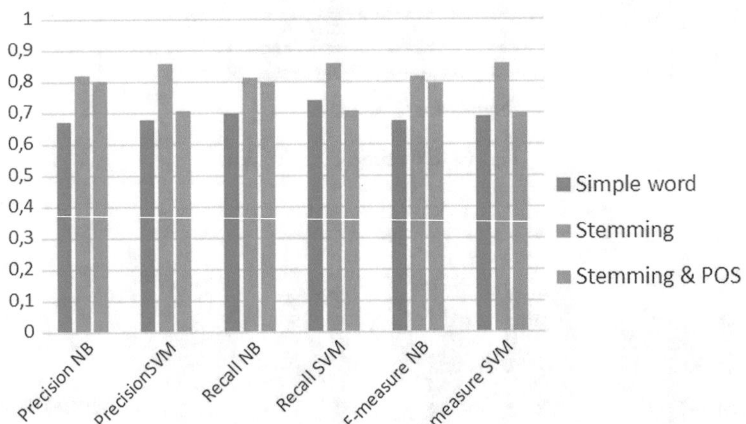

Fig. 8. Comparison.

* Both the NB and SVM classifiers performed best when trained on sufficiently large corpus.
* The SVM performs the best when using simple words or stemming. When considering a corpus of 500 tweets, SVM slightly improves the performance when using simple words of 1% in terms of precision, 4% in terms of recall, 1.5% in terms of F-measure, and when using stemming SVM exceeds NB of 4.9% in terms of precision, 4.6% in terms of recall, and 4.1% in terms of F-measure.
* Stemming combined with POS adversely affected the performance of classification.

- NB performs better than SVM in the case of stemming combined with the grammatical filtering of 9.7% in terms of precision, 9.2% in terms of recall, and 9.6% in terms of F-measure.
- By comparing the performance results of both classifiers, we see that using stem words as features give better classifier performance compared with other types of features (simple words and POS).

5 Conclusion

In this paper, we propose a method to extract opinion target from Arabic tweets. For this goal, we employed SVM and NB classifiers. The feature vectors of the tweets are preprocessed in several ways and the effects of these features on the classifiers' accuracy were investigated. The comparison is based on standard accuracy measures such as precision, recall, and F-measure. The results showed that, with 500 tweets collected and manually tagged, stemming combined with stop words removal improved the performance of classification.

We cast the problem of Opinion Target Extraction as a machine learning classification task. Hence to obtain the best results, this method requires a large corpus to allow better learning.

In the future work, a larger corpus will be used and we intend to use deep learning approaches.

References

1. Liu, B., Zhang, L.: A survey of opinion mining and sentiment analysis. Mining Text Data, pp. 415–463. Springer, Boston (2012)
2. Liu, B.: Sentiment analysis and opinion mining. Synth. Lect. Hum. Lang. Technol. 1–167 (2012)
3. Elarnaoty, M.: A machine learning approach for opinion holder extraction in Arabic language. Int. J. Artif. Intell. Appl. **3**(2), 45–63 (2012)
4. Alkadri, A.M., ElKorany, A.M.: Semantic feature based Arabic opinion mining using ontology. Int. J. Adv. Comput. Sci. Appl. **7**(5), 577–583 (2016)
5. Ismail, S., Alsammak, A., Elshishtawy, T.: A generic approach for extracting aspects and opinions of Arabic reviews. In: Proceedings of the 10th International Conference on Informatics and Systems, pp 173–179. ACM (2016)
6. Farra, N., McKeown, K., Habash, N.: Annotating targets of opinions in Arabic using crowd sourcing. In: 2015 Proceedings of the Second Workshop on Arabic Natural Language Processing, pp. 89–98 (2015)
7. Ibrahim, M.A., Salim, N.: Aspect oriented sentiment analysis model of Arabic tweets. Int. J. Comput. Sci. Trends Technol. **4**(4), 342–353 (2016)
8. Khoja, S., Garside, R.: Stemming Arabic Text, Lancaster, UK, Computing Department, Lancaster University (1999). http://www.comp.lancs.ac.uk/computing/users/khoja/stemmer.ps
9. Behdenna, S., Barigou, F., Belalem, G.: Sentiment analysis at document level. In: International Conference on Smart Trends for Information Technology and Computer Communications, pp 159–168. Springer, Singapore (2016)

Shifted 1D-LBP Based ECG Recognition System

Meryem Regouid[1(✉)] and Mohamed Benouis[2]

[1] Computer Science Department, University of Ferhat Abbas, Setif 1, Pôle 2 - El Bez, 19000 Setif, Algeria
mregouid@yahoo.com

[2] Computer Science Department, University of M'sila BP, 28000 M'sila, Algeria
benouis.mohamed@univ.msila.dz

Abstract. ECG analysis has been investigated as promising biometric in many fields especially in medical science and cardiovascular disease for last decades in order to exploit the discriminative capability provided by these liveness measures developing a robust ECG based recognition system. In this paper, an ECG biometric recognition system was proposed based on shifted 1D-LBP. Shifted 1D-LBP was applied to extract the representative non-fiducial features from preprocessed and segmented ECG heartbeats. For matching step, K Nearest Neighbors (KNN) was adopted. Two benchmark databases namely MIT-BIH/Normal Sinus Rhythm and ECG-ID database were used to validate the proposed approach. A Correct Recognition Rate (CRR) of 100% and 97% was achieved with MIT-BIH/Normal Sinus Rhythm and ECG-ID databases, respectively.

Keywords: ECG · Shifted 1D-LBP · KNN · MIT-BIH/normal sinus rhythm ECG-ID · CRR

1 Introduction

There are different attacks that can be launched against authentication systems based on passwords and tokens. So it is necessary to find other techniques to solve these problems. Biometric recognition is considered as a real alternative to passwords, signatures and other identifiers. Biometrics can be defined as a set of measurable, robust and distinctive traits. These biological, behavioral or morphological traits can be used to identify or verify the claimed identity of an individual. Biometric system can perform in verification or identification mode depending on the application contexts. In verification mode, the biometric systems validate the user's identity by comparing the input traits with one template exist in the database. Whereas, in identification mode, the biometric systems validate the user's identity by comparing the input traits with all templates exist in the database. There are many biometric characteristics used in various applications. Each biometric technology characteristic has itself proprieties and it depends on various factors such the nature, requirements, variety of issues and the matching performance of an application [1].

S. Chikhi et al. (Eds.): MISC 2018, LNNS 64, pp. 168–179, 2019.
https://doi.org/10.1007/978-3-030-05481-6_13

Biometric systems demonstrate its effectiveness and robustness to various types of malicious attacks in various areas in the modern society. The most used biometric characteristics are face, fingerprint, Palm print, iris, Keystroke, Signature, Voice, Gait…etc. The choice of a biometric modality depends on its strength and weakness besides the application requirements. Many properties can determine the suitability of biometric traits: universality (user's application must have the proposed trait), uniqueness (each user have its unique traits comparing with other individuals), permanence (the stability of the biometric trait over a period of time), measurability (the acquisition of the biometric trait using adequate device in order to extract the representative feature sets), performance (the constraints imposed by the application should be accepted when calculating the recognition accuracy), acceptability (users do not disturb presenting their biometric trait to the system) and circumvention (the possibility of imitating trait of an individual using fake traits) [2].

Analysis of electrocardiogram (ECG) describes the electrical activity of the heart, is considered as new biometrics measure for human identity recognition. The activity recorded by the ECG comes from extracellular currents related to the propagation of a depolarization front (atrial P wave, then ventricular QRS complex). Nowadays, ECG biometric has been gaining more attention from many research laboratory because of its strengths especially in the security domain, e-health medical, diseases prevention…etc. [3–5].

The ECG is known to contain characteristics originating from the geometrical features of the individual body and heart that might be utilized for biometrical applications [6]. ECG can be used as a liveness detector, universality, remote login process by ECG signal captured from a finger which improves security and privacy [7]. Emotion, diet, physical exercises, diseases or position of the electrodes have the ability to change the human ECG signal [8].

An ECG recognition system consists of three main steps: ECG preprocessing, features extraction and classification. The preprocessing step aims to reduce the effect of several types of noise. This stage includes also the segmentation of the filtered ECG signal thereby to locate the different ECG waveforms by using an appropriate algorithm. The feature extraction step, several techniques have been proposed in the literature to extract the most significant features from the ECG signal, which are divided on three class, fiducial, non-fiducial features and hybrid approach.

In biometric filed, feature extraction plays a major role in the biometric process, so far, this paper presents a feature extraction approach, a shifted 1D-Local Binary Patterns (shifted 1D-LBP). Shifted 1D-LBP is built on the local binary pattern (LBP), which is an image processing method, and one-dimensional local binary pattern (1D-LBP). In LBP, each pixel is compared with its neighbors. Similarly, in 1D-LBP, each sample in the signal raw is compared against its neighbors. 1D-LBP extracts feature based on local changes in the signal. Therefore, it has high a potential to be employed in medical purposes. Since, each heart beat activity, which is recorded in ECG signals, has its own pattern, and via the 1D-LBP these (hidden) patterns may be detected. But, the positions of the neighbors in 1D-LBP are constant depending on the position of the sample in the ECG signal. Also, both LBP and 1D-LBP are very sensitive to noise. Therefore, its capacity in detecting hidden patterns is limited. To overcome these drawbacks, shifted 1DLBP was proposed. So, the positions of the neighbors and their values can be simply shifted in order to increase the potential of getting different micro and macro patterns in an adaptable way. Therefore, the transformation codes are obtained for all the signal points, the bin histogram of these codes formed the feature vector of the ECG signal and then fed to a classifier to perform the classification. To validate the proposed feature extraction approach, ECG-ID and Normal Sinus Rhythm datasets were employed.

In this paper, a shifted 1D-Local Binary Patterns (shifted 1D-LBP) was implemented and applied on the segmented ECG waveforms to extract the non-fiducial features. KNN (K Nearest Neighbors) classifier was used in the classification step.

The rest of paper is organized as follow: in Sect. 2 reviews of relevant literature on ECG—based biometric system. Section 3 presents the mechanism of the basic LBP and the proposed shifted 1D-LBP. The experimental results were discussed in Sect. 4. The last Sect. 5 concludes the paper.

2 Literature Review

ECG analysis, or the heart beat variability, has been studied in many fields such as medical science, psychology, robotic, and Cardiovascular disease for decades [3–5]. Recently, numerous research works have been published tackling the problem of ECG based recognition as the new key parameter to biometric identification. For example, in [9, 10] we can find a survey on this problem recapitulating some of the most popular approaches. Some of them use explicit morphological models of heartbeat variability (P-QRS-T), whereas others use either spatio-temporal or frequency domain features of the ECG signal.

Biel et al. in [3] has proposed a new approach by using 12-lead electrocardiogram (ECG) recorded for human identification where was carried out on ECG-ID database. Fiducial features mainly capture the holistic patterns from the ECG waveform signal and may be represented as the distances, amplitudes and angle deviation and so on. The experimental results demonstrate the possibility of identifying a person using features extracted from only one lead achieving 98%.

Wübbeler et al. in [6]: an ECG biometric verification and identification system was developed based on fiducial features. The developed work aims to study the long-term stability of the individual ECGs; a short recording time beside the usage of easily applicable ECG leads hardly affected by actual positioning of the electrodes. ECG-ID database was used for testing and validating the proposed approaches. An EER of 2.8% was achieved for verification mode and recognition rate of 98% was achieved for identification mode.

Louis et al. in [8]: One Dimensional Multi-Resolution Local Binary Patterns Features (1DMRLBP) was proposed for the purpose of extracting regular from irregular ECG waveforms. The achieved result demonstrates the efficiency of 1DMRLBP in tolerating noise and preserving morphology of ECG waveforms. The extracted regular ECG waveforms are passed through to a biometric system, and the irregular ECG waveforms are filtered out. There proposed approach are validate using PTB benchmark database. An EER and accuracy of 0.09% and 91% respectively was obtained.

Dar et al. in [11]: Discrete Wavelet Transform (DWT) of cardiac cycle and heart rate variability (HRV) based features were proposed for the purpose of developing an ECG recognition system. The developed system was validated using publicly available databases like MIT-BIH/Arrhythmia (MITDB), MIT-BIH/Normal Sinus Rhythm (NSRDB) and ECG-ID database (ECG-IDDB) including all subjects. Random Forests technique was applied for classification stage achieving an accuracy of 100% using NSRDB database and 83.88% for a challenging ECG-ID database.

Chun in [12]: the use of a single pulse ECG for authentication assuming that there is no access to others ECG signals. For the enrollment phase, a guided filter (GF) method was applied to reduce the noise of a single ECG pulse. Two simple similarity measures, Euclidean distance and DWT, between the enrolled ECG and the input signal were employed that they may be applicable to both small scale and large scale authentication systems. A comparison with PCA based authentication system was performed achieving an EER of 2.4%.

Barra et al. in [13]: an ECG recognition system based on fiducial features (normalized local maxima and minima) using simple peak detection technique. A dataset has been created, by composing the signals of PTB database. An EER of 1.33% was reached up.

Bassiouni et al. in [14]: an ECG biometric identification system was performed three kind of features extraction step. The first approach, non-fiducial approach is applied the autocorrelation/discrete cosine transform on the signal ECG and achieved an overall accuracy of 70%. The second approach was based on fiducial features involving duration, amplitude differences, along with angles between (PQRST…). Thus, fiducial points were detected from each ECG waveform. An EER of 0.4% and accuracy of 52% was obtained. The third approach fused both the P-QRS-T fragments and ECG features decomposed by the wavelet transform as a unique feature. The reported total classification accuracy was 98%.

In this paper, we propose an ECG-based recognition system. A shifted 1D-LBP algorithm was implemented to extract the desired features from the segmented ECG waveforms. The use of the shifted 1D-LBP technique to extract the different patterns existing in the preprocessed ECG signal increase the potential of getting different micro and macro patterns in an adaptable way which improves the correct recognition accuracy and decreases the Equal Error Rate beside to the ease of programming and a complexity of O(n) was obtained. We aim that our proposed approach will be able to outperform the performance, achieve a high-security level and reduce the complexity which minimizes the execution time of the system.

3 Proposed Methodology

Local descriptors such as Local Binary Patterns (LBPs) have proven its effectiveness in real world condition than the global descriptor. LBPs have largely used for texture analysis for 2D image processing [15, 16] LBPs methods can solve various computer vision issues, moreover to its invariance against monotonic gray level changes caused. e.g., by illumination variations, furthermore, their simplicity of computation allows analyzing images in challenging real-time settings. [17]. Owing to its numerous advantages on a 2D image, the application of LBPs in 1D processing signal has investigated. Shifted 1D-LBP method was extended from a 1D-LBP method which was introduced by Chatlani et al. [18]. Both methods are applied to sensor signals for the purpose of extracting features or segmentation.

3.1 Basic LBP

The local binary pattern, proposed by Ojala in 1996 [15], is considered one of the famous approaches used for the feature extraction task, therefore, it has been successfully applied in several fields of machine vision. The original version of the local binary pattern operator is applied on a 3 × 3 pixel block of an image which means 8 neighborhoods. The neighborhoods of each block are threshold by its center pixel value to generate an LBP-code for the center pixel [19]. The formulation of LBP for a center pixel x_c is given as:

$$LBP(x_c) = \sum\nolimits_{p=0}^{7} s(x_p - x_c) * 2^p \tag{1}$$

Where the S() indicates the sign function and is defined as

$$S(x) = \begin{cases} 1 & for \;\; x \geq 0 \\ 0 & for \;\; x < 0 \end{cases} \tag{2}$$

A value of 1 was assigned to the S() function if the input parameter are grater or equal to 0, Otherwise, a value of 0 was assigned. Figure 1 illustrates the basic LBP operator.

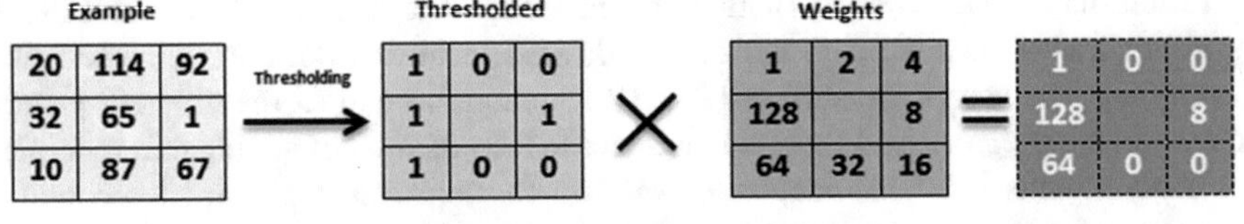

Fig. 1. The basic LBP operator.

3.2 Shifted 1D LBP

Shifted 1D-LBP method was extended from a 1D-LBP method which was introduced by Chatlani et al. [18]. The Shifted 1D-LBP and the 1D-LBP operators have both the same process. The main difference was consisting the selection of neighbors so that a fixed number of neighbors were selected for both left and right of the central point in which a limited number of detectable micro and macro patterns were generated. While, shifted 1D-LBP solve this problem by shifting the number of neighbors in left and right sides in order to increase the potential of generating more micro and macro patterns

[20]. The choice of left (Pl) and right neighbors (Pr) still considered as problem. Once the selection of neighbors was performed, each sample of the processed signal was compared with its left and right neighbors. Then, a binary code was generated for the center point, which named a Shifted 1D-LBP code, by thresholding each value of left and right neighborhood with the value of each center samples from a signal.

The formulation of Shifted 1D-LBP of center sample Pc of processed signal x is given as

$$\text{Shifted 1DLBP}(P_c) = \sum\nolimits_{j=1}^{P} S\left(P_j - P_c\right)2^{j-1} \tag{3}$$

Where the S() indicates the sign function as mentioned in Eq. 2 and where Pc and Pj represent the center sample and number of neighbors respectively. An example of Shifted-1D-LBP operator is given in Fig. 2 where PL = 6 and PR = 2, the center sample Pc is mentioned.

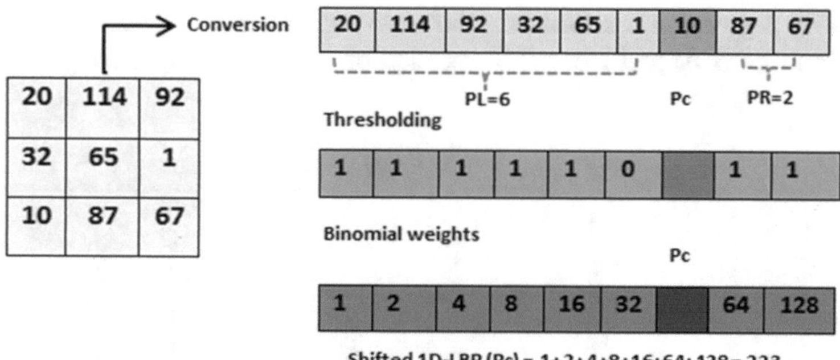

Fig. 2. An example of Shifted-1D-LBP operator.

3.3 Proposed System

The proposed approach consists of three stages

In the preprocessing step, SG-FIR (Savitzky-Golay Finite impulse response) technique was applied to reduce the noise from the ECG signal and removes various artifacts and improves the signal equality. Figures 3 and 4 shows about 10 s from the original ECG signal and it's preprocessed one from ECG-ID and Normal Sinus Rhythm databases, respectively.

The second step consist of segmenting the preprocessed ECG signal by locating the fiducial point QRS adopting Pan-Tompkins algorithm [4] to isolate the fiducial point (P, Q, R) for each beat segment. Each detected R-peak determines the center of the QRS complex.

Then, to isolate the heartbeat, we take 94 samples before the R-peak and 150 samples after the R-peak which means that each ECG heartbeat has 245 samples and 490 ms segment duration for ECG-ID database and we take 50 samples before and

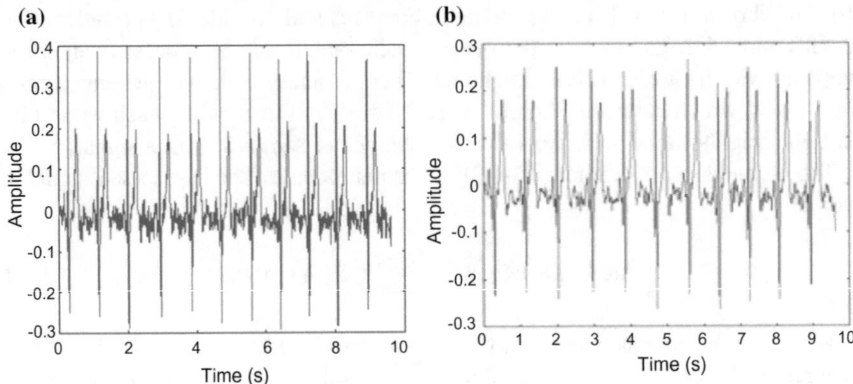

Fig. 3. The original (a) and the preprocessed (b) ECG signal from ECG-ID database

after the R-peak obtaining 101 samples for Normal Sinus Rhythm database. Figures 5 and 6 shows segmented heartbeats for the same and different subjects aligned with the R peak from ECG-ID and Normal Sinus Rhythm databases, respectively.

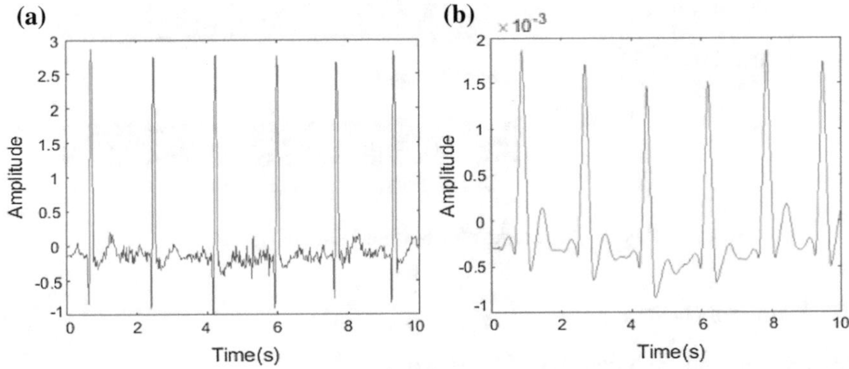

Fig. 4. The original (a) and the preprocessed (b) ECG signal from Normal Sinus Rhythm database

For features extraction step, Shifted 1d-LBP was applied to extract the non-fiducial features from the ECG heartbeats. After many experiences to select the best combination of parameters, a number of 5 neighbors from the left side and 3 neighbors from the right side (PL = 5, PR = 3). Figure 7 shows Shifted 1D-LBP features extraction diagram from an ECG heartbeats with ECG-ID and Normal Sinus Rhythm databases.

In the matching step, we have assessed the performance of our proposed approach by using KNN (K Nearest Neighbors) classifier. The performances of the classifiers were evaluated in terms of accuracy by Eq. 4.

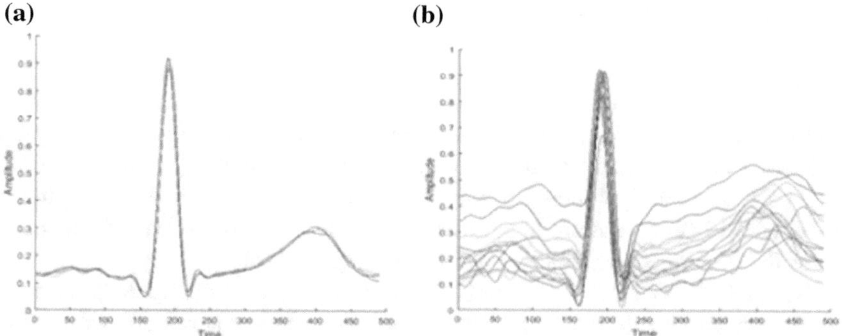

Fig. 5. Segmented heartbeats for the same (a) and different (b) subjects aligned with the R peak from ECG-ID database

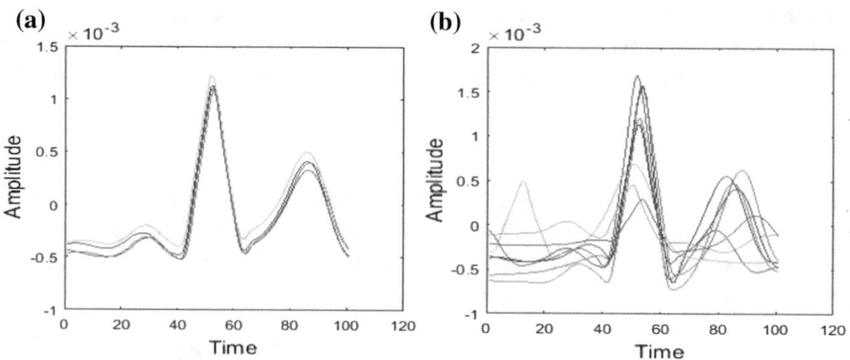

Fig. 6. A segmented heartbeats for the same (a) and different (b) subjects aligned with the R peak from Normal Sinus Rhythm database

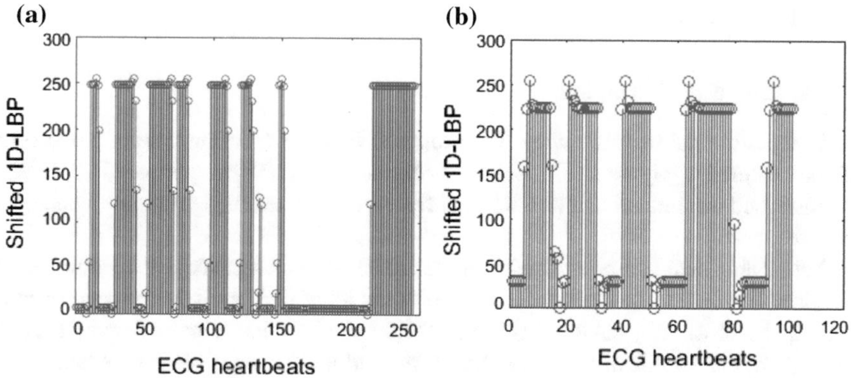

Fig. 7. Shifted 1D-LBP features extraction diagram of an ECG heartbeats from (a) ECG-ID database, (b) Normal Sinus Rhythm database

4 Experimental Result

4.1 Database

Two ECG databases were employed for testing the developed ECG biometric systems:

ECG-ID database contains different signals from 90 subjects, each subject has at least two records and at most 20 records acquired for about 20 s and digitized at 500 Hz with 12-bit resolution. In our experiments, two records were used per subjects, one for the training set and the second as a testing set [21].

MIT-BIH Normal Sinus Rhythm Database contains different long-term ECG recordings from 18 subjects. This dataset has been used in many studies focusing on different problems. One record for each of 18 persons was required for about 20 h referred to the Arrhythmia Laboratory at Boston's Beth Israel Hospital. The included subjects have no diagnosed cardiac abnormalities, with beat annotations. In our experiments, each ECG signal was divided into two sub-signals, about 20 s were used from the first sub signal for the training set and 20 s were used from the second sub signal for testing set [21]. We notice here that the Nearest Neighbors (KNN) classifier was made in overall our experiences to assess the accuracy (CRR) of our approach.

4.2 Evaluation Metrics

Over both databases, the various results are compared through Receiver Operating Characteristic (ROC), Equal Error Rates (EER), False Accept Rate (FAR), False Reject Rate (FRR) and Correct Recognition Rate (CRR) which was defined in Eq. 4 to evaluate the performance of a biometric system:

$$\text{Correcte Recognition Rate} = 100 - \frac{\text{FRR} + \text{FAR}}{2} \tag{4}$$

Where: FRR indicates that genuine person was considered as an imposter and FAR indicates that imposter was considered as a genuine person. ERR indicates the point where: FRR-FAR = 0.

4.3 Results and Discussion

In this experimental research work, the proposed ECG recognition system based on no fiducial features applying the implement Shifted 1D-LBP was compared to existing systems based on fiducial and non-fiducial features validated using different benchmark databases.

From Table 1, it can be noticed that the achieved results prove that the most of the ECG heartbeats in the test dataset can be correctly identified. As it can be seen from the table, an ERR of 2%, FAR of 0.7 and 97% for CRR has obtained with ECG-ID database. While a FAR of 0%, an ERR of 2% and a CRR of 100% has obtained with Normal Sinus Rhythm database. Figure 8 shows the ROC curve for the proposed approach using ECG-ID and Normal Sinus Rhythm databases which allows us to visualize easier the accuracy of our results.

Table 1. Comparison of the performance evaluation of the proposed approach to the other existing systems

Athors	Extraction method	Database	CRR %	EER %	FAR %	FRR %
Biel et al. [3]	Fiducial features	ECG-ID	98	–	–	–
Wübeler et al. [6]	Fiducial features	ECG-ID	98	2.8	–	–
Louis et al. [8]	1D-MR-LBP	PTB	91	**0.09**	0.09	**0.09**
Dar et al. [11]	DWT and HRV	ECG-ID	83.8	–	16.1	0.3
Chun [12]	Guided filter	ECG-ID	99	2.4	–	–
Barra et al. [13]	Simple peak detection	PTB	96.1	1.33	–	–
Bassiouni et al. [14]	Fiducial features +DWT	ECG-ID	98	–	–	–
Nemirko et al. [22]	Fiducial features	ECG-ID	96	–	–	–
Our proposed approach	Shifted 1D-LBP	ECG-ID	97	2.16	0. 71	5.56
		Normal Sinus Rhythm	**100**	2.61	**0**	5.56

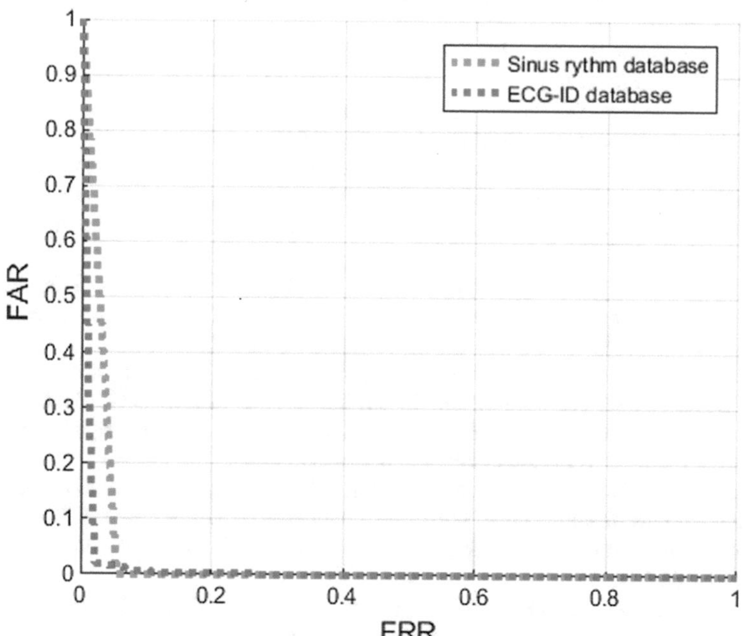

Fig. 8. ROC curve for the proposed approach using ECG-ID and Normal Sinus Rhythm databases

5 Conclusion

In this paper, an ECG biometric recognition system was proposed based on shifted 1D-LBP. After preprocessing the ECG signal, the implemented Shifted 1D-LBP extract the important features from each ECG heartbeat to generate a unique template which presents the signature of the subject. The proposed technique is easy to implement and computationally simple. Nether less, the time complexity of the proposed techniques was not assessed against some of state-of-the-art ear recognition systems. Two public databases, namely ECG-ID and Normal Sinus Rhythm are utilized to validate and evaluate our achieved results. In the future, the use of regular ECG heartbeats instead of using all regular and irregular heartbeats which degrade its performance. Applying a hybrid approach for extraction features which fuse our extracted non-fiducial traits with a fiducial one will be interesting. The effectiveness of these feature extraction techniques may also be verified with a larger dataset.

References

1. Woodward, J.D., Webb, K.W., Newton, E.M., Bradley, M.A., Rubenson, D.: Army Biometric Applications: Identifying and Addressing Sociocultural Concerns. Rand Corporation (2001)
2. Jain, A.K., Flynn, P., Ross, A.A.: Handbook of Biometrics. Springer Science \& Business Media, New York (2007)
3. Biel, L., Pettersson, O., Philipson, L., Wide, P.: ECG analysis: a new approach in human identification. IEEE Trans. Instrum. Meas. **50**(3), 808–812 (2001)
4. Shen, T.-W., Tompkins, W. J., Hu, Y.H.: One-lead ECG for identity verification. In: Engineering in medicine and biology, 2002. Proceedings of the Second Joint 24th Annual Conference and the Annual Fallmeeting of the Biomedical Engineering Society Embs/Bmes Conference, vol. 1, pp. 62–63 (2002)
5. Belgacem, N., Nait-ali, A., Fournier, R., Bereksi Reguig, F.: ECG based human identification using random forests. In: The International Conference on E-Technologies and Business on the Web (EBW2013), Bangkok, Thailand (2013)
6. Wübbeler, G., Stavridis, M., Kreiseler, D., Bousseljot, R.-D., Elster, C.: Verification of humans using the electrocardiogram. Pattern Recogn. Lett. **28**(10), 1172–1175 (2007)
7. Islam, M.S., Alajlan, N.: Biometric template extraction from a heartbeat signal captured from fingers. Multimed. Tools Appl. **76**(10), 12709–12733 (2017)
8. Louis, W., Hatzinakos, D., Venetsanopoulos, A.: One dimensional multi-resolution local binary patterns features (1DMRLBP) for regular electrocardiogram (ECG) waveform detection. In: 2014 19th International Conference on Digital Signal Processing (DSP), pp. 601–606 (2014)
9. Fratini, A., Sansone, M., Bifulco, P., Cesarelli, M.: Individual identification via electrocardiogram analysis. BioMedical Eng. Online **14**(1) (2015)
10. Pereira Coutinho, D., Figueiredo, M., Fred, A., Gamboa, H., Silva, H.: Novel fiducial and non-fiducial approaches to electrocardiogram-based biometric systems. IET Biom. **2**(2), 64–75 (2013)

11. Dar, M.N., Akram, M.U., Shaukat, A., Khan, M.A.: ECG based biometric identification for population with normal and cardiac anomalies using hybrid HRV and DWT features. In: 2015 5th International Conference on IT Convergence and Security (ICITCS), pp. 1–5 (2015)
12. Chun, S.Y.: Single pulse ECG-based small scale user authentication using guided filtering. In: 2016 International Conference on Biometrics (ICB), pp. 1–7
13. Barra, S., Casanova, A., Fraschini, M., Nappi, M.: Fusion of physiological measures for multimodal biometric systems. Multimed. Tools Appl. **76**(4), 4835–4847 (2017)
14. Bassiouni, M.M., El-Dahshan, E.-S.A., Khalefa, W., Salem, A.M.: Intelligent hybrid approaches for human ECG signals identification. SIViP **12**(5), 941–949 (2018)
15. Ojala, T., Pietikäinen, M.: Unsupervised texture segmentation using feature distributions. Pattern Recogn. **32**(3), 477–486 (1999)
16. He, S., Soraghan, J.J., O'Reilly, B.F., Xing, D.: Quantitative analysis of facial paralysis using local binary patterns in biomedical videos. IEEE Trans. Biomed. Eng. **56**(7), 1864–1870 (2009)
17. Pietikäinen, M., Hadid, A., Zhao, G., Ahonen, T.: Computer Vision Using Local Binary Patterns, vol. 40. Springer Science \& Business Media, London (2011)
18. Chatlani, N., Soraghan, J.J.: Local binary patterns for 1-D signal processing. In: 2010 18th European Signal Processing Conference, pp. 95–99 (2010)
19. Boodoo-Jahangeer, N.B., Baichoo, S.: LBP-based ear recognition. In: 2013 IEEE 13th International Conference on Bioinformatics and Bioengineering (BIBE), pp. 1–4 (2013)
20. Ertuğrul, F., Kaya, Y., Tekin, R., Almali, M.N.: Detection of Parkinson's disease by shifted one dimensional local binary patterns from gait. Expert. Syst. Appl. **56**, 156–163 (2016)
21. Goldberger, A., et al.: PhysioBank, PhysioToolkit, and PhysioNet: components of a new research resource for complex physiologic signals. Circulation **101**(23), e215–e220 (2000). Circulation Electronic Pages. http://circ.ahajournals.org/cgi/content/full/101/23/e215. Accessed 13 June 2000
22. Nemirko, A.P., Lugovaya, T.S.: Biometric human identification based on electrocardiogram. In: Proceedings of the 8th Russian Conference on Mathematical Methods of Pattern Recognition, Moscow, Russian, pp. 20–26 (2005)

Dynamic Time Warping Inside a Genetic Algorithm for Automatic Speech Recognition

Fadila Maouche[1(✉)] and Mohamed Benmohammed[2(✉)]

[1] Emir Abd El Kader University, Constantine, Algeria
mifad_5@yahoo.fr
[2] Abd El Hamid Mehri University, Constantine, Algeria
Mohamed.benmohammed@univ-constantine2.dz

Abstract. The technology of the automatic speech recognition is in full grow, a multitude of algorithms have been developed to improve the performance and robustness of ASR (Automatic Speech Recognition) systems. The most studied methods in recent years are those inspired by nature as genetic algorithms. In this article, we will introduce a system that uses a genetic algorithm and dynamic time warping for automatic recognition of isolated Arabic words, tested in noisy environment and in isolated environment.

Keywords: Automatic speech recognition · Genetic algorithm
Dynamic time warping · Arabic language
Mel frequency cepstral coefficients (MFCC) · Oral corpus

1 Introduction

Nowadays, automatic speech recognition's system are increasingly widespread and used in very different acoustic conditions, and by very different speakers, but despite the spectacular progress, the ideal system does not exist yet, and to overcome the current performance of ASR systems, many works are carried out in various laboratories in the world. The most studied methods in recent years are those inspired by nature as genetic algorithms.

The speech signal conveys linguistic and extra-linguistic information; it must be filtered and reduced to retain only the relevant ones. In addition, this signal is extremely variable even for the same speaker. Therefore, a direct comparison treatment on this kind of signal is impossible. To counteract these problems, the Mel Frequency Cepstral Coefficients (MFCC) and the Dynamic Time Warping (DTW) are used in this work.

The Arab States have at least 300 million people, in addition to Arabic minorities scattered throughout the world. Despite this large number, the research on Arabic's speech processing did not reach the level of research on other language as English and French. This is due to several reasons including the lack of standard Arabic sound corpora.

In this paper, we propose a new system for automatic speech recognition of isolated Arabic words, intended for voice control of a wheelchair. This system uses genetic algorithm and dynamic time warping for recognition and MFCC coefficients to model the speech signal.

S. Chikhi et al. (Eds.): MISC 2018, LNNS 64, pp. 180–192, 2019.
https://doi.org/10.1007/978-3-030-05481-6_14

2 Architecture of the Suggested System

The proposed system is a multi-speakers ASR system, and even independent of the speaker, designed to be embedded in a wheelchair. It uses the global approach. when a word is pronounced, the acoustic image of this word is facing all references in the dictionary and the most resembles word, by calculating a distance, is then chosen [1].

Our system consists of several processes; its scheme is represented in the Fig. 1.

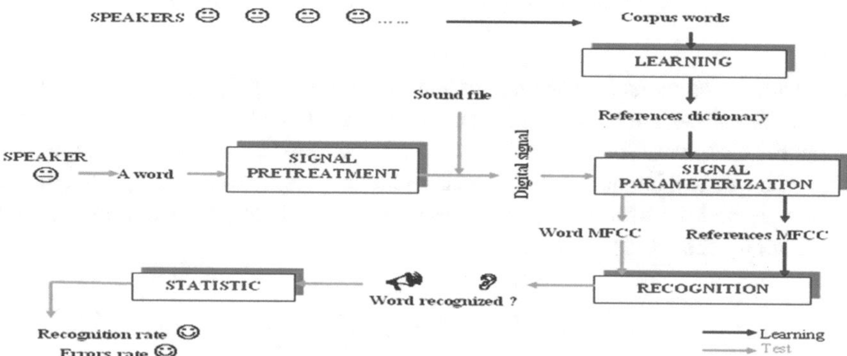

Fig. 1. System architecture

2.1 The Learning Process

The realization of ASR system requires the provision of oral corpora for both learning and test. Several corpora were built for different languages as TIMIT for English language and BDSONS for French language, but there is big lack of standard Arabic oral corpora. To evaluate our system, we were obliged to make two new corpora.

The creation of corpora is a difficult task and takes a great time. First, we have to define the content of the corpora that can depend on the application referred to or not. Then, we must limit the corpora size considered sufficient. After this, we choose the registration conditions, noisy environment or calm. Finally, select speakers and make the recordings, this stage is the most difficult.

The first corpus used to evaluate the proposed system is corpus A, it's composed of 30 words where all phonetics' characteristics of Arabic language were taken into account, without binding to a specific field. The words of corpus A are grouped in Table 1.

The recording environment of corpus A was very calm (no echo and no noise). The digitization of the signal was made by the professional software "Sound Forge version 8", which is well known in the digital audio editing [2].

Four women and four men pronounced the vocabulary words. Their ages are between 22 years and 55 years. Each speaker repeats 3 times each word, so every word has 24 different occurrences. The entire corpus contains 720 sound files.

The second corpus, Corpus B, is recorded in real environment with the open source software Audacity, it contains 33 words, which represent 11 different movements of the

Table 1. Vocabulary of Corpus A

Type of sub-corpus	Words of sub-corpus
Corpus simple	شحذ – دحرج – عرف – كتب – وزن
Corpus gemination	رَدَ – كرَس – فسَر – الشَمس – قيَد
Corpus emphasis	نظر - قرب - ضرب - صرف - طبع
Corpus duration	نادى ، مماليك ، يؤول ، غروب ، هتاف
Corpus tanwine	مكتب ، لون ، منزل ، فرش ، عزف
Corpus mixture	مشتط ، قض ، خزاعة ، اضطر ، مواد

wheelchair, 3 synonyms for each movement, pronounced by 22 speakers of both sexes and of different age categories, children, young, adults and old person. Each speaker pronounces each word five times, the three first occurrences with the same rhythm, the fourth with a strong intonation that reflects suddenness to avoid an unexpected obstacle for example and the fifth reflects weakness. The total number of sound files is 3630. Table 2 group the words of corpus B.

Table 2. Vocabulary of Corpus B

Movements	Words
Turn on	توكل – باشر - هيا
Turn off	انتهى – استرح – تم
Start	امشي - انطلق– تقدم - سر
Stop	قف – توقف – وقوف
Speed	اسرع – عجل – هرول
Delay	تمهل – تكاسل – تريث
Ahead	امام – قدام - قبالة
back	خلف – وراء – تأخر
Right	يمين – تيمن – يمن
Left	يسار – شمال – تيسر
Return	التف – استدر

2.2 The Signal Pre-treatment Process

To be usable by a computer, the speech signal must first be digitized. The transition process from analogue sound to the digital sound is called “sampling”. This consists of measuring the voltage of analogue signal at regular intervals. The value obtained is finally encoded in binary. Another very important parameter of sampling is the precision in which the voltage of analogue signal is read and coded (2nbits) [3].

In the suggested system, speech is recorded at a sampling rate of 44100 Hz and coded in 16-bit.

2.3 The Signal Parameterization Process

The speech signal can't be used directly in ASR systems, it contains many other elements than the linguistic message targeted by these systems, hence the need for pre-processing the signal.

The parameterization step consists in representing the signal by a reduced, relevant, discriminant and robust set of parameters, such as: Energy, the zero crossing rate, the Mel Frequency Cepstral Coefficients (MFCC), the Linear Prediction Coefficients (LPC), the Perceptual Linear Prediction (PLP), etc. This list is not exhaustive but provides an overview of the various parameters that can be extracted from a speech signal.

All these techniques are reasonable solutions for speech signal parameterization, but the famous MFCC are better than the other candidates. The MFCC paradigm has maintained its dominance since its introduction in 1980 by Davis and Mermelstein. The purpose of MFCC is to reduce the number of data characterizing the signal and shows a limited number of parameters or coefficients, discriminating and robust, even in noisy environment [3–6].

To file transform an audio in MFCC cepster several steps are necessary [4, 5, 7], see Fig. 2.

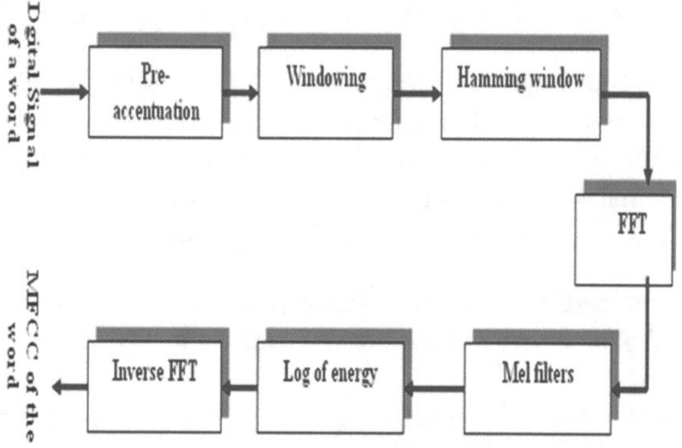

Fig. 2. From digital signal to MFCC

The Signal Pre-accentuation. In speech recognition by MFCC, the signal must first undergo pre-accentuation to remedy the fact that high frequencies are less powerful than the low frequencies. The signal pre-accentuation formula is [3, 4]:

$$h_n = 1 - \alpha * Z_n^{-1} \tag{1}$$

is the pre-accentuation factor, commonly taken to 0,970. h is the pre-accentuation of the signal Z.

The Windowing. The audio signal can't be treated as a whole because this would require many calculations for the machine, so we cut the signal into slices called windows that have the particularity of overlap in half with the aim of have a better treatment for FFT (Fast Fourier Transform). It typically uses a window of N samples, N is a number that is a power of two because the FFT algorithm is much faster for these numbers [3].

Applying Hamming Window. A Hamming window is applied to each window in order to decrease the spectral distortion created by the overlap, and minimize errors produced by FFT. The Hamming window improves the sharpness of harmonics and removes discontinuities on the edges. To create the Hamming signal, we use the following formula:

$$0.54 - 0.46 * \cos\left(\frac{2\pi * n}{N - 1}\right), n \in [0, N - 1] \quad (2)$$

N is the size of the signal (the number of samples) [3, 8].

Application of FFT (Fast Fourier Transform). To transform the signal from time domain to frequency domain we must calculate the discrete Fourier transform (DFT). The Fast Fourier Transform (FFT) is a very powerful algorithm for calculating the discrete Fourier transform. The FFT calculation time is about 10 times lower than a classic DFT [3]. The result of this step is the signal spectre, horizontal axis represent time, vertical axis represent frequency and intensity is represented by the color [9].

Mel Filters Bank. The frequencies range in the FFT spectrum is very wide, so much data to process, we must use a filter bank in the Mel scale. We passes the speech signal through a filter bank, the Mel filter bank is built from triangular filters, each filter will give a cepstral coefficient, commonly we use 12 factors, so we use 13 filters because the 0ème coefficient is not needed for speech recognition [3, 10].

The Cepstrals Coefficients. This is the final step, we transforms data from Mel scale to time scale. We make the inverse of the Fourier transform. The result of this step will be the MFCC itself [3].

The most popular MFCC implementation is written by Malcolm Slaney in the "Auditory toolbox" of MATLAB toolbox.

2.4 The Recognition Process

The recognition is made by a genetic algorithm that compares MFCCs references and MFCC of the word to recognize, the outcome of this process is the sound recognized as mentioned in Fig. 3. In the test phase, the user must confirm the recognition to avoid false recognitions.

The main processes of our genetic algorithm are cited in Fig. 4.

Initialization Of The Population. The reference dictionary (learning corpus) is the population managed by our genetic algorithm. The number of individuals is the number of sound file in this dictionary. The population is divided into sub-population according to the number of words in the dictionary. The choice of the initial population is random

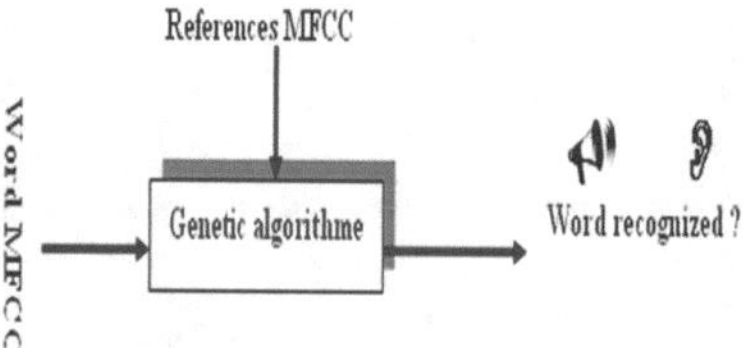

Fig. 3. Recognition process

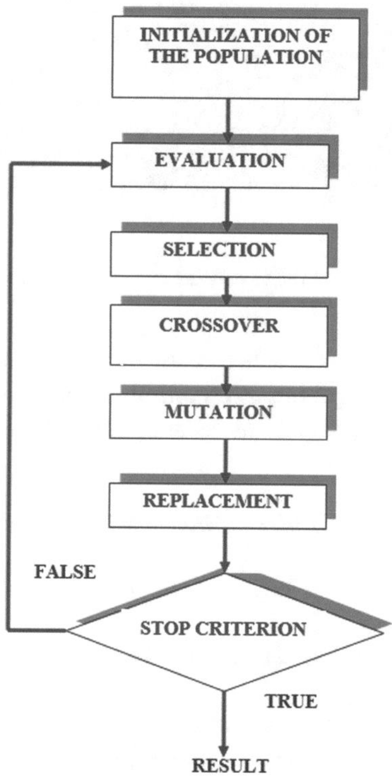

Fig. 4. Genetic algorithm processes

for each word to recognize. An initial population is made up of all occurrences of a word. If we do not reach the recognition threshold in this sub population after 5 generations, we change initial population, and so on, until we reach the recognition threshold of the word to recognize. Each individual of this population contains the word it represents, its MFCC cepster and its fitness.

A MFCC cepster is a matrix, the number of lines is the number of MFCC coefficients. The number of columns differs from one word to another. Figure 5 represents the MFCC cepster of the word 'وزن':

Evaluation Of The Population. To evaluate the population, we calculate the fitness of each individual, it is the distance between the cepster of the word to recognize and the

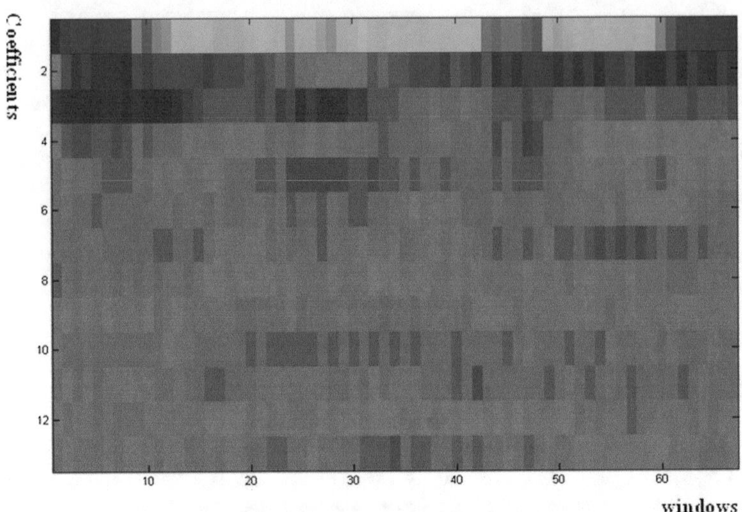

Fig. 5. MFCC cepster of the word 'وزن'

cepster of each individual, if this distance is less than the recognition threshold then the word is recognized. The recognition threshold of a word is defined as a function that relates the maximum and the minimum distance between the different occurrences of this word in the learning corpus.

$$Threshold\_World_i = (Max_i - Min_i) * Fact + Min_i \tag{3}$$

Max_i: The maximum distance between the $word_i$ and the different occurrences of this word in the learning corpus.

Min_i: The minimum distance between the $Word_i$ and the different occurrences of this word in the learning corpus.

Fact: Adjustment factor, a number between 0 and 1.

Two types of distance were used separately to evaluate individuals.

Euclidean Distance. Its formula is [11]:

$$Distance(A, B) = \frac{1}{N * M} \sum \sum (A - B)^2 \tag{4}$$

A and B are two matrices of size (N * M).

Dynamic Time Warping Distance. The speech signal is characterized by high variability inter-speaker and even intra-speaker. The temporal scales of occurrences of the same word don't coincide. To counter this problem, several recognition systems use Dynamic Time Warping (DTW). It is based on the work of R. Bellman in 1957 and Sakoë and Shiba in 1970 [5].

Given two acoustic forms $T = (T_i,\ 1 <= i <= I)$ and $R = (R_j,\ 1 <= j <= J)$ of respective lengths I and J, the distance between T and R consists of accumulation of the distances d (i, j) between the events T_{ik} and R_{jk} along a path C of length K. The comparison between T and R amounts to searching the optimal path C that minimizes this cumulative sum. The cost of a path is obtained by the recurrent formula of Sakoe and Shiba [5].

In this work, the proposed system uses the DTW as a distance measure only, but does not traverse all the set of references to have the minimum distance, which is the major disadvantage of systems based on DTW.

The Stop Criterion. The stop criterion of this genetic algorithm is either the word is recognized either all sub populations have been covered.

The Selection. After evaluating all individuals of the population, we apply the elitist selection method (Many researchers have found that elitism improves the performance of the AG). This method allows the genetic algorithm to retain a number of best individuals for the next generation. These individuals may be lost if they are not selected to reproduce [12, 13].

The crossover. Its fundamental role is to enable the recombination of information contained in the genetic heritage of the population. We applied the one point cross with the probability 0.80 [12, 14].

The mutation. A mutation is simply a change of a gene found in a locus randomly determined. The altered gene may causes an increase or a weakening of the solution value that represents the individual [15].

The principle of the mutation used in our system is new, we inject a gene of the word to recognize directly in the reference dictionary, instead of using a random gene (a gene is a column of MFCC cepster). This principle will accelerate the convergence of the algorithm without influence the result [16]. The probability of mutation is 0.01 [12, 14].

The replacement. The elitist replacement is the most suitable in our case, it keeps individuals with the best performance from one generation to the next. In general, a new child takes place within the population if it is more efficient than the less powerful individuals of the previous population, so we replace the worst parents [15].

2.5 The Statistical Process

The recognition's system errors can be classified into three basic types, which have not the same weight [17]

- Substitution: a word is confused with another word (false recognition),
- Elision: a word is not recognized (non-recognition),
- Insertion: a word that does not belong to the vocabulary has been recognized.

We considered two types of errors, substitution and elision. The statistical process gives several results: the recognition rate, the substitutions rate, the elision rate and the number of tested words.

To evaluate our system, we carried out five tests on corpus B with each type of distance

- Test1: Occurrences used to test the system are taken from the learning corpus, the recognition rate reach 100% regardless the corpus used. This test proved that the search mechanism is very efficient.

- Test2: Each word in the corpus B has 110 occurrences. We removed some occurrences of this corpus to use them in this test. Therefore, any speaker who participated to the learning can use this system. In other words, the system is multi-speakers.
- Test3: The same principle of test2 is used, but we eliminate occurrences that represent unusual elocutions from the corpus. The purpose of this test is to compare its results with the test2 results to show the influence of unusual elocutions on the recognition rate in multi-speakers systems.
- Test4: Speakers, who haven't participated to learning, pronounce the occurrences of the test. This test has proved that the system is independent of speaker.
- Test5: The aim of this test is to show the influence of unusual elocutions on the independent of speaker systems.

The corpus A is recorded in an isolated environment. We conducted two tests on this corpus with each type of distance. The aim is to show the impact of noise on the system.

- Test1: is the same of corpus B.
- Test2: prove that the system is independent of speakers.

3 The Experiments Results

The purpose of the tests performed on corpus A and corpus B is to compare the impact of dynamic time warping on the behavior of the genetic algorithm in an isolated environment and in a noisy environment, either for multi-speaker recognition or for independent of speaker recognition. Corpus B is recorded in real conditions because it is intended for an embedded system on a wheelchair. The results of tests are mentioned in the following figures:

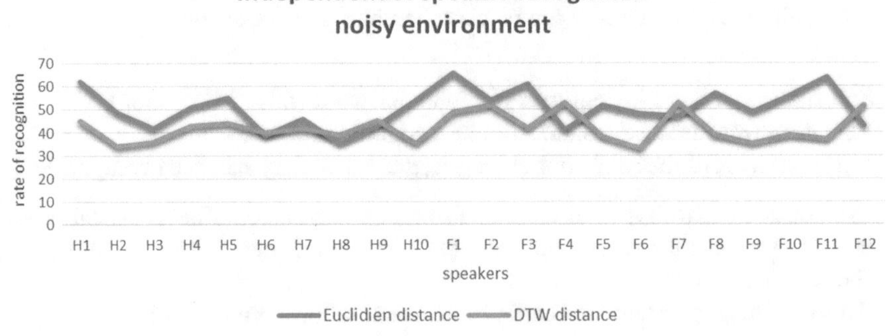

Fig. 6. Results of test4

- **Result1**. According to Figs. 6 and 7, the best recognition rate given by the system in noisy environment, with both distances, does not exceed 66% for independent of

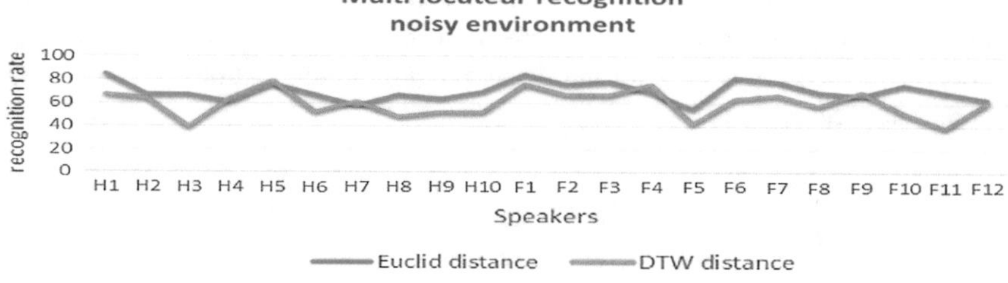

Fig. 7. Results of test2

speaker recognition and 85% for multi speaker recognition. These rates are acceptable in noisy environment. It is noticeable also, that Euclidean distance is slightly better than Dtw distance for both type of recognition.

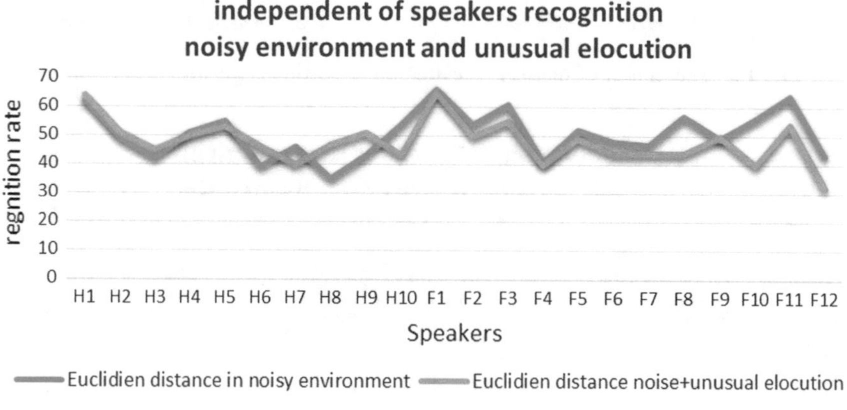

Fig. 8. The impact of unusual elocution on Euclidian distance.

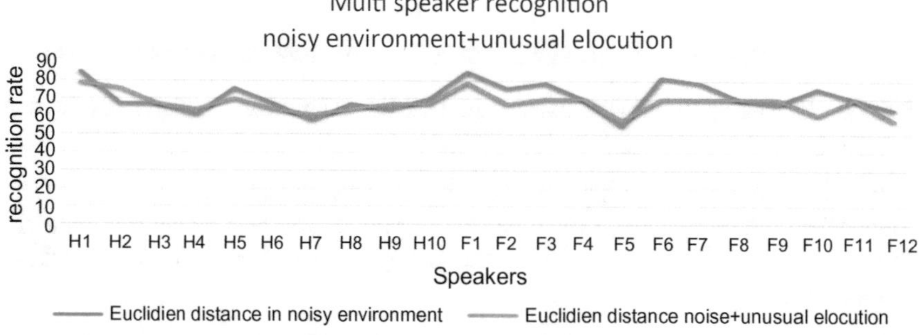

Fig. 9. The impact of unusual elocution on recognition rate with Euclidian distance

- **Result2**. According to Figs. 8 and 9, uunusual elocution has almost no effects on Euclidian distance regardless the type of recognition. The best recognition rate given by the system is 79% for multi speaker recognition and 65% for independent of speaker recognition.

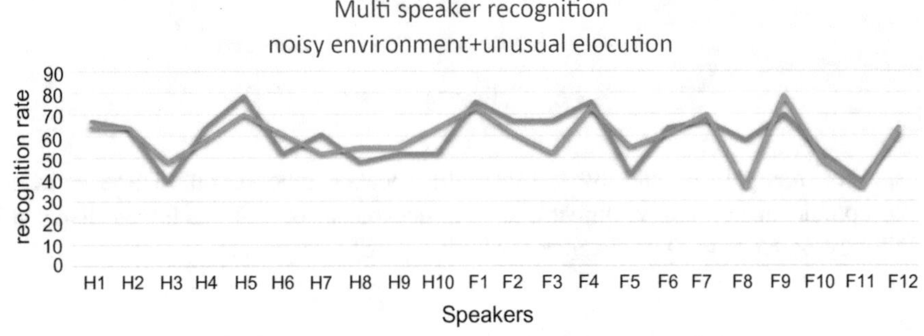

Fig. 10. The impact of unusual elocution on recognition rate with Dtw distance

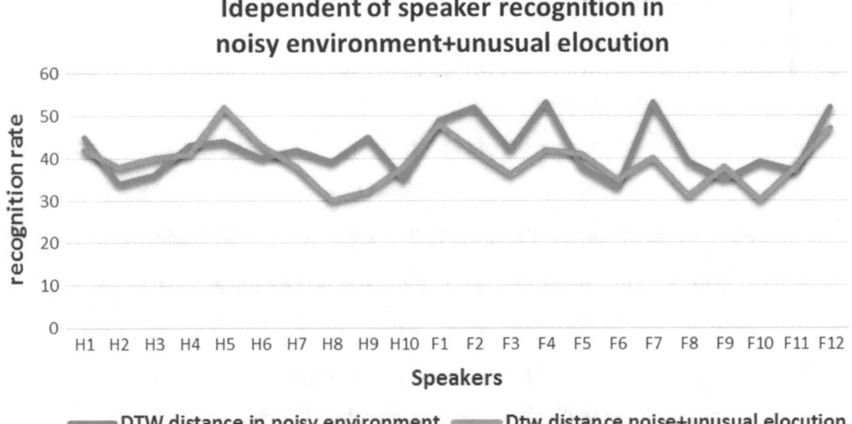

Fig. 11. The impact of unusual elocution on Dtw for independent of speaker recognition

- **Result3**. According to Figs. 10 and 11, the Dtw distance is not responsive to unusual elocution regardless the type of recognition. The best recognition rate given by the system is 79% for multi speaker recognition and 65% for independent of speaker recognition.

Corpus A is recorded in an isolated environment, the aim is to evaluate the real performances of the system and to show the impact of noise on the two distance used. The results are reported in Fig. 12.

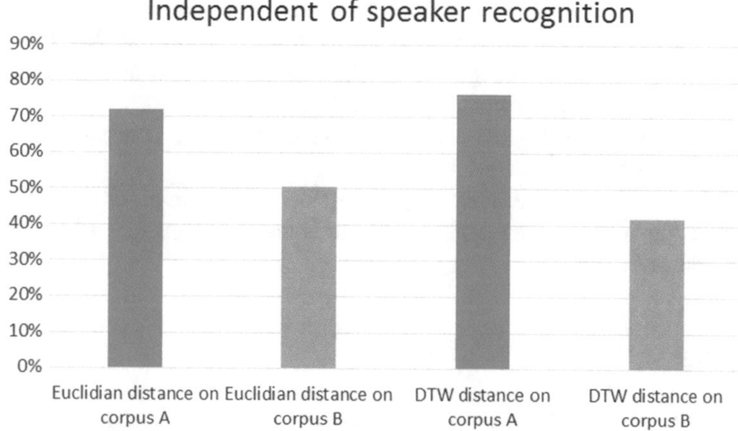

Fig. 12. Impact of DTW in noisy and isolated environment

Tests on corpus A will also proved, if the system is influenced by the different characteristics of the Arabic language.

- **Result4**. According to Fig. 12, the Dtw distance is more suitable for isolated environment, while Euclidian distance is more relevant for noisy environment.
- **Result5**. The recognition rate of the tests on corpus A differs from one sub-corpus to another according to the characteristics of the Arabic language. So the type of words affects the recognition rate and good choice of vocabulary can yield better results.
- **Result6**. The system does not make the distinction between man voice and woman voice, if a woman pronounces a word of the vocabulary; the recognized sound may be a man voice, because our system works on the word's signal itself without taking any consideration of speaker's personality.

4 Conclusion and Perspectives

In this paper, we propose a new system for automatic speech recognition of isolated Arabic words, intended for voice control of a wheelchair. This system uses genetic algorithm and dynamic time warping for recognition and MFCC coefficients for the speech signal parameterization.

To evaluate the performance of this system, we have made two oral corpora, which could be used by other researchers to test and validate their systems working on Arabic language. The words of these corpora have been carefully selected. The first corpus is recorded in an isolated environment. The second one is recorded in noisy environment.

The results acquired by our system confirm that the use of genetic algorithms joined with MFCC parameterization is a very promising method in the ASR field. In the other hand, the various tests carried out proved that the use of dynamic time warping within the genetic algorithm gives results slightly less than the use of Euclidian distance in noisy environment. We can also conclude that unusual elocution does not really affect recognition rates in noisy environment. These tests also proved that the different characteristics of Arabic language affects the recognition rate and good choice of vocabulary can yield better results.

In addition, we have suggested a new mutation method that accelerated the convergence of genetic algorithm

In the future, it would be interesting to test our system in an embedded environment, extend our corpora size and experiment other technics for automatic speech recognition as ant colonies.

References

1. Haton, J.-P., Pierrel, J.M., Perennou, G., Callean, J., Gauvain, J.-L.: Automatic speech recognition, Dunod edition (1999)
2. Sound Forge 8. https://www.clubic.com/telecharger-fiche10882-sound-forge.html
3. Robot guidé par reconnaissance vocale, PPE (2004). www.tigen.org/perso/geogeo/TPE_Vocale/Dossier%20(Reconnaissance%20vocale).pdf
4. Ganchev, T.D.: Speaker recognition. Dissertation subject at the Patras University for the Doctor degree of philosophy, Greece (2005)
5. Jacob, B.: Un outil informatique de gestion de modéles de markov caches: expérimentations en reconnaissance de la parole, Ph.D. thesis at the Paul Sabatier University in Toulouse III (1995)
6. Ellis, D.: « PLP et RASTA (et MFCC, et inversion) dans Matlab employer melfcc.m et invmelfcc.m » (2006). http://www.ee.columbia.edu/~dpwe/resources/matlab/rastamat/
7. Mami, Y.: Reconnaissance de locuteurs par localisation dans un espace de locuteurs de references, doctoral thesis, signal and image speciality, telecommunication national school, Paris (2003)
8. Foukou, A., Henaff, S., Lagacherie, M., Rouget, P.: Final presentation of Marvin project (Modest-encoding AlgoRIthm with Vocal IdentificatioN), EPITA (2002)
9. Hernandez, J.: Algorithmes d'acquisition, compression et restitution de la parole à vitesse variable, étude et mise en place » , graduation project (1995)
10. Moraru, D.: « Segmentation en locuteurs de documents audios et audiovisuels: application à la recherche d'information multimédia » , thesis for doctor degree of inpg, speciality: signal image parole télécoms, national polytechnic institute Grenoble (2004)
11. Mathematic distance, Wikipédia, https://fr.wikipedia.org/wiki/Distance_(math%C3%A9matiques). Accessed 8 Sept 2018
12. Obitko, M.: Genetic algorithms, University of Applied Sciences Dresden (FH) (1998)
13. Melanie, Mitchell: An Introduction to Genetic Algorithms, A Bradford Book the MIT Press, Fifth Printing. Massachusetts, London, England, Cambridge (1999)
14. Dumeur, R.: Synthèse de comportements animaux individuels et collectives par algorithms génétiques, science department, artificial intelligence institute, Paris_8 university (1995)
15. Amédée, S., Francois-Gérard, R.: Genetic algorithms, thesis of the end of the year, Mr Philippe Tutoring Audebaud, supported on 21/06/2004
16. Maouche, F., Benmohammed, M.: Evolutionist approach and MFCC modeling for Arabic automatic recognition. RIST, vol 18 (2010)
17. Néel, F., Chollet, G., Lamel, L., Minker, W.: Reconnaissance et compréhension de la parole: evaluation et applications. The LIMSI and CNRS laboratories in France

Enhancing the Sales Diversity Using a Two-Stage Improved KNN Algorithm

ChemsEddine Berbague[1(✉)], Nour El islem Karabadji[1,2], and Hassina Seridi[1]

[1] Electronic Document Management Laboratory (LabGED), Badji Mokhtar-Annaba University, BP 12, Annaba, Algeria
{berbague,Karabadji,Seridi}@labged.net

[2] Higher School of Industrial Technologies, P.O. Box 218, 23000 Annaba, Algeria

Abstract. In recommender systems field (RS), considering a commercial system perspective involves covering in an accurate manner most items available in the market. However, in the memory based collaborative filtering (CF),the recommendation ability is limited because of the huge size of users and items. For this reason, the clustering algorithms were employed to improve the scalability of the system by partitioning users data into clusters then performing computations on each cluster separately. We propose in this paper a recommendation approach that targets two well-known issues: the scalability problem and the recommendation diversity. Our contribution consists of two successive stages: a) K-nearest neighbor (KNN) algorithm based on the use of an adapted similarity measure. b) An adjusted neighborhood selection performed by a genetic algorithm. The approach aims to improve the quality of the neighborhood set by exploring the reduced search space obtained in the first step, to choose among them the best ones who can enhance the quality of the recommendations. The proposed algorithm was compared to baseline recommender systems and showed competitive results in terms of the diversity and the precision of the recommendations.

Keywords: Memory-based collaborative filtering · Genetic algorithm Diversity

1 Introduction

E-commerce is considered a hot business field that attracts interesting investments in many domains such as online services, movies, books, and tourism [1]. The objective of a recommender system is to increase the mutual benefits of the clients and the system by satisfying users needs and promoting most products under the trade equivalently. This objective is a hard task because of the opposite directions of the accuracy and the recommendation coverage as long as the accuracy goes positively with the recommendation popularity which means less covered items. This phenomenon is known in the literature by the long tail problem [2].

S. Chikhi et al. (Eds.): MISC 2018, LNNS 64, pp. 193–203, 2019.
https://doi.org/10.1007/978-3-030-05481-6_15

Long tail problem is explained by the trend of recommender systems at only recommending popular items. In effect, items highly rated have more chances to be recommended. In more words, recommendation algorithms have more reliable prediction values toward known items (i.e.in term of the expressed preferences). In the case of a newly inserted item, predicting a rating value for a user who has not seen it yet involves few participant neighbors and the predicted value might be unreliable and consequently being ignored.

Moreover, the long tail issue affects negatively the novelty and the diversity of the recommendations because of the limited recommendation coverage in most memory-based collaborative filtering algorithms. Popular items are usually of a limited number and their diversity may be limited and pushed by the popularity factor. These issues may strongly decrease the commercial revenues because of the low covered items and the redundant recommendations.

The purchasing habits tend to be more frequent for a limited amount of items which have the highest consumption frequency and less with the long tail items. This latter came with a decreasing consumption and are usually unknown items and are probably newly inserted. CF algorithms aim to exploit the social phenomenon of users to offer accurate recommendations. Furthermore, CF was widely used in the recommendation field due to their simplicity and its considerable results. In effect, the collaboration concept came from the fact that similar users tend to act similarly in commercial markets. Thus, the user similarity computation has a great importance at accurately predicting the user's ratings. In the literature, there are many similarity metrics which were explored, improved and examined experimentally for different purposes and quality objectives. Indeed, the similarity efficiency of two given users depends on their status: cold users or warm users. Furthermore, the similarity metric could be applied on two levels: the user's demographic information and the user's behavioral information (i.e ratings given by users to items).

The based rating similarity metrics include a wide range of measures such as Euclidean distance, Cosine, Spearman...etc. The similarity value of two users explains how much they share the same preferences. High similarity value means that probably a liked item by a first user would be also liked by the second user. In opposite, low similarity value reflects the independence of the considered users. For a given user, the CF process looks for a set of neighbors by comparing the similarity values. In this algorithm, K nearest neighbors are selected according to the similarity values. Then, for a given unrated item, a prediction formula is used and consists of aggregating the neighbor's ratings. The CF's efficiency depends on two critical factors. From one hand, the size of the neighborhood (i.e. a high neighborhood size leads to a better prediction accuracy). From the other hand, the sparsity level (i.e. less sparse user profile leads to a better similarity measurement). However, the CF approach suffers from high computation complexity. In fact, the neighborhood selection involves exploring the whole users set, to locate the most suitable neighbors. This problem was treated in the literature by proposing based clustering collaborative algorithms. The clustering techniques concern the statistical and bio-inspired based algorithms. The idea consists of avoiding a general view of the users set by dividing them into groups of similar profiles. Then, applying a collaborative algorithm on the level of each cluster separately. However, this approach considers only one side of the recommendation quality which is the accuracy.

We propose in this paper a two-stage neighborhood selection approach. In a first step, we seek to reduce the search space extension by performing an adapted KNN algorithm. In this stage, we modified a similarity measure to combine a pairwise user diversity measure and a similarity-based rating measure. Then, in a second step, we employed a based genetic algorithm to improve the neighborhood selection. During this step, we explore for each user the possible neighborhood sets obtained from the first step to select among them the best ones which may improve the recommendation quality. The objective of the proposition is to enhance the scalability of the system and to improve the recommendation quality.

In the next section, we cite related works that discuss the recommendation diversity in recommender systems and the different proposed solutions. We pass later, in Sect. 3, to explain in details our proposed approach. Next, we show, in Sect. 4, the obtained results with different configurations and comparisons against the state of the art recommender systems. Finally, we discuss and conclude in the last section.

2 Related Works

A recommender system efficiency is subject to a number of factors. In fact, machine learning tools took an important position in modeling and understanding user behavior. Techniques such as hidden Markov model (HMM) [3], association rules(AR) [4], neural networks (ANN) [5], decision trees (DT) [6] and the clustering [7] was proposed for targeting both the top N recommendation and the rating prediction.

In this context, the data structure plays a crucial role in indicating which algorithm to use. Additionally, The emersion of new evaluations metrics has oriented the search to a more interesting quality criterion. On the one hand, users expect from the system relevant recommendations to those ones which they may like. On the other hand, the system aims to promote most items in the trading market.

The research in [8] made on eBay data has analyzed the recommendation field from different perspectives through the issue of the long tail and the improvement possibilities. In fact, the authors have discussed five different recommendation aspects: the definition of the target clients, the recommendation approach choice, the recommendation context, and the recommendation explanation.

In [9] the authors prove that in item-based top N recommendation, item similarity measures are strongly correlated to the recommendation popularity. In the aim of controlling this effect, a two steps process was proposed. Firstly, the rating matrix was standardized to completely remove the correlation between popularity and the similarity. Secondly, a ranking function was used to balance both accuracy and popularity.

Additionally, a modified cosine similarity measure was proposed in [10,11]. The new formulas use the user/item profile size as a weight to decrease the popularity effect, said a user with large profile size has probably different preference range.

[2]'s authors have proposed an item clustering technique to decrease the rating prediction error by dividing the items into two sets according to the size of the items profile. Then, for long tail items, an item clustering is used to leverage the long tail problem. Moreover, the researchers analyzed the clusters number parameters and related scalability improvements.

Model-based collaborative filtering approaches were equally subject to the diversification adjustments. As an example, in [12] the researchers have proposed an adapted matrix factorization technique to address the recommendation ranking problem.

From a different side, in [13] a graph based recommender was proposed. In this work, the item-based recommender system was analyzed in terms of the recommendation coverage. The presented analyze showed that some items are not reachable during the recommendations process. In details, the authors have modeled the recommendation network in form of a graph. Then, they adopted a probabilistic model to minimize the number of isolated items.

The evolutionary algorithms (EAs) were used for clustering problems, optimizing user/item features' weights as well as recommender hybridization [14]. In this context, the authors of [15] have discussed the hard balance between accuracy and diversity. Thus, they proposed to take benefit of combining different recommendation algorithms since each algorithm has different recommendation features.

In [16], the researchers have proposed an evolutionary algorithm that optimizes a ranking function. The algorithm takes as parameters primitive measures: precision, recall, diversity, and novelty. The resulted function was used to rank the recommendations of a collaborative filtering algorithm.

In [17], a genetic algorithm was proposed to refine the recommendation list. In the first stage, an item based collaborative algorithm was used to generate K recommendations for each user. Then, a genetic algorithm was applied to refine the recommendations using a fitness function that combines both of the item popularity and the accuracy. The algorithm generates a set of valid solutions for every user.

In an earlier work [18] we proposed the use of a genetic algorithm to tackle the initial seeds selection in K-means algorithm and concluded by mentioning the need for the recommendation diversification in RS. On the contrary of the works presented in [14,19] that aim to improve only the accuracy of the recommendations by improving the user's similarities using a genetic algorithm. We propose in this paper, a two-stage diversification approach. In the first step, we address the wide extent of the search space by applying an adjusted KNN algorithm. We addressed the limited view of the similarity metrics by proposing a new metric that combines a pairwise diversity measure with a conventional similarity measure. However, in the second stage, we made a neighbor selection based on the use of a genetic algorithm. The details of our proposition are explained more in the next section.

3 Proposed Approach

Recommender systems occupy a considerable place like a machine learning application in e-commerce websites. This bag of tools facilitates online access to products and services as well as increasing business affairs.

We present in this article an algorithm that involves two successive stages each of which has its high considerable role in delivering diverse recommendations. The first step concerns reducing the search space by applying an adjusted KNN. For this reason, we used a linear combination of a diversity measure with a conventional similarity measure to control the diversity quality of the formed neighborhood. In the second stage, we used a genetic algorithm to optimize a well-defined fitness function. The objective of our fitness choice is to ensure that for each user, the selected neighborhood set would offer a maximum item coverage. We adopted in this paper a linear combination of the precision and the popularity of recommendation as a fitness function to evolute the genetic algorithm. The chosen fitness function was taken for its simplicity and efficiency.

3.1 An Adjusted Similarity Based Neighborhood Selection

KNN algorithm is a well-known algorithm used for classification and regression purposes [20]. This algorithm was used in the memory based collaborative filtering to predict the user's ratings. In this type of approach, a similarity measure is used to select for each user a set of neighbors. The selected neighbors are used to predict the ratings according to the closeness of each neighbor to the target user. A close neighbor has a strong effect on the rating value while a far neighbor has a weaker effect. The formula describes the neighbors rating aggregation to predict the user's preference:

$$Pr(u,i) = \bar{r_u} + \frac{\sum_{j \in N(u)} sim(u,j) \times R_{j,i}}{\sum_{j \in N(u)} sim(u,j)} \tag{1}$$

Where $N(u)$ is u's selected neighborhood, sim(u,j) is the considered similarity between the target user and the neighbor $\bar{r}$ u's average rating while $R_{j,i}$ is j's rating to the item i.

The use of the conventional similarity metric allows to select very similar neighbors and consequently limiting the diversity of the recommendations. However, the addition of the diversity inside the similarity metrics allows a dual understanding of the users set to select similar and diverse ones at the same time. Following this conception, we propose the use of the next formula:

$$new\_sim(u1, u2) = \alpha \times sim(u1, u2) + (1 - \alpha) \times div(u1, u2) \tag{2}$$

For the similarity calculation, we can use whatever measure from the state of the art proposed metrics. However, the diversity calculation involves new metrics. The quality of the recommendation is a result of the whole contributions from every member of the neighborhood. In the aim of selecting neighbors with potential diversification abilities we propose the next diversity measure:

- Popularity: the users who can decrease the popularity effect are the ones who tend to select rare items from the market. A good measure of the user preference toward the less popular items could be the next:

$$div(u_1, u_2) = \sum_{i \in I_2 - I_1} 1 - \frac{P(i)}{P} \tag{3}$$

Where P(i) is the item's profile length and P is the number of users.

The adjusted similarity metric allows only a one to one selection and do not observe the whole scene of the neighbors together. For this end, we propose the use of a genetic algorithm as a complementary step that allows optimizing the quality of the neighborhood.

3.2 An Optimized Neighborhood Using Genetic Algorithm

Genetic algorithms (GA) were initially proposed as a metaheuristic for discrete optimization problems. GA was used for both numerical function approximation as well as searching problems. The algorithm is inspired from the natural concept of survive for the fittest. GA involves an initialization step whereas a set of chromosomes are generated randomly. Additionally, the solutions are encoded on each chromosome. The optimization of the quality of the solution is performed by means of the fitness function and done during the evolution iterations. In each iteration, a number of genetic operations are applied such as the crossover, the mutation, and the selection strategy.

3.3 Description

The objective of the proposed scheme is to address two well-known issues in the recommendation field: the scalability of the system and the recommendation diversity. In the first step, we performed a KNN algorithm as described in the previous section. In fact, each user gets a set of possible neighbors using an adapted similarity measure. The assignment of the neighbors allows for each user to have a view over the whole set of users and to increase the neighbors' possibilities to a range of choices that are similar and hold the diversity.

In the second step, we have for each user a set of possible neighbors with a size larger than the required minimum neighborhood size which we adopted. The genetic algorithm takes as parameters the set of clusters and the users' membership. The exploration of the search space lasts 200 iterations. The details of the genetic algorithm are set in the next elements.

Initialization. The initialization step importance depends on the complexity and nature of the problem. We adopted a random chromosome generation. Also, we put as parameters the number of chromosomes equal to 20, the mutation probability equal to 0.2 and the crossover probability equal to 0.7. Finally, we adopted a binary tournament selection strategy.

Fitness Function. The fitness function choice has a crucial role at accelerating the approximation as well as improving the desired recommendation quality. We used in our case a linear combination of the precision and the diversity as a function to guide the exploration process.

We believe that the restricted use of the similarity alone does not allow the enrichment of the neighborhood by users who can improve the diversity. However, conserving a certain level of similarity during the neighbors' selection helps to get relevant recommendations. Furthermore, the diversity term ensured the inclusion of users with potential diversification ability. The balance between the recommendation relevancy and the recommendations diversity is controlled during the evolutionary process.

Our intuition supposes that the users are differentiated according to their personalities to users with trends to consume diverse products and others are not. We seek during the genetic-based selection to only keep neighbors candidates who can maximize the diversity. Targeting the binomial fitness function allows maximizing both the diversity and the accuracy of the recommendations. We propose the use of a fitness function in the form of:

$$fitness(ch) = \beta \times (1 - precision) + (1 - \beta) \times (1 - diversity) \tag{4}$$

Encoding. The chromosome encoding, as illustrated in Fig. 1, holds for each user his possible neighbors. Thus, each user occupies a length of K binary bits. The K bits represents the candidate neighbors existing in the selected neighborhood set. A bit equals to one means that the neighbor could participate in the collaborative process while a value of zero means that the neighbor is excluded.

The search space has a direct link to the size of the neighborhood. In the case of movielens 100K dataset. A reduced search space of only 300 possible neighbor decreases the number of the possibilities exponentially. In comparison to an encoding which uses for each user a specific identifier, we still find that the required bits are higher and depends on $10 \times 50bit$ in the case of considering a search space composed of 943 user (the whole set of the available users) and $9 \times 50bit$ while considering a search space composed of only 300 user which is respectively higher by 150% and 130%. Moreover, the decoding step consumes time during the convention from the binary to real values while the proposed encoding is directly interpretable. We get a total size of the chromosome as the sum of the initial neighborhood size of each user:

$$Number\_of\_bits = \sum_{u \in U} K \tag{5}$$

Where K is the initial number of neighbors.

Decoding. The main advantage of the encoding which we proposed is its easy and fast interpretation. The chromosome encoding scheme used as the users'

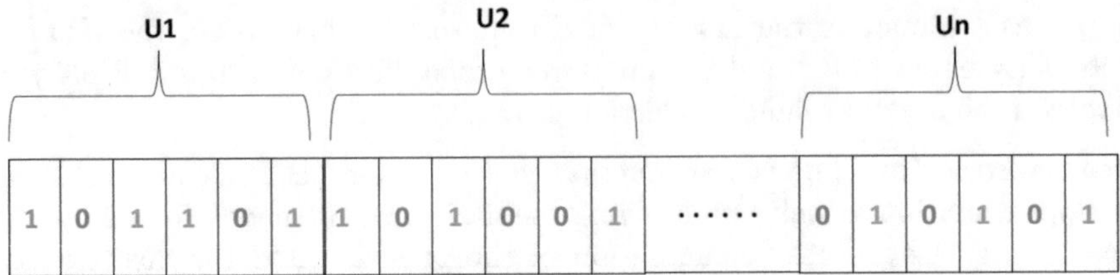

Fig. 1. The chromosome encoding scheme

approval/exclusion vector. The weighting of similarities by zero or to consider the users during the neighborhood selection.

We have adopted during the similarity calculation an additional criterion. So, each user can get recommendations from a possible neighbor only if this neighbor can also get recommendations from the current user. The algorithm 1 explains how similarity values are calculated.

Data: weights, U_1, U_2, indexer, neighborhood, similarity;
Result: weighted similarity
if *$U_2 \notin$ neighborhood(U_1) OR $U_1 \notin$ neighborhood(U_2)* **then**
  return 0;
else
  i ← indexer(U_1, U_2);
  j ← indexer(U_2, U_1);
  if *weights(i) =1 AND weights(j) =1* **then**
    return similarity(U_1, U_2);
  else
    return 0;
  end
end

Algorithm 1. Similarity calculation

4 Experimental Design

We have applied our algorithm on movielens dataset which is formed of 943 users who have given ratings to 1682 item movie. Each rating is in the range of 1 to 5. Our results were compared to two of baseline recommender algorithms. The first one is a memory based collaborative filtering algorithm while the second one is a model based collaborative algorithm. Specifically, we employed user-based collaborative filtering by changing neighbor size in a range of values between 10 and 50 gotten by incrementally adding 10 at every test. The memory collaborative filtering was applied using a basic KNN algorithm. During this latter, we used the conventional similarity metric and the adjusted one. Also, we made a comparison against K-means as a reference clustering algorithm. However, for the model-based algorithm, we used LDA [21] recommender.

For each recommender system, we choose the best results gotten for the targeted quality metrics which are the accuracy and the diversity detailed in the next section.

4.1 Evaluation Metrics

The desired recommendation quality which we adopted in this contribution consists of making a balance between the users and the system benefits. Thus, we used both precision and coverage as accuracy and diversity evaluators. These metrics are defined in the next elements:

- Precision: this metric indicates the portion of relevant recommendations among the set of delivered recommendations. A rating value upper than 3 reflects a positive preference.

$$Precision = \frac{1}{|U|} \sum_{u \in U} \frac{|\{i \in I_u | R_{ui} > \beta\}|}{|I_u|} \tag{6}$$

- Aggregate diversity: this metric computes the amount of promoted products among the the recommendations sets and is defined by:

$$Aggregate\ diversity = \frac{1}{|I|} \bigcup_{u \in U} I_u \tag{7}$$

4.2 Results and Discussion

The tables presented nextly show the results obtained using the memory user-based collaborative filtering with different configurations on a recommendations size of 10. We selected the best results which address the desired quality metrics. These latter were proportional to the highest neighborhood size of the range we studied and it was equal to 50.

	Normal similarity	Adjusted similarity	K-means best precision	K-means best coverage	Proposed approach	LDA
Precision	0.6106	0.6130	0.6379	0.6321	0.6524	0.3368

The results show that for the precision measure, the adjusted similarity measure has a negligible improvement in comparison to the conventional similarity metric. However, the optimization of the neighborhood has given a better result than the rest of the algorithms.

The coverage results, shown in Fig. 2, are equally close in case of using both similarity measures and are better than the LDA algorithm. However, the proposed approach gave the highest recommendation coverage and is better than K-means algorithm. In general, the results prove the potential of decreasing the search space extent using the KNN algorithm and also optimizing the neighborhood quality which may enhance the recommendation quality.

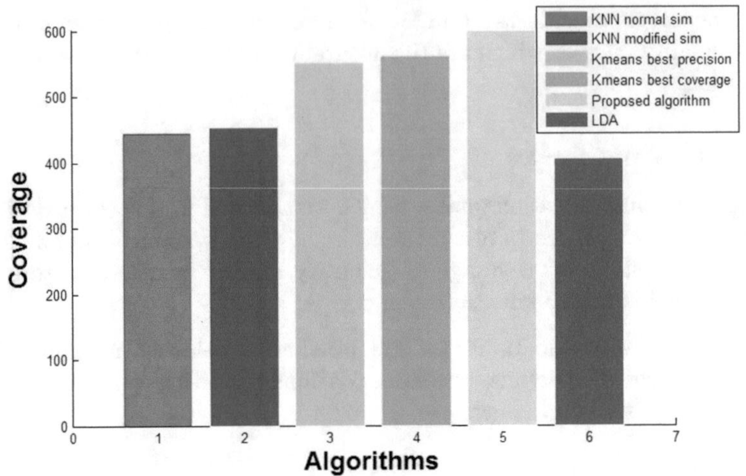

Fig. 2. Coverage results

5 Conclusion

The long tail problem influences negatively the financial revenues since focusing only on the popularity limits the size of promoted products. We have proposed in this paper a two-stage recommendation algorithm to alleviate the popularity effect on the recommendation process by keeping an acceptable balance between accuracy and diversity. The proposed genetic algorithm has a simple fitness function while its chromosome encoding scheme is easily interpretable and less space taking than other encodings. The binary user similarity weighting limits the calculation complexity but it entirely ignores a possibly valuable information in the excluded users. Our next intent is to analyze formally the users and the items profiles in the aim of enhancing our fitness function as well as improving the chromosome encoding to mainly increase the genetic algorithm efficiency.

References

1. Lu, J., Wu, D., Mao, M., Wang, W., Zhang, G.: Recommender system application developments: a survey. Decis. Support Syst. **74**, 12–32 (2015)
2. Park, Y.J., Tuzhilin, A.: The long tail of recommender systems and how to leverage it. In: Proceedings of the 2008 ACM Conference on Recommender Systems - RecSys 2008. ACM Press (2008)
3. Chen, L., Guo, H., Lv, H., Wu, S.: Intelligent recommendation algorithm based on hidden markov chain model. In: 2016 International Conference on Machine Learning and Cybernetics (ICMLC), IEEE (2016)
4. Cakir, O., Aras, M.E.: A recommendation engine by using association rules. Procedia Soc. Behav. Sci. **62**, 452–456 (2012)
5. Paradarami, T.K., Bastian, N.D., Wightman, J.L.: A hybrid recommender system using artificial neural networks. Expert Syst. Appl. **83**, 300–313 (2017)

6. Thiengburanathum, P., Cang, S., Yu, H.: A decision tree based recommendation system for tourists. In: 2015 21st International Conference on Automation and Computing (ICAC), IEEE (2015)
7. Liao, C.L., Lee, S.J.: A clustering based approach to improving the efficiency of collaborative filtering recommendation. Electron. Commer. Res. Appl. **18**, 1–9 (2016)
8. Sundaresan, N.: Recommender systems at the long tail. In: Proceedings of the Fifth ACM Conference on Recommender Systems - RecSys 2011, ACM Press (2011)
9. Zhang, M., Hurley, N., Li, W., Xue, X.: A double-ranking strategy for long-tail product recommendation. In: 2012 IEEE/WIC/ACM International Conferences on Web Intelligence and Intelligent Agent Technology, IEEE (2012)
10. Adomavicius, G., Kwon, Y.: Improving aggregate recommendation diversity using ranking-based techniques. IEEE Trans. Knowl. Data Eng. **24**(5), 896–911 (2012)
11. Liu, R.R., Jia, C.X., Zhou, T., Sun, D., Wang, B.H.: Personal recommendation via modified collaborative filtering. Phys. A Stat. Mech. Appl. **388**(4), 462–468 (2009)
12. Abdollahpouri, H., Burke, R., Mobasher, B.: Controlling popularity bias in learning-to-rank recommendation. In: Proceedings of the Eleventh ACM Conference on Recommender Systems - RecSys 2017. ACM Press (2017)
13. Seyerlehner, K., Flexer, A., Widmer, G.: On the limitations of browsing top-n recommender systems. In: Proceedings of the Third ACM Conference on Recommender Systems - RecSys 2009. ACM Press (2009)
14. Alhijawi, B., Kilani, Y.: Using genetic algorithms for measuring the similarity values between users in collaborative filtering recommender systems. In: 2016 IEEE/ACIS 15th International Conference on Computer and Information Science (ICIS), pp. 1–6 (2016)
15. Ribeiro, M.T., Lacerda, A., Veloso, A., Ziviani, N.: Pareto-efficient hybridization for multi-objective recommender systems. In: Proceedings of the Sixth ACM Conference on Recommender Systems - RecSys 2012. ACM Press (2012)
16. Guimarães, A.P., Costa, T.F., Lacerda, A., Pappa, G.L., Ziviani, N.: GUARD: a genetic unified approach for recommendation. JIDM **4**(3), 295–310 (2013)
17. Wang, S., Gong, M., Ma, L., Cai, Q., Jiao, L.: Decomposition based multiobjective evolutionary algorithm for collaborative filtering recommender systems. In: 2014 IEEE Congress on Evolutionary Computation (CEC), IEEE (2014)
18. Berbague, C., Karabadji, N.E.I., Seridi, H.: An evolutionary scheme for improving recommender system using clustering. In: CIIA. Volume 522 of IFIP. Advances in Information and Communication Technology, pp. 290–301. Springer (2018)
19. Verma, D.A., Virk, H.K.: A hybrid recommender system using genetic algorithm and knn approach (2015)
20. Imandoust, S.B., Bolandraftar, M.: Application of k-nearest neighbor (knn) approach for predicting economic events: Theoretical background (2013)
21. Xie, S., Feng, Y.: A recommendation system combining LDA and collaborative filtering method for scenic spot. In: 2015 2nd International Conference on Information Science and Control Engineering, IEEE (2015)

A Statistical Approach for the Best Deep Neural Network Configuration for Arabic Language Processing

Abdelhalim Saadi[1(✉)] and Hacene Belhadef[2]

[1] Faculty of Technology, Department of Basic Education in Technology, University Ferhat ABBAS Setif 1, 19000 Setif, Algeria
halim.saadi@gmail.com

[2] MISC Laboratory, NTIC Faculty, Abdelhamid Mehri, Constantine 2 University, Constantine, Algeria
hacene.belhadef@univ-constantine2.dz

Abstract. The widespread of the computer technology and the Internet lead to a massive amount of textual information being available in written Arabic. This that more is available, it becomes more difficult to extract the relevant information. To meet this challenge, many researchers are directed to the development of information retrieval systems based on syntactic and semantic parsing. In Arabic, this field is restricted by the lack of labeled datasets. Thus, it is important to build systems for part-of-speech tagging and language modeling and use their results for further syntactic and semantic parsing in fields like chunking, semantic role labeling, information extraction, named entity recognition and statistical machine translation. Deep neural networks have proved efficient in fields like imaging or acoustics and recently in natural language processing. In this study, we used the Taguchi method to find the optimal parameter combination for a deep neural network architecture. Therefore, the neural network obtained the most accurate results. The main use of the Taguchi method in our work is to help us to choose the best context which is the number of words before and after the word on which the training is made.

Keywords: Deep neural networks · Arabic language processing
Taguchi Method · Statistical Machine Translation · Parallel Computing

1 Introduction

Part-of-speech tagging is the process by which a specific tag is assigned to each word of a sentence to indicate the function of that word in its specific context. It demonstrates the word's syntactic category such as noun, verb, pronoun, adverb, adjective or other tags for the purposes of resolving lexical ambiguity [1]. The process of part-of-speech tagging is a very important task in text parsing. It is considered as one of the fundamental tools in natural language processing and it

S. Chikhi et al. (Eds.): MISC 2018, LNNS 64, pp. 204–218, 2019.
https://doi.org/10.1007/978-3-030-05481-6_16

is often part of an higher-level application such as machine translation, speech recognition, and information retrieval.

A language model is a probabilistic model that assigns probabilities to any sequence of words $p(w_1, w_2, \ldots w_T)$. Language modeling is the task of learning a language model that assigns high probabilities to well formed sentences and it plays a crucial role in speech recognition and machine translation systems. Language models (LM) can be classified into two categories: count-based and continuous-space LM. The count-based methods, such as traditional statistical models, usually involve making an n-th order Markov assumption and estimating n-gram probabilities via counting and subsequent smoothing. In continuous-space LM is in general feed-forward neural probabilistic language models (NPLMs) or recurrent neural network language models (RNNs). This Neural Language Models (NLM) solves the problem of data sparsity of the n-gram model, by representing words as vectors (word embeddings) and using them as inputs to a NLM.

In this paper, we focus on the use of part-of-speech tagging and language modeling in Arabic. This task is very hard especially in the Arabic language because of the complexity of the modern Arabic morphology where the absence of diacritics influences the meaning of the words. Therefore there is a need for a disambiguation process. The main problem in Arabic language is that it is very rich in inflectional variation. Proclitics and enclitics can be added one after the other to the basic form of the word, in order to form new words with different meanings. In our work, we have used the MADA toolkit which is a system for morphological analysis and disambiguation for Arabic to annotate our corpus and to calculate the probabilities of each word that belongs to a tag from the set of possible tags. MADA also offers us the features for each word: The gender, the number, the person, and the basic form of the word and the affixes.

To solve our problem we focus on one of the machine learning techniques, called deep neural networks. The idea of using deep neural networks in natural language processing (NLP) began by the architecture proposed by [2]. It is a unified architecture for the six standard NLP tasks: Part-of-speech tagging, chunking, named entity recognition, semantic role labeling, language modeling and semantically related words.

Choosing good parameters for the architecture is very important in order to train deep neural networks. In our paper, we used two strategies for the learning rate: Adaptive learning rate and a fixed value learning rate. Different values for the momentum. And different sizes for the context which is the number of words in the input layer. We used the Taguchi method as a method of parameter optimization which could significantly reduce the cost of time in simulation process.

Natural language processing involves analyzing large volumes of data. Therefore, parallel computing like multithreading and utilization of the processing capacity of graphics cards can help to achieve the above requirements. In our architecture, we have implemented both multithreading on a multicore multi-threaded central processing unit (CPU) and multithreading using the capacity of a graphics processing unit(GPU) by the use of Nvidia CUDA technology.

2 Related Work

Many taggers have been built for Arabic. In general, there are three approaches to deal with tagging problem:

- **The rule-Based approach:** This kind of tagger is based on grammatical or morphological rules, either to assign a tag to a word or to define the possible transitions between different tags. In [3] we can find a rule-based tagger using the ideas of the Brill tagger for Arabic.
- **The statistical approach:** This kind of tagger use frequencies and probabilities. The use of probabilities in tags is quite old, probabilities in tagging were first used by Stolz in [1]. For Arabic, we can find a statistical tagger developed in [4] using support vector machines. Another kind of statistical tagger for Arabic is based on hidden Markov models (HMMs) built for the Egyptian dialect [5], where a performance of a 68.48% was reported.
- **The hybrid approach:** This approach consists of combining a rule-based approach with a statistical-based one. In [6] we can find the hybrid tagger for Arabic that combines statistical and rule-based techniques.

For the language modeling in Arabic we can find many attempts by using the n-gram models, in [7] we can find the use of an approach based on the finite state transducers. In [8] we can find the use of the factored language models.

3 Arabic Morphology

Morphology is a very important task in Arabic language processing because Arabic is a morphologically rich language, which presents big problems in part of speech tagging. Arabic is an inflectional language with a very rich vocabulary. Words in Arabic consist of a base-word or a stem surrounded by affixes. These affixes show the grammatical classes such as: The number, the gender for the nouns and the tense for the verbs. The prefix is added to the beginning of the word and the suffix is attached to the end of the word. Morphology is the branch of linguistics and a computational process that analyzes natural words by considering their internal structures. It studies the word formation, including affixation behavior, roots, and pattern properties. Morphology can be classified as either inflectional or derivational.

Derivational or lexical morphology: In this kind of morphology. We focus on how the words are formed, from a root and a patterns we get lexemes. Therefore, we call this kind of morphology **templatic**, it concatenates a set of morphemes to the word and may affects its grammatical class.

Inflectional morphology: Words are created by lexemes and some features such as part of speech like: Verb, noun and particle, in computational field we find more classes such as: N, V, Adv, Pron, Num, Conj, Det, Pun, and others. In Arabic, nouns are characterized by:

1. **The number:** "مُفْرَدْ": mufrad: singular, "مُثَنّى" : muṯanā: dual, "جَمْعْ": jam`: plural.
2. **The gender:** "مُذَكَّرْ": muḏakkar: masculine, "مُؤَنَّثْ": mu'annaṯ: feminine.
3. **Definiteness:** "مَعْرِفَة": ma`rifat: definite, "نَكِرَة": nakirat: indefinite.
4. **Case:** "الرَّفع": ar-raf`: nominative, "النّصْب": an-nasb: accusative, "الجَّر": aljjar: genitive.
5. **Possessive clitics.**

Whereas verbs are characterized by:

1. **Aspect:** "مَاضِي": māDiy: perfective, "مُضَارِع": muDāri`: imperfective, "أَمْر": 'amr: imperative.
2. **Voice:** "مَبْنِّي لِلْمَعْلُوم": mabnniy lilma`luwm: active, "مَبْنِّي لِلْمَجْهُول": mabnniy lilomajohuwl : passive.
3. **Tense:** "مَاضِي": māDiy: past, "حَاضِر": HāDir: present, "مُسْتَقْبَل": musotaqobal: future.
4. **Mood:** "مَرْفُوع": marofuw`: indicative, "مَنْصُوب" : manoSuwb: subjunctive, "مَجْزُوم" : majozuwm: jussive.
5. **Subject:** "المُتَكَلِم ، المُخَاطِب ، الغَائِب" : almutakalim, almuxāTib, algaa'ib: 1st, 2nd and 3rd person, "مُفْرَد ، مُثَنّى ، جَمْع": mufrad, muṯanā, jam`: singular, dual, plural, "مُؤَنَّثْ ، مُذَكَّرْ" : mu'annaṯ, muḏakkar: feminine, masculine.
6. **Object clitic.**

Particles known in Arabic grammar as ("إِنَّ وَ أَخَوَاتُهَا" : 'nna wa akhawātuhā: that and its sisters), these are ("إِنَّ" : 'nna: indeed); ("أَنَّ" : anna: that); ("كَأَنَّ": ka'anna: it seems that); ("لَكِنَّ": lakinna: but); ("لَيْتَ": layyota: i wish) and ("لَعَلَّ": la``ala: perhaps), these come before a noun and change its case from nominative to accusative, Arabic particles include also interrogation, conjunction, negation and preposition particles.

Inflectional morphology does not affect the word's grammatical category because if the word is a noun, even we change the number or the gender, the word remains a noun, and the same case for a verb. If we change the person or the number the word stills a verb too. **Stemming** for Arabic is the process of extracting roots [1] or stems [2] by removing all affixes from words.

[1] The root is the original form of the word before any transformation process, and it plays an important role in language studies

[2] A stem is a morpheme or a set of morphemes which expresses some central idea or meaning

In computational stemming algorithm is a clustering process that gathers all words that share the same stem and have some semantic relation.

Lemmatization: Is the mapping of a word form to its corresponding lemma. The canonical representative of its lexeme. It is a specific instantiation of the more general task of lexeme identification in which ambiguous lemmas are further resolved. It is not a stemming process, but it is about mapping the word into its stem like in the stem مُحَمَّد: `muHammad: a proper noun`. From the word الْمُحَمَّدُون: `AlmuHammadwun: those worthy of praise`.

Root extraction: Is the process which focuses on finding the root of the word like the root حَمَدَ: `Hamada: he praised`. From الْمُحَمَّدُون: `AlmuHammadwun: those worthy of praise`.

Diacritization: Is a word sense disambiguation process since many Arabic words may have the same constituent letters with different meanings. For instance the word (قدم : `qdm`) may be (قَدَم: `qadam: foot`) or (قَدِمَ: `qadima: he came`).

Word structure: In Arabic a word may represent a proposal. The word in Arabic is composed of a base-word surrounded by affixes, proclitics and enclitics.

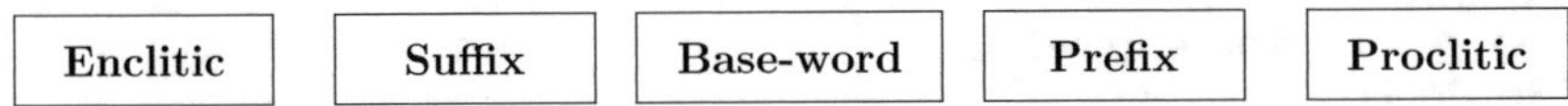

Fig. 1. Arabic word structure

4 Architecture of the Neural Network

There are many possible architectures, for example: Multilayer perceptron, convolutional neural networks [10], restricted Boltzmann machine [11] and another kind of recurrent neural networks [12]. Our architecture is described in Figure 1. It is an architecture inspired from the works of [2]. In this section, we will discuss the different layers the input layer, the hidden layers, and the output layer. As we use the word2vec [9] to represent the words instead of the one hot encoding we don't need to use a layer before the input layer to map the words into real-valued vectors by using the same matrix of weights. The input layer represents the window of words, which means the word on which the training is made, surrounded by a number of words. Each word represented by the continues bag of words CBOW model concatenated by the word features. For the surrounding words, we add the probabilities for which grammatical class the word belongs to. Next, we will discuss the mechanisms of propagation and backpropagation.

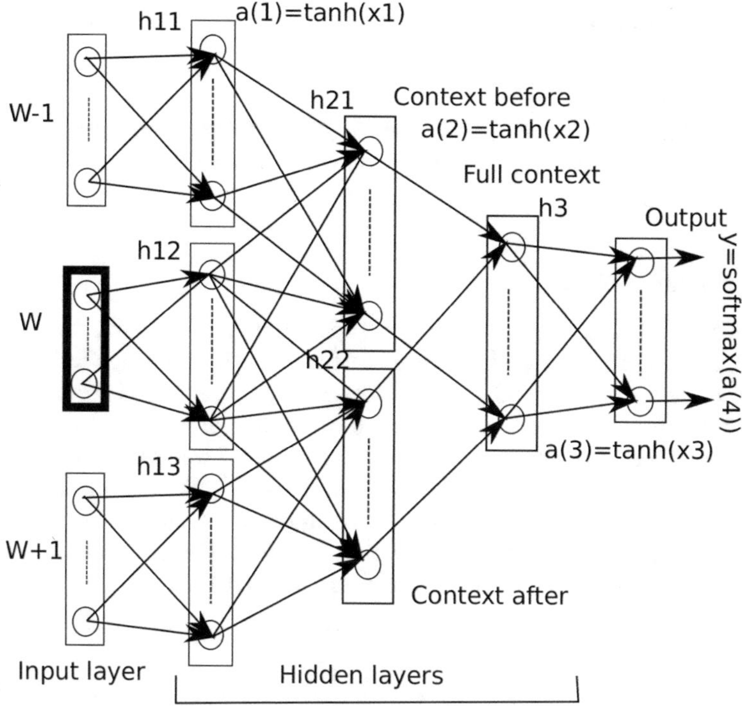

Fig. 2. The neural network architecture

4.1 Preprocessing of Datasets

The preprocessing of datasets is the first step to prepare words in a manageable representation to use in our neural network. It is divided into three parts: firstly, the extraction of word features. Secondly, the calculation of probabilities and finally the CBOW representation for each word. We used MADA[3] as the main toolkit for the preprocessing of data. The features in MADA are `Aspect, Case, Gender, Mood, Number, Person, State, Voice, Stem, POS, and affixes` . The word structure in MADA follows this pattern:

$$[\text{prc3}\ [\text{prc2}\ [\text{prc1}\ [\text{prc0} - \text{Baseword} - \text{enc0}]]]]$$

prc3 is question proclitic or `QUES`, prc2 is a conjunction proclitic or `CONJ`, prc1 is preposition proclitic or `PREP`, prc0 is an article proclitic or `ART`, enc0 are pronominals enclitics or `PRON`. By extracting the features we created the first Look-up-table LW_1. To accomplish the word representation we run word2vec in order to get the word representation in the vector space, to get the second Look-up-table LW_2. The final step in the data preparation to feed the neural network is the computing of probabilities by using the output file of MADA and

[3] MADA+TOKAN version 3.2 using Aramorph version 1.2.1.

the following formula:

$$\text{p}(w_j \mid t_i) = \frac{\text{Count}(w_j, t_i)}{\text{Count}(t_i)}. \tag{1}$$

The probabilities are listed in LW_3 which is the 3rd look-up-table. The vector input of the word w is the concatenation of all the values extracted from the three look-up-tables.

4.2 Propagation

To learn the network, we use the backpropagation method with the stochastic gradient descent algorithm. Therefore the first step is the propagation and calculation of the error, as well as backpropagate to update weights to minimize the error E. In our experiments, we used a window of 3, 5, 7, 9 and 11 words in the input layer. Firstly, we initialize the matrix of weights for each input $\rightarrow$ hidden 1, hidden 1 $\rightarrow$ hidden 2 and hidden 3 $\rightarrow$ output. The initial values for the weights of the hidden layer i should be uniformly sampled from a symmetric interval that depends on the activation function. For the tanh function the results obtained in [13] shows that the interval should be:

$$w_{ij} \in Uniform\left[-\sqrt{\frac{6}{fan_{in} + fan_{out}}}, \sqrt{\frac{6}{fan_{in} + fan_{out}}}\right] \tag{2}$$

In the hidden layers we applied the *tanh* function as the activation function, in fact we used the classic formula:

$$x_j^{(1)} = 1,716 \times \tanh(\frac{2}{3} a_j^{(1)}) \tag{3}$$

to limit the phenomena of saturation. This is because of the function tanh tends rapidly to 1 or -1, and its derivative is then close to 0. Therefore the backpropagation of the error stops. The phenomenon is known under the name of saturation.

Then we use soft-max, a log-linear classification model, to obtain the probabilities in the output:

$$y_k = p(T \mid W_C) = \frac{\exp(a_k^{(3)})}{\sum_{j'=1}^{n_2} \exp(a_{j'}^{(3)})} \tag{4}$$

Where T is the set of tags and W_C is the input words. we can write y_k as:

$$y_k = \frac{\exp(\sum_{j=1}^{n_1} w_{jk} x_j^{(2)})}{\sum_{j'=1}^{n_2} exp(\sum_{j=1}^{n_1} w_{jj'} x_j^{(2)})}. \tag{5}$$

4.3 Backpropagation

The training objective is to maximize $p(T \mid W_C)$, the conditional probability of observing the actual output, the set of tags for the word w_I (denote its index in the output layer as y_{k^*}), surrounded by the words of the context (all the explanation based on Fig. 1).

$$max(p(T \mid W_C)) = max(y_{k^*}) = max \log y_{k^*} \tag{6}$$

We define our loss function E as:

$$E = -\log p(T \mid W_C) \tag{7}$$

$$E = -a_{k^*}^{(3)} + \sum_{j'=1}^{n_2} \exp(a_{j'}^{(3)}) \tag{8}$$

The loss function E for an example is:

$$E_{example} = \frac{1}{2} \sum_{k=1}^{n_2} e_k^2 \tag{9}$$

For all examples we have:

$$E = \sum_{l=1}^{N} E_l \tag{10}$$

Where N is the number of all examples.

5 Parallel Computing

5.1 Parallel Computing Using CPU

Neural POS tagging is a long-running and time-consuming process. In order to accelerate the execution, we split the execution over multiple threads on multi-core processors, because our architecture of the neural network supports the parallel aggregation pattern, and the following conditions are satisfying:

1. We can split the architecture into multiple independent parts. For each hidden layer, we can compute h^l (which is the product of the input by the matrix of weights w_{ij}) independently. Each produces a partial result which will be used in the last hidden layer.
2. In our architecture, the result of each part is a vector h^l and it is not related to the results of the other parts.
3. The partial results from each worker thread will be aggregated when all the threads have finished. In our case the vector of the last hidden layer which is the sum of the vectors of the hidden layers, $conv = \sum_C h^l$.

We have made our experiments on an Intel Core i7-2670QM CPU 2.20GHz $\times$ 8 with 8 GB of RAM, and GeForce GT 525M/PCIe/SSE2 Nvidia graphic card. The number of logical cores in the machine is 8-core processors between physical and hyper-threading enabled. In general, the operating system does a 1-to-1 mapping of the application threads to cores. Therefore we can calculate the hidden layers at the same time, which accelerates the process. In general, in NLP tasks most computations are done between the output layer and the last hidden layer. Therefore using GPU to calculate the output will speed up the process of training. The kernel here is a matrix multiplication function.

5.2 Parallel Computing Using GPU

In neural networks, most calculations are a multiplication between matrix. To use CUDA in our work we created a kernel which calculates the multiplication of a matrix by a vector. We define first the number of blocks needed to make the operation of the multiplication. In general, each block containing 512 threads. The number of threads in general equal to the number of the elements of the result of the multiplication. When we call the kernel all the threads will be launched at the same time, and the result will be obtained very fast.

We tested both versions for 100000 examples from the KALIMAT which is an Arabic natural language resource, the data collection articles fall into six categories: culture, economy, local-news, international-news, religion, and sports [14].

We find that the average time for an epoch using CPU is 120 s and the average time using GPUs is 90 s. From the two kinds of parallel computing, we can see that in deep learning projects, using GPUs is very requested especially when the size of the tensors is important. Because of this, we find that most deep learning frameworks like: Tensorflow, PyTorch, Keras, Caffe ...etc, use GPUs in deep learning applications.

6 Taguchi Method

The Taguchi method is a structured approach for determining the best combination of the inputs to produce a product or service, it was proposed in the 1950s by Dr. Taguchi in Japan [17] it is based on a Design of Experiments (DOE) methodology for determining parameter levels. The main aim of this method is to use a few amount of experimental data for a systematic analysis. instead of using Full-Factorial experiments which is a time-consuming approach, in the Taguchi method we use the orthogonal array, in this case, there is no need to implement each experiment and we can reach accuracy within a short period of time. In our study, we selected four factors: The context which is the number of words in the input layer, the learning rate, the momentum and the epoch. We selected four levels without considering the interaction among the factors when choosing the four-level standard. In our study we selected $L_{16}(4^4)$. Our experiment is based on the mean square error (MSE). Therefore signal-to-noise smaller is better were chosen. The idea here is to gain the smaller error and to get the higher accuracy (Table 1).

6.1 Levels for Factor Design

We selected four factors as the main configuration for our neural network as follows:

1. **Context**: We used four values: 5, 7, 9 and 11. In fact, we used a context with 3 words but in this case, we find that the model did not converge.
2. **Epoch**: Four values are used for this factor which are: 3000, 5000, 7500, 10000.
3. **Learning rate**: We used two fixed values: 0.01 and 0.001, and our two adaptive learning rate methods: The first one based on an Exponential Decay (MED), second one based on a Linear Decay (MLD).
4. **Momentum**: We used four values: 0.25, 0.50, 0.75 and 1.00.

Table 1. Factors and levels for a Taguchi plan

Level/ Factor	Context	Epoch	Learning rate	Momentum
Level 1	5	3000	MED	0.25
Level 2	7	5000	MLD	0.50
Level 3	9	7500	0.01	0.75
Level 4	11	10000	0.001	1.00

Table 2. Differences of all factors among the different levels

Level	Context	Epoch	Learning rate	Momentum
1	17.975	17.556	7.499	13.882
2	17.243	15.961	18.070	20.004
3	−2.406	17.618	22.639	11.285
4	28.746	10.421	13.348	16.386
Delta	31.152	7.197	15.140	8.720
Rank	1	4	2	3

Table 2 illustrates the S/N ratios among factors using the smaller is better method. Using the Minitab[4] software we calculated for each factor in each level the S/N ratio. We can see also the Dela value which is the difference between the highest and the lowest ratio for the same factor. For all factors, we ranked the

[4] Minitab 18: Statistical Software Free Trial. http://www.minitab.com.

Delta value from the highest to the smallest. In our case, the context factor is ranked 1 and epoch is ranked 4. The factor ranked one has the most significant effect in building the neural network. Figure 2 shows the optimized configuration for the neural network in our case our model will take the following values for our factors: Context (9 words); Learning rate (MED adaptive LR with exponential decay); Epoch (10000) and Momentum (0.75).

6.2 Optimal Configuration Choice

Figure 3 demonstrates The optimal configuration for our model which are: Context (9 words); Learning rate (MED adaptive LR with exponential decay); Epoch (10000) and Momentum(0.75). To predict the correspondent theoretical rate of defect T (Theo), we need just add to the average defect rate (15.39) the different effects as follows:
Effect(Context = 9 words) = −2.406 − 15.39 = −17.80
Effect(LR = MED) = 7.499 − 15.39 = −7.89
Effect(Epoch = 10000) = 10.421 − 15.39 = −4.97
Effect(Momentum = 0.75) = 11.285 − 15.39 = −4.10
Ttheo = 15.39 − 17.80 −7.89 − 4.97 − 4.10 = −19.37

This negative result may surprise, but it is a theoretical calculation and a 0 rate will be very satisfied.

To validate our model we made 10 trials using the optimal configuration and the results have confirmed the theoretical outcome which means that the experimental design is a success.

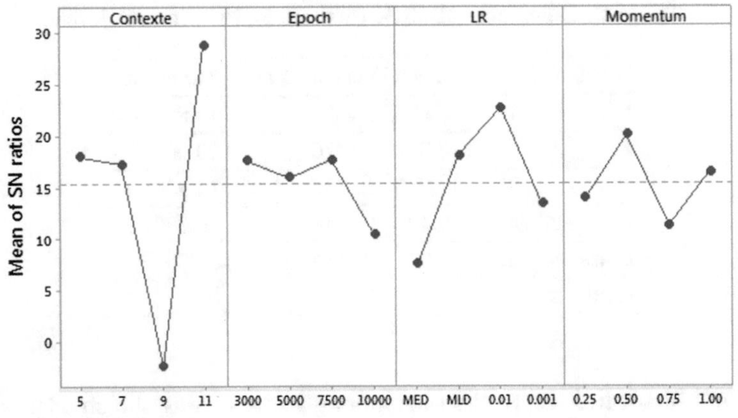

Fig. 3. The optimal parameters for the factors

7 Experimental Results

We tested our model on a window with size 3, 5, 7, 9 and 11 words. In the case of a context with 3 words, the model does not converge. For the rest of the contexts, the Taguchi method helped us to find the best context (9 words in our case). To compare our results with the modern approaches, we built a Hidden Markov POS tagger (HMM-POS), and for the same testing set we have applied an SVM-based tagger [4], we have used the Amira toolkit which is a set of tools for processing Arabic text from raw to base phrase chunks, and we compared our results also with the results given by the modern taggers like the Maximum Entropy Tagging approach exactly the Stanford tagger [16].
For the HMM-POS tagger, we learned a model from the same tagged corpus used in the neural tagger. Our model has two parts:

- The model of the generation of tags with the probabilities $P(w|t)$ that is the probability of generating a word w from a tag t.
- The model of tags sequence with the transition probabilities $P(t_i|t_{i-1})$

To calculate the tag sequence that maximizes the probability $P(w|t) \times P(t_i|t_{i-1})$, we used the Viterbi algorithm.

We applied the four taggers: Neural network, SVM, HMM and maximum entropy tagger on five categories: Culture, economy, local-news, international news and sports. Taken from the KALIMAT corpus [14]. Table 3 illustrates the accuracy found for the 4 methods.

Table 3. Results of the different methods of Arabic part-of-speech taggers

Method	Culture accuracy %	Economy accuracy %	Local-news accuracy %	International accuracy %	Sports accuracy %
Neural network	95.24	96.12	96.89	96.75	94.21
HMM	85.50	85.78	85.33	81.28	81.90
SVM	75.88	80.39	81.89	80.40	80.95
Max entropy	95.10	95.22	95.84	96.00	92.97

From Table 3 we can see that the weakest results found in the case of the SVM tagger. The errors encountered in the in the SVM-based tagger occurred due to the confusion between nouns and adjectives. A comparison study between an HMM model and an SVM model in [15] showed that the SVM model is weak in estimating a classification for very low-frequency words. Best results found are those of the neural network tagger we can see that they are even better than the maximum entropy tagger. From the results, we found that the main advantage of the deep neural networks is their generalization capability.

The generalization here is the ability to handle unseen data. For the known words in the training set the maximum entropy tagger gives better results than deep neural networks, but for the unknown words (words that exist in the training set and don't exist in the testing set), the deep neural networks are much better than the maximum entropy taggers.

For the Arabic language models, we have used the n-gram models with three techniques of smoothing: linear, written-bell and good-turing. We compared our results by using the neural language models and the other methods by calculating the perplexity in four cases with one, two, three and four grams. Perplexity represents the average branching factor of a model; that is, at each point in the test set, we calculate the entropy of the model. Therefore, a lower perplexity is desired. Table 4 demonstrates that good results found with the neural-based approach and in the cases of three and four grams This is due to the fact that this segmentation makes the corpus statistically viable. Indeed, the decomposition decreases the variety of bigrams and increase the frequency of tree and four grams.

Table 4. Results of the different methods of Arabic language models

n-gram	Neural LM PP/E	Linear PP/E	Written-bell PP/E	Good-turing PP/E
Unigram	150.23/7.23	250.14/7.97	210.21/7.71	227.56/7.83
Bigram	132.15/7.05	238.24/7.90	202.78/7.66	225.89/7.82
Trigram	120.88/6.92	238.12/7.90	201.50/7.65	225.89/7.82
Four-gram	120.88 /6.92	249.84/7.96	200.12/7.64	225.89/7.82

8 Conclusions and Future Work

In this work, we proposed an architecture based on deep neural networks for part-of-speech tagging and language models. This architecture is based on the automatic extraction of vector representations of words and deep learning theories. The most important task in this work is the data preprocessing, the words are converted into a manageable representation using word2vec. To the words, we added their features and the probabilities of belonging to a certain grammatical class and well formed sentences. The architecture contains: the input layer, the hidden layers, and the output layer. Training this kind of architecture is a time-consuming process. Therefore, using parallel computing is a good solution to reduce the time of training.

To find the best configuration of the neural network we used the Taguchi method which is based on a design of experiments methodology. This method allowed to us to determine the best context, in our work a context with 9 words which means 4 words before and 4 words after the word on which the training is made is the best context. Combining Taguchi and deep neural networks lead getting strong models by using a few amount of experimental data for a systematic analysis. After the validation of the model, we used it to compare the results given by the neural network tagger and very recent approaches in this field. We built an HMM part-of-speech tagger, and we used an SVM-based tagger in addition to modern taggers like Stanford tagger. From the four approaches, we can see that for the same dataset the tagger based on neural networks gives the best results. Four Arabic language models, we compared our results with the n-gram models with three techniques of smoothing the results found of our approach were very better than the n-gram method. The Arabic part-of-speech tagging and the Arabic language models will be the main components for our project which is a comparative study between the Arabic statistical machine translation and the Arabic neural machine translation.

References

1. Jurafsky, D., Martin, J.H.: Speech and language processing: an introduction to speech recognition. In: Computational Linguistics and Natural Language Processing. Prentice Hall (2008)
2. Collobert, R., Weston, J.: A unified architecture for natural language processing: deep neural networks with multitask learning. In: Proceedings of the 25th International Conference on Machine learning, pp. 160–167. ACM (2008)
3. Freeman, A.: Brill's POS Tagger and a Morphology Parser for Arabic (2004)
4. Diab, M., Hacioglu, K., Jurafsky, D.: Automatic tagging of Arabic text: from raw text to base phrase chunks. In: Proceedings of HLT-NAACL 2004: Short papers, Association for Computational Linguistics, pp. 149–152 (2004)
5. Duh, K., Kirchhoff, K.: POS tagging of dialectal Arabic: a minimally supervised approach. In: Proceedings of the ACL Workshop on Computational Approaches to Semitic Languages, Association for Computational Linguistics, pp. 55–62 (2005)
6. Khoja, S.: APT: Arabic part-of-speech tagger. In: Proceedings of the Student Workshop at NAACL, pp. 20–25 (2001)
7. Heintz, I.: Arabic language modeling with finite state transducers. In: Proceedings of the ACL-08: HLT Student Research Workshop (Companion Volume), Association for Computational Linguistics, Columbus, pp. 37–42 (2008)
8. Vergyri, D., Kirchhoff, K., Duh, K., Stolcke, A.: Morphology-Based Language Modeling for Arabic Speech Recognition (2004)
9. Mikolov, T., Sutskever, I., Chen, K., Corrado, G.S., Dean, J.: Distributed representations of words and phrases and their compositionality. In: Advances in Neural Information Processing Systems, pp. 3111–3119 (2013)
10. Collobert, R., Weston, J., Bottou, L., Karlen, M., Kavukcuoglu, K., Kuksa, P.: Natural language processing (almost) from scratch. J. Mach. Learn. Res. **volume 12, Aug**, 2493–2537 (2011)
11. Fischer, A., Igel, C.: An introduction to restricted Boltzmann machines. In: Iberoamerican Congress on Pattern Recognition, pp. 14–36. Springer (2012)
12. Toutanova, K., Klein, D., Manning, C.D., Singer, Y.: Feature-rich part-of-speech tagging with a cyclic dependency network. In: Proceedings of the 2003 Conference of the North American Chapter of the Association for Computational Linguistics on Human Language Technology, Association for Computational Linguistics, vol. 1, pp. 173–180 (2003)
13. Glorot, X., Bengio, Y.: Understanding the difficulty of training deep feedforward neural networks. Aistats **9**, 249–256 (2010)
14. Mahmoud, E.: KALIMAT a multipurpose Arabic corpus. https://sourceforge.net/projects/kalimat/files/kalimat/Part%20of%20Speech%20Tagged%20Corpus (2013)

15. Takeuchi, K., Collier, N.: Use of support vector machines in extended named entity. In: Proceedings on Computational Natural Language Learning, Taiwan, pp. 119–125 (2002)
16. Toutanova, K.: Stanford log-linear part-of-speech tagger. https://nlp.stanford.edu/software/tagger.shtml (2011)
17. Lee, H.H.: Principles and Practices of Quality Design, 4th edn. Gaulih Book Publishing (2013)

Implementation of Real Time Reconfigurable Embedded Architecture for People Counting in a Crowd Area

Gong Songchenchen(✉) and El-Bay Bourennane

Université Bourgogne Franche-Comté, Laboratoire LE2I, Dijon, France
gscc19@hotmail.com, ebourenn@u-bourgogne.fr

Abstract. We propose a feature fusion method for crowd counting. By image feature extraction and texture feature analysis methods, data obtained from multiple sources are used to count the crowd. We count people in high density static images. Most of the existed people counting methods only work in small areas, such as office corridors, parks, subways and so on. Our method uses only static images to estimate the count in high density images (hundreds or even thousands of people), for example, large concerts, National Day parade. At this scale, we can't rely on only one set of features for counting estimation. Therefore, we use multiple sources of information, namely, HOG and LBP. These sources provide separate estimates and other combinations of statistical measurements. Using the support vector machine (SVM) classification technique, and regression analysis, we count the crowd with high density. The method gives good results in crowded scenes.

Keywords: HOG · LBP · SVM

1 Introduction

In recent years, due to the awareness of social security, much more attention has been paid on people counting and human detection area. With the improvement of living standard, there are various recreational activities, some of which will bring potential dangers to human safety. In fact, it is important to accurately estimate the number of people in public areas to ensure safety and prevent the overcrowding. A stampede occurred in a "love parade" electronic music festival in Germany in 2010, and another accident occurred at the Bund of Shanghai China in 2014. In order to prevent the occurrence of such accidents in overcrowded areas, it is important to accurately estimate the number of people.

2 Related Work

The HOG function is used to detect human heads. Dalal et al. firstly used this function for pedestrian detection in static images. The main idea is to estimate

S. Chikhi et al. (Eds.): MISC 2018, LNNS 64, pp. 219–229, 2019.
https://doi.org/10.1007/978-3-030-05481-6_17

the gradient histogram of the local region, which is used to describe the target characteristic. Then, the HOG function is combined with Support Vector Machine (SVM) classifier to detect heads. SVM classifier and head detection are divided into two stages [1] (Fig. 1).

Fig. 1. This figure shows four arbitrary images from the dataset used in this paper

A method extracted texture features using Gabor filters and Least Squares vector support machine. The local texture characteristics of each cell on the grid mask are more efficient than the overall characteristics with the same multichannel Gabor filters [2].

A method is proposed that can be used to detect and count people. For feature extraction, local binary texture descriptors are applied and classified using a support vector machine (SVM) [3].

A recognition method for head and shoulders in a complex background. It starts with Head and shoulder contour features like 'Ω', gose through the coarse filtering using Haar classifiers as well as precise verification using SVM cascade classifiers based on Joint HOG features,and detection and identification of head-shoulders.It has a good detection rate and is suitable for population statistics and video surveillance, especially in public places [4].

Dense crowds can be thought as a texture that corresponds to a harmonic pattern at fine scales. Counts from texture analysis methods: crowds are repetitive in nature since all human appear similar from a distance. We will adopt three different texture analysis methods which separately give an estimate of the count [5,6].

Detect the crowd, use the point of interest detector to specify the image of the head area, and then count the number of people in the crowd [7,8].

A novel integrated framework for the analysis of crowds including all relevant aspects as simulation, detection and tracking of pedestrians and dense crowds and event detection. It utilizes an appearance-based approach for object detection since this method has been successfully applied for very small objects. Track the pedestrian using the Bayesian tracking method. Graph-based event detection using hidden markov models (HMM) [9].

A deep-learning approach to estimate the number of individuals in mid-level crowd visible in a still image. The proposed deep-learning framework is used to learn a feature-count regressor, which can estimate the number of people within each part, and the crowd estimation in whole image is therefore the sum of that in all parts [10,11].

To use local information at pixel level substitutes a global crowd level or a number of people per-frame. The proposed approach consists of generating fully automatic and crowd density maps using local features as an observation of a probabilistic crowd function [12,13].

According to the people counting method based on head detection and tracking to evaluate the number of people who move under an over-head camera. The main methods are: foreground extraction, head detection, head tracking, and crossing-line judgment [14–16].

Based on MID foreground segmentation module, it provides active areas for the head-shoulder detection module to detect head and count the number of people [17,18].

Human detection in crowded scenes using a Bayesian 3D model based method [19].

3 Proposed System

We propose a feature fusion method of crowd counting. By image feature extraction and texture feature analysis methods, data obtained from multiple sources are used to count the crowd. We count people in high density static images.

3.1 HOG Based Head Detections

HOG is used for head detection. Locally normalized HOG descriptors perform better than existing feature sets (including wavelets). Compared to the edge direction histograms and SIFT descriptors, the HOG is calculated on grid-sized uniform cell units, and in order to improve performance, overlapping local contrast normalization is also used.

In a static image, the representation and shape of the local target can be well described by the gradient or edge direction density distribution. Its basic is the gradient of statistical information, and the gradient mainly exists on the edge of the space. In order to reduce the effect of light factors, we firstly normalize the image. For the texture intensity of the image, the proportion of the local surface exposure is larger. Therefore, the compression process can effectively reduce the image of the local shadow and light changes. Because the function of color information does not work well, images are usually converted to grayscale. We calculate the gradient of the image abscissa and the ordinate direction, as well as the gradient direction value for each pixel position. The guidance operation captures the edge, silhouette and some texture information, further weakening the lighting effects. The gradient of the pixel (x, y) in the image is:

$$G_x(x,y) = H(x+1,y) - H(x-1,y) \tag{1}$$

$$G_y(x,y) = H(x,y+1) - H(x,y-1) \tag{2}$$

Where G_x, G_y represents the horizontal gradient, vertical gradient, and $H(x,y)$ pixel values at the pixel (x, y) in the input image.The gradient and gradient directions at the pixel (x, y) are:

$$G(x,y) = \sqrt{G_x(x,y)^2 + G_y(x,y)^2} \tag{3}$$

$$\partial(x,y) = tan^{-1}(\frac{G_x(x,y)}{G_y(x,y)}) \tag{4}$$

A gradient orientation histogram is constructed for each cell unit.We divide the image into several "cells", each cell consisting of 8*8 pixels and each block consisting of 2*2 cells. Suppose we use nine bin histograms to count the 8*8 pixel gradient information as shown in Fig. 2:

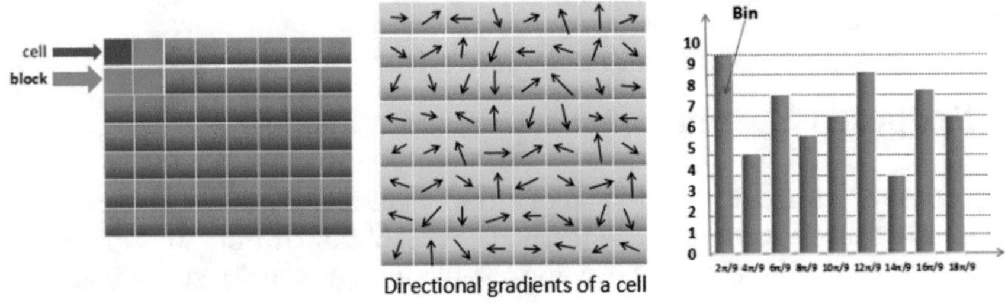

Fig. 2. The definition of a cell, a block and a bin.

We divide the gradient of the cell by 360 degrees into nine directions.Due to changes in local illumination and changes in foreground and background contrast, the range of gradient intensity is very large. This requires normalizing the gradient intensity. Normalization can further illuminate light, shadows, and edge compression. The approach we take is to combine the individual cell units into large blocks and spatially connected intervals. Normalizing the HOG feature vector in the block and introducing v to represent a vector that has not been normalized. It contains all histogram information for a given block. According to the $||V_k||$ standard, where k is 1 or 2, and e is a small constant. At this point, the normalization factor can be expressed as follows:

$$L2 - norm, f = \frac{v}{\sqrt{\|V\|_2^2 + e^2}} \tag{5}$$

We refer to the normalized block descriptors (vectors) as HOG descriptors. In this way, the HOG descriptor becomes a vector consisting of the histogram components of all cell units in each interval. A vector with HOG feature dimensions.

3.2 LBP Feature

The Local Binary Patterns (LBP) were first proposed by T. Ojala, M.PietikÃd'inen, and D. Harwood. LBP is a simple but very effective texture operator. It compares each pixel with its nearby pixels and saves the result as a binary number. The most important advantage of LBP is its robustness to changes in grayscale such as illumination changes. Its other important feature is its simple calculation, which makes it possible to analyze the image in real time. The basic LBP operator is defined as the 3*3 window. Using the value of the center pixel of the window as threshold, the gray value of the adjacent 8 pixels is compared with it. If the surrounding pixel value is greater than the central pixel value, the pixel value is of the location is marked as '1'. Otherwise it is '0'. In this way, the 8 points in the 3*3 neighborhood can be compared to produce 8-bit binary numbers (usually converted to decimal numbers, is 256 types of LBP codes), which is to get the LBP value of the pixel in the center of the window, and use this value to reflect the texture information of the area, for example: 00010011. Each pixel has 8 adjacent pixels, and 2^8 possibilities.

The basic LBP feature for a given pixel is formed by thresholding the 3×3 neighbourhood with the centre pixel value as the threshold, where (Xc, Yc) is the center pixel,ic be the intensity of the centre pixel and in (n = 0, 1, 2 ... 7) pixel intensities from the neighborhood. The LBP is given by:

$$LBP\left(X_c, Y_c\right) = \sum_{n=o}^{P-1} 2^n s\left(i_n - i_c\right) \tag{6}$$

Where P is the number of sample points and:

$$s\left(x\right) = \begin{cases} 1 \;\; if \;\; x \geqslant 0 \\ 0 \; else \end{cases} \tag{7}$$

The LBP could be interpreted as an 8-bit integer. The basic LBP concept is presented in Fig. 3:

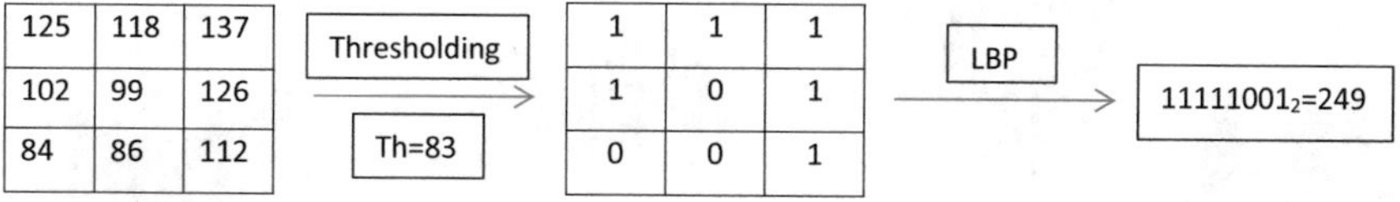

Fig. 3. Illustration of the standard LBP operator. Image taken from the Daimler pedestrian dataset.

When the LBP operator is used for texture classification or face recognition, the statistical histogram of the LBP mode is often used to express the image information, and more pattern types will make the data volume too large and the histogram too sparse. Therefore, it is necessary to reduce the dimension of the original LBP mode so as to best represent the image information in the case of a reduced data amount. In order to solve the problem of too many binary patterns and improve statistics, we proposed a "Uniform Pattern" to reduce the dimension of LBP operator's pattern. In an actual image, most LBP patterns only contain at most two transitions from '1' to '0' or from '0' to '1'. When a loop binary number corresponding to an LBP changes from '0' to '1' or from '1' to '0' at most twice, the binary corresponding to the LBP is called an equivalent pattern class. For example, 00000000 (0 jumps), 00000111 (only one transition from '0' to '1'), and 10001111 (first jump from '1' to '0', then '0' to '1' and two jumps in total) are all equivalent modes class. The modes other than the equivalent mode class are classified as another class, called a mixed mode class, such as 10010111 (a total of four transitions). With this improvement, the variety of binary patterns is greatly reduced without losing any information. The number of modes is reduced from the original 2P to P (P-1)+2, where P represents the number of sampling points in the neighborhood set. For the 8 sampling points in the 3*3 neighborhood, the binary pattern is reduced from the original 256 to 58. This allows the feature vector to have fewer dimensions and can reduce the effects of high-frequency noise.

3.3 Training of Joint HOG-LBP Classifiers

Feature extraction is one of the most critical aspects of human head detection. Extracting features with distinguishing significance plays an important role in the accurate detection of the human head. Our work integrates the features of HOG and LBP, which not only combines the effective identification information of multiple features, but also eliminates most of the redundant information, there by realizing effective compression of information, saving information storage space, and facilitating the acceleration of operations and real-time processing of information. Here we use a serial fusion approach, as shown in Fig. 4:

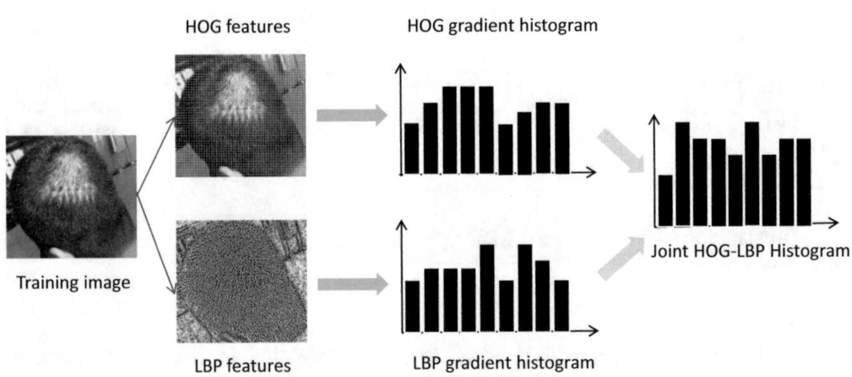

Fig. 4. Joint HOG-LBP histogram

Here, we use support vector machines to achieve optimal classification of linearly separable data. For a linear SVM with the training samples $((x_i, y_i)), 1 \leqslant$

$i \leqslant N\}$, where x_i is the ith instance sample, y_i is the corresponding category labels (i.e., the expected response), its decision surface equation can be expressed as:

$$\omega \cdot x + b = 0 \tag{8}$$

Where x is the input vector, ω is the dynamically variable weight vector,and b is the offset. In essence to find an optimal classifier is to find an optimal hyper plane according to formula(8), which can not only separate two classes correctly but also maximize the between-class distance. Accordingly, support vectors refer to the training sample points located in the classification boundaries, which are the key elements of the training sample set. Based on these theories and concepts, the following formula is used to classify the input samples:

$$f(x) = sgn\left\{\sum_{i=1}^{k} a_i y_i (x_i \cdot x) + b\right\} \tag{9}$$

where a_i is the weight coefficient corresponding to the support vector x_i.

The next step, we connect the sample HOG feature vector and the LBP feature vector in series to form a joint feature vector input SVM. Here, in the classifier process, the linear inseparable low-dimensional space is converted into a linearly separable high-dimensional space mainly through SVM kernel functions and use the cross-validation method to select the SVM optimal parameters, so that the classifier has the highest classification accuracy for the input training samples. In accordance with the above method, the training process for joint HOG-LBP SVM classifiers is shown in Fig. 5:

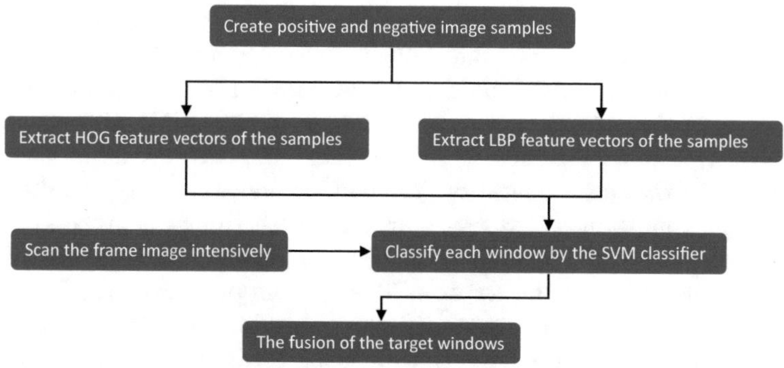

Fig. 5. Training process for joint HOG-LBP SVM classifiers

4 Experimental Results

In order to objectively verify the performance of the algorithm proposed, some experiments were made for it and some frequently-used algorithms. Our experiment is mainly divided into two parts: the first part is the pedestrian detection, and the second part is the counting of the crowd pictures.

The first part of the experiment: The training set we used contains 500 head face images clipped manually and enough non-head face images from some sample sets including INRIA, PETS2000, and MIT. During training negative samples can be selected as needed automatically from the background images. The test set contains 500 images with or without pedestrians, including about 1,500 apparent head faces and covering various scenes, angles, postures, and clothing, etc.The algorithm is embodied in a program developed with Matlab 2017a function library.

Extract sample HOG and LBP features:

Sample HOG feature calculation steps: For each positive and negative sample set, each size of 32*32 grayscale pictures (in this case, the grayscale picture is used to consider the effect of the size of the calculation, and the final detection result has little effect), and the rectangular HOG feature is calculated. Descriptor: The set cell size is 8*8, the size of the block is 16*16, and the slide step size is the width of a Cell. The specific process of HOG feature calculation is as follows: To reduce the influence of lighting, the sample is first Gamma standardization of images. Then calculate the gradient of x and y directions for each pixel in the grayscale image, and use the [-1,0,1] template to calculate the direction and amplitude of the gradient. In each Cell, set the projection direction to 9 bins, and use the gradient magnitude of each pixel as the weight, and vote to count the weighted histograms of gradient directions of each Cell. The dimension of this histogram is 9. Four cells in a Block (with overlaps between blocks) are normalized using L2-norm, and the gradient histograms of four cells are counted, and the dimension is 36. Finally, all blocks in the image are concatenated, and the dimension of the obtained HOG feature vector is calculated.

Sample LBP feature calculation steps: For each positive and negative sample set, each size is a 32*32 grayscale image, and LBP feature extraction based on a sliding window is used. The general description of the sliding window for the image algorithm is as follows: In an image of size W*H, the w*h window (W¿¿w, H¿¿h) is moved according to a certain rule, and a pixel in the window is performed. In the series operation, the window moves one step to the right or down after the operation is completed, until the entire image is processed. Set the size of the window to 16*16, and set the window's horizontal and vertical sliding steps to half the width of the window. The specific process of the LBP feature calculation is as follows: For one pixel in each window, an LBP feature value is calculated using an operator LBP (representing a radius 1, a ring containing 8 neighborhoods, a uniform mode). According to the LBP eigenvalue calculated in the window, the histogram of each window is calculated, that is, the number of occurrences of each LBP eigenvalue, and then L2-norm is used for normalization. The statistical histograms of all windows in the tandem image, the dimensions of the resulting LBP eigenvectors.

Finally, the sample HOG feature vector and the LBP feature vector are connected in series to form a joint feature vector. We use the SVM classifier to transform the linearly indivisible, low-dimensional space into a linearly separable, high-dimensional space by using a kernel function. The cross-validation method is used to select the optimal SVM parameters so that the classifier pairs the input training samples. The highest classification accuracy. The experimental test image size is 384*288. The algorithm is modified based on Matlab 2017a and runs on an Inter Core i5-5250 (1.60 GHz), 4 GB RAM computer. The experimental results are shown in Table 1.

Table 1. Algorithm performances shown by 3 experiments.

Detection algorithm	Test sequence	False number	Detection rate
Dalal HOG	1	39	92.2%
T.Ojala LBP	2	32	93.6%
Joint HOG+LBP	3	17	96.6%

The second part of the experiment: we will process the image of the crowd, and divide it into multiple small pieces. For example, a picture of a crowd of 256*256 pixels, the size of each small piece is defined as: 1*1 pixel, 2*2 pixels, 4*4 pixels, 8*8 pixels. Through the feature fusion of HOG-LBP, combined with the SVM classifier, population density estimation and counting are performed on each small block image. The training set contains 500 head face images that were manually cut from the INRIA sample set. The experimental test image size is 595*350, and the experimental results are as Table 2 (Fig. 6).

Fig. 6. Crowd counting

We have tried many times and got the results. In the table, number of people detected, number of people actually present in the scene, difference between detected number of people and actual number of people and time.Based on the above results, precision calculated is: 91.923%.

Table 2. Result summary.

Test sequence	Number of people detected	Actual	Difference	Time
1	236	260	24	94.32 s
2	231	260	29	92.56 s
3	220	260	40	89.41 s
4	228	260	32	90.28 s
5	239	260	21	96.21 s
6	226	260	34	90.03 s

5 Conclusion

In this paper, we propose a feature fusion method for crowd counting. By image feature extraction and texture feature analysis methods, data obtained from multiple sources are used to count the crowd. Therefore, we use multiple sources of information, namely, HOG and LBP. These sources provide separate estimates and other combinations of statistical measurements. Using the support vector machine (SVM) classification technique, and regression analysis, we count the crowd with high density. The approach adopted is easy and fast in processing. Our experiments showed the method gives good results in crowded scenes.

6 Future Work

In order to improve the detection efficiency and apply it to the crowd testing in the real-world, we plan to use the FPGA cards which is famous for its performance in real-time crowd image processing.

References

1. Li, M., Zhang, Z.: Estimating the number of people in crowded scenes by MID based foreground segmentation and head-shoulder detection. In: 19th International Conference on Pattern Recognition, pp. 1–4. Tampa, FL (2008)
2. Le, T.S., Huynh, C.K.: Human-crowd density estimation based on gabor filter and cell division. In: International Conference on Advanced Computing and Applications (ACOMP 2015), pp. 157–161 (2015)
3. Kryjak, T., Komorkiewicz, M., Gorgon, M.: FPGA implementation of real-time head-shoulder detection using local binary patterns, SVM and foreground object detection. In: Proceedings of the 2012 Conference on Design and Architectures for Signal and Image Processing, Karlsruhe, pp. 1–8 (2012)
4. Chen, L., Wu, H., Zhao, S., Gu, J.: Head-shoulder detection using joint HOG features for people counting and video surveillance in library. In: IEEE Workshop on Electronics, Computer and Applications, Ottawa, ON, pp. 429–432 (2014)

5. Rohit, Chauhan, V., Kumar, S., Singh, S.K.: Human count estimation in high density crowd images and videos. In: 2016 Fourth International Conference on Parallel, Distributed and Grid Computing (PDGC), Waknaghat, pp. 343–347 (2016)
6. Idrees, H., Saleemi, I., Seibert, C., Shah, M.: Multi-source multi-scale counting in extremely dense crowd. In: 2013 IEEE Conference on Computer Vision and Pattern Recognition, Portland, OR, pp. 2547–2554 (2013)
7. Subburaman, V.B., Descamps, A., Carincotte, C.: Counting people in the crowd using a generic head detector. In: 2012 IEEE Ninth International Conference on Advanced Video and Signal-Based Surveillance, Beijing, pp. 470–475 (2012)
8. Conte, D., Foggia, P., Percannella, G., Vento, M.: A method based on the indirect approach for counting people in crowded scenes. In: 2010 7th IEEE International Conference on Advanced Video and Signal Based Surveillance, Boston, MA, pp. 111–118 (2010)
9. Butenuth, M., et al.: Integrating pedestrian simulation, tracking and event detection for crowd analysis. In: 2011 IEEE International Conference on Computer Vision Workshops (ICCV Workshops), Barcelona, pp. 150–157 (2011)
10. Hu, Y., Chang, H., Nian, F., Wang, Y., Li, T.: Dense crowd counting from still images with convolutional neural networks. In: 2016 Journal of Visual Communication and Image Representation, pp. 530–539 (2016)
11. Zhang, C., Li, H., Wang, X., Yang, X.: Cross-scene crowd counting via deep convolutional neural networks. In: 2015 IEEE Conference on Computer Vision and Pattern Recognition (CVPR), Boston, MA, pp. 833–841 (2015)
12. Fradi, H., Dugelay, J.: Crowd density map estimation based on feature tracks. In: IEEE 15th International Workshop on Multimedia Signal Processing (MMSP). Pula, pp. 040–045 (2013)
13. Wang, Z., Liu, H., Qian, Y., Xu, v: Crowd density estimation based on local binary pattern co-occurrence matrix. In: 2012 IEEE International Conference on Multimedia and Expo Workshops, Melbourne, VIC, pp. 372–377 (2012)
14. Krishna, A.N., et al.: A people counting method based on head detection and tracking. In: 2016 International Journal of Innovative Research in Computer and Communication Engineering, pp. 2320–9801 (2016)
15. Yang, H., Zhao, H.: A novel method for crowd density estimations. In: IET International Conference on Information Science and Control Engineering 2012 (ICISCE 2012), Shenzhen, pp. 1–4 (2012)
16. Ma, Z., Chan, A.B.: Crossing the line: crowd counting by integer programming with local. In: 2013 IEEE Conference on Computer Vision and Pattern Recognition, Portland, OR, pp. 2539–2546 (2013)
17. Li, M., Zhang, Z., Huang, K., Tan, T.: Estimating the number of people in crowded scenes by MID based foreground segmentation and head-shoulder detection. In: 2008 19th International Conference on Pattern Recognition, Tampa, FL, pp. 1–4 (2008)
18. Rodriguez, M., Laptev, I., Sivic, J., Audibert, J.: Density-aware person detection and tracking in crowds. In: 2011 International Conference on Computer Vision, Barcelona, pp. 2423–2430 (2011)
19. Wang, L., Yung, N.H.C.: Bayesian 3D model based human detection in crowded scenes using efficient optimization. In: IEEE Workshop on Applications of Computer Vision (WACV). Kona, HI, pp. 557–563 (2011)

Towards a Better Training for Siamese CNNs on Kinship Verefication

Sellam Abdellah(✉) and Azzoune Hamid

Laboratory for Research in Artificial Intelligence,
Computer Science Department,
University of Science and Technology Houari Boumediene (USTHB),
BP 32, 16111 El-Alia, Bab Ezzouar, Algiers, Algeria
asellam@usthb.dz, azzoune@yahoo.fr

Abstract. Kinship verification from facial images in the wild is a recent problem that received an increasing interest from the computer vision research community. Due to the limited size of the existing datasets, applying Deep Learning approaches results in a model that overfits to the training data, therefore, the purpose of this study is to reduce the degree of overfitting when training a Deep Learning model on kinship datasets. To this end, we propose a new training mechanism for siamese convnets, in which we train the model on all images from all types of kinship relations instead of training on each of these subsets separately, then we evaluate the model on each subset individually. Experimental results demonstrated that using this training method resulted in better performance compared to training on each subset separately, and allowed to achieve results comparable to the most recent state of the art approaches. This paper focuses on the impact of adding more data over adding the gender information by separating kinship relation types in different subsets.

Keywords: Kinship verification · Siamese CNN · Deep learning
Feature learning

1 Introduction

The verification of parental relationship between two persons identified by their facial images; aka kinship verification; has attracted a lot of attention from the computer vision research community in these last years, as it has so many potential applications, ranging from family album creation and organization to finding missing children and suspects' family members.

Despite the importance of the problem and the efforts made by researchers in this field, current results suggest further improvements. One of the reasons for which results are still not good enough is the small size of the existing datasets, especially for deep learning approaches, as a small set of training examples causes

S. Chikhi et al. (Eds.): MISC 2018, LNNS 64, pp. 230–242, 2019.
https://doi.org/10.1007/978-3-030-05481-6_18

the model to overfit to the training data making it unable to generalize well. Other difficulties include: age and gender variations and pose and illumination changes.

In this study, we will investigate the effect of training a Deep Siamese Convnet on all the facial images regardless of the gender of the subjects.

The rest of this paper is structured as follows, the next section presents some related work in the field of kinship verification, Sect. 3 explains in details the proposed siamese architecture and training method, Sect. 4 shows the obtained results and compares them to other state of the art approaches, In Sect. 5 we discuss the obtained results and their meaning, and the last section is a conclusion that lists possible future improvements.

2 Related Work

The problem of kinship verification from facial images was first tackled by Fang et al. [1], the authors proposed the use of a set of engineered features to represent facial images of the subjects and K Nearest Neighbors algorithm for kinship verification. Zhou et al. [2] proposed the SPLE descriptor as a facial feature, vectors of SPLE where then classified using SVM to predict kinship.

Lu et al. [3] proposed Neighborhood Repulsed Metric Learning (NRML), face images were represented by vectores of LBP, HOG or SIFT features, then a distance metric is learned by making pairs of feature vectors with kinship relations as close as possible, while negative pairs are pushed away. The authors also presented a multiview version of NRML (MNRML) that enable the use of more than one type of features in the distance metric learning.

Motivated by the fact that different face descriptors deliver complementary information, [4–8] proposed to learn multiple similarity metrics from multiple feature representations and combine them all to have a more precise and discriminative kinship recognition.

Yan et al. [9] proposed a feature learning approach in which they learn mid-level discriminative features from low-level descriptors, each one of the mid-level features is a SVM hyperplane decision.

Bottino et al. [10] combined geometrical and textural features to represent facial images. Moreover they reduced the size of the feature-vectors using a two stage feature selection. Finally, the SVM classifier was used for kinship verification.

Zhou et al. [11] proposed Scalable Similarity Learning that learns a diagonal bi-linear similarity matrix instead of a full matrix resulting in faster and more scalable learning.

Qin et al. [12] proposed to use two matrices to model the kinship similarity, one is shared among all kinship subsets (without the gender information) and the second one is specific to each subset.

Lan et al. [13] prensented a new descriptor called QMD (Quaternion-Michelson descriptor) to extract local features from facial images using Michelson contrast law on the quaternion representation of facial images. Lan et al. [14] proposed another quaternion representation based descriptor (QWLD) that uses Weber law.

Yan et al. [15] replaced euclidean distance metric used in most metric learning with the correlation distance. Patel et al. [16] proposed a Block-based extension to the NRML [3] approache (BNRML) where multiple local distance metrics are learned from different blocks of facial images.

Lu et al. [17] proposed Descriminative Deep Metric Learning (DDML) in which a Deep Neural Network is used to learn facial images' features, the euclidean distance between the feature vectors will then be compared to determine kinship.

Zhang et al. [18] proposed a basic CNN architecture that takes an input with 6 channels corresponding to the RGB channels from the two input images, then a sequence of three convolutional layers with increasing number of 5×5 filters is used for feature extraction followed by a fully connected layer that outputs 640 features, finally, a softmax layers outputs two probabilities corresponding to the two kinship classes (negative/positive).

In this paper we present a new Siamese CNN architecture and a new training method that reduces the amount of overfitting of the learned Deep Model. The next section will describe in details the proposed approach.

3 The Proposed Approach

In this section we will discuss the details of the proposed siamese architecture and the training method.

3.1 The Proposed Siamese Architecture

Siamese Neural Networks proved to be very efficient in tasks involving finding similarity include signature verification [19], face verification [20], one-shot image recognition [21] ...etc.

A Siamese CNN is a pair of Convolutional Neural Networks that share the same set of weights, which means that any change to the weights of one CNN will be applied to the second one as well. The two CNNs serve as a Feature Learning mechanism which extract features from their input images, providing two 1D feature vectors as a result, the feature vectors from the two CNNs will then be merged in a single vector using a mathematical function (for example, L2 which is simply the element wise squared difference). At last, a stack of Dense (Fully connected) layers will compute the binary output (Same/Different) from the merged feature vector. (See Fig. 1 for a graphical illustration).

The Siamese architecture proposed in this paper consists of two convolutional neural networks with a stack of three convolution-pooling sequences (see Table 1 for the implementation details of this convnet).

Each CNN takes as input a RGB image (3D volume), which is then processed by being passed trough a sequence of convolutional and max pooling layers,

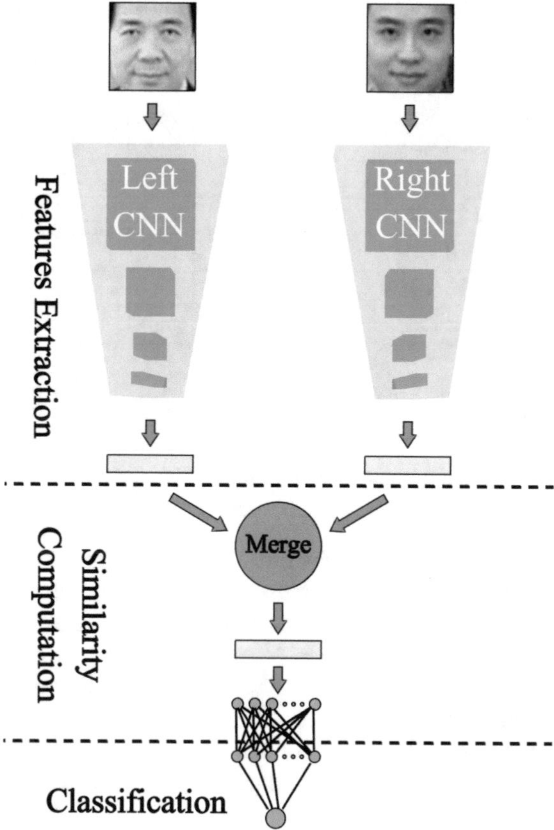

Fig. 1. An illustration of a Siamese CNN. The upper part serves as a feature extraction mechanism and consists of two convolutional neural networks sharing the same set of weights. The lower part serves as a similarity computational unit and a classification unit.

the max pooling layers reduce the width and height of the volume while the convolutional layers increase its depth (number of channels), Each convolution operation represents a filter that learns how extract a specific facial feature by filtering out the information irrelevant to the concerned feature. The last volume containing the extracted features is flattened into a 1D vector of features.

The two vectors extracted using the convolutional neural network will be merged into a single feature vector using the L1 function. Each element in the resulting vector is the absolute value of the difference between the two elements at the same position in the original vectors. (See Eq. (1) for the definition of this operation)

$$L_1(V_1, V_2) = (|V_1(1) - V_2(1)|, |V_1(2) - V_2(2)|, ..., |V_1(N) - V_2(N)|) \quad (1)$$

Table 1. The implementation details of the convolutional neural network.

Layer	Output volume size	# Filters	Filter size	Stride	Activation
Input	$64 \times 64 \times 3$	–	–	–	–
Convolutional	$64 \times 64 \times 16$	16	13×13	1×1	ReLU
Max Pooling	$32 \times 32 \times 16$	–	2×2	2×2	–
Convolutional	$32 \times 32 \times 64$	64	5×5	1×1	ReLU
Max Pooling	$16 \times 16 \times 64$	–	2×2	2×2	–
Convolutional	$16 \times 16 \times 128$	128	3×3	1×1	ReLU
Max Pooling	$8 \times 8 \times 128$	–	2×2	2×2	–

Where:

- **V1** and **V2** are the two vectors being merged
- **N** is the number of elements in each vector

The choice to use of the simple L1 norm over L2 norm (or more complicated distance similarity metrics) is mainly experimental, using L2 merge on our architecture caused the model to underfit the training data. Moreover, the weighted L1 distance was used in other Siamese Neural Network papers [20,21].

A final dense layer (fully connected layer) with a sigmoid activation takes all the values in the merged vector and outputs the final kinship answer. The dense layer computes a weighted sum of the vector's elements (see Eq. (2)), then it adds a bias value to it and apply the sigmoid (σ) function (see Eq. (3)) to this value.

$$y = \sum_{i=1}^{N}(W(i) \times V(i)) + b = W(1) \times V(1) + W(2) \times V(2) + ... + W(N) \times V(N) \quad (2)$$

Where:

- **W(1), W(i), ... W(N)** are the weights of the synapses of the dense layer
- **V(1), V(i), ... V(N)** are the elements of the vector resulting from the L1 merge function
- **N** is the number of elements in the merged vector

$$\sigma(y) = \frac{e^y}{e^y + 1} \quad (3)$$

For an overall view of the architecture of the siamese cnn see Fig. 2.

3.2 Interpretation of the Layers and their Outputs

The first part of the siamese neural network consists of convolutional and max pooling layers that modify input images and encode them into feature vectors. Figure 3 shows some of the outputs of different convolutional layers as grayscale images. We can clearly see that each convolution highlights a specific facial

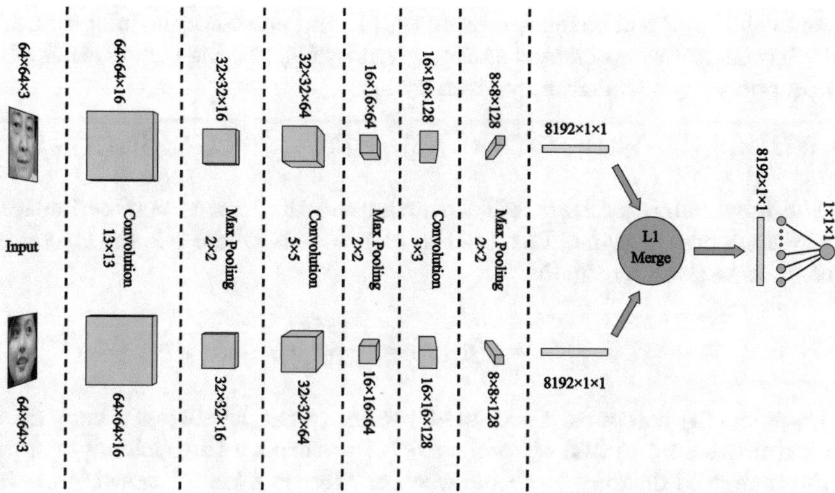

Fig. 2. The proposed Siamese CNN architecture. The two input images are passed through a series of convolutions and pooling layers in which the depth grows bigger while the width and height shrink. The volume of the last pooling layer get flattened into a 1D vector. The two vectors corresponding to the input images will be merged using L1 norm yielding a single vector that is passed to a Dense Layer to compute the kinship output.

features by filtering out pixels irrelevant to it. The size of the images from subsequent convolutional layers shrinks as a result of the pooling layers between them which enables a compact representation of features.

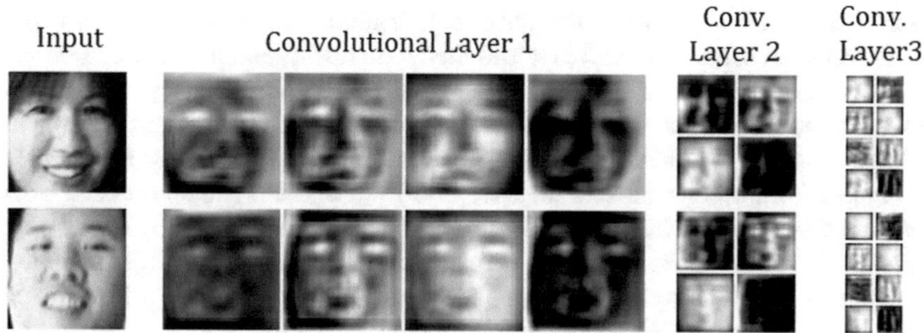

Fig. 3. Hidden Convolutional Layers' outputs for a positive pair of images visualized as gray-scale images.

The last pooling layer outputs a 3D volume of $128 \times 8 \times 8$ values, it is a list of 128 features extracted using convolutions, these features will then be organized in a 1D vector. The two vectors V_1 and V_2 corresponding to the images from the (parent,child) pair will be merged using Eq. (1) and passed to the fully connected layer that computes a weighted sum as given by Eq. (2), the combination of the two equations is equivalent to **y** given by Eq. (4).

$$y = W(1) \times |V_1(1) - V_2(1)| + W(2) \times |V_1(2) - V_2(2)| + ... + W(N) \times |V_1(N) - V_2(N)| \tag{4}$$

The fully connected layer will then compute the kinship verification result (**K**) by applying the sigmoid (σ, see Eq. (3)) to y+b, where **b** is the layer's bias value. This is given by Eq. (5).

$$K = \sigma(y + b) = \frac{e^{y+b}}{e^{y+b} + 1} \tag{5}$$

Equation (4) computes a weighted version of the L1 distance between the two feature vectors V_1 and V_2, while Eq. (5) performs a thresholding operation on the computed distance to decide whether there is a kin relationship between the two persons identified by the two facial images in the input layer.

This alongside with the convolutional layers consists a system of deep metric/feature learning that learns the kinship features automatically from raw image data and learns the distance metric to evaluate thier similarity and the thresholding operation to decide kinship (as illustrated in Fig. 1).

3.3 The Siamese CNN Training

A major problem with deep learning approaches like this one is the lack of training data, having so little data will cause the deep learning model to overfit to the training samples, this will make the model unable to generalize well to new data unseen in the training phase.

Most kinship datasets have a small number of sample for each type of kinship relation (Father-Son, Mother-Daughter ...), training the model on each type of kinship relation separately will give the model an additional information describing the gender of the two subjects being compared, unfortunately this information is not as important as the size of the dataset for deep learning approaches.

Motivated by this idea we propose to train the deep learning model on all the images from all the subsets we have and then test the model on each subset separately, this is equivalent to training on each set separately in terms of the confidence of the evaluation as the total number of training and test samples is the same in both cases, but in terms of overfitting we observed a huge impact in the positive way, which is shown in the next section.

The following parameters were used to train the model:

- **Optimizer:** Stochastic Gradient Descent (SGD).
- **Loss Function:** Mean Squared Error (MSE).
- **Learning Rate:** 0.1

4 Experimental Results

In this section we will present in details the experimental study and its results, the source code of our implementation is publically available on Github (See Reference [22])

4.1 Data Collection

To evaluate our approach we used the famous KinfaceW dataset that contains facial images collected in Lu et al. [23]. (See Fig. 4 for examples of images in this dataset).

Fig. 4. Example of Images from the KinfaceW dataset (Left: KinFaceW-I, Right: KinFaceW-II

There are two versions of the KinfaceW dataset: KinfaceW-I and KinfaceW-II, each of these two datasets consists of four subsets corresponding to specific types of kin relations: Father-Daughter (FD), Father-Son (FS), Mother-Daughter (MD) and Mother-Son (MS). (Table 2 show the number of pairs of facial images in each subset of KinfaceW).

Table 2. The number of pairs in each subset of KinfaceW.

	KinfaceW-I	KinfaceW-II
FS	156	250
FD	134	250
MS	116	250
MD	127	250
Total	533	1000

4.2 Experimental Protocol

The KinFaceW dataset contains a predefined set of 5-fold cross validation train/test pairs. For the evaluation of our approach we used the same set of train/test pairs, but instead of training eight different classifiers for the eight subsets available, we grouped all the training pairs from all the datasets together and feed them to a single deep learning model. Using all the training samples (pairs) from all subsets allowed us to reduce the overfitting and increase the 5-fold cross validation accuracy while the proportion of test samples to train samples is the same as other existing approaches in the literature for the sake of a fair comparison.

4.3 Results

Table 3 shows the accuracies of our approach and some of state of the art approaches for kinship verification on the KinfaceW-I and KinfaceW-II datasets. We were able to achieve a better accuracy on the Mother-Daughter subset of KinfaceW-II, results on other subsets where competitive with other approaches.

Table 4 compares the accuracy of the model trained on all kinship relations (A-SCNN) and the accuracy of the model trained on each subset separately (SCNN). Training the model on all samples from every subset improves the test accuracy on all subsets with a difference of +4% to +7% compared to training on each subset separately.

The amount of time needed to complete the training depends on several hyperparameters (Learning Rate, Accuracy Goal, ...). For this reason, in order to compare the two training methods in terms of execution time, we computed the total time needed to complete **30** epochs of training for each of the **5** folds, results are given in Table 5, of course, the same architecture and the same batch size were used for this evaluation. The experiment to measure the execution time was carried out on a computer with an Intel i7 6700HQ 2.6GHz CPU and an NVidia Geforce GTX 1060.

Table 3. Comparison with other state of the art approaches on the KinfaceW dataset.

Approach	KinfaceW-I				KinfaceW-II			
	FS	FD	MS	MD	FS	FD	MS	MD
NRML [3]	70.5%	65.2%	65.2%	71.0%	76.8%	73.1%	76.8%	77.0%
MNRML [3]	72.5%	66.5%	66.2%	72.0%	76.9%	74.3%	77.4%	77.6%
DMML [7]	74.5%	69.5%	69.5%	75.5%	78.5%	76.5%	78.5%	79.5%
MPDFL [9]	73.5%	67.5%	66.1%	73.1%	77.3%	74.7%	77.8%	78.0%
ESL [8]	83.9%	76.0%	73.5%	81.5%	81.2%	73.0%	75.6%	73.0%
Multiview SSL [11]	82.8%	75.4%	72.6%	81.3%	81.8%	74.0%	75.3%	72.5%
S^3L [6]	82.4%	72.8%	74.6%	79.1%	82.6%	73.8%	74.1%	73.6%
QMCBP [13]	74.4%	69.6%	68.6%	77.8%	77.2%	71.6%	79.0%	75.6%
Chen et al. [24]	**88.5%**	81.0%	81.0%	82.6%	88.0%	82.8%	84.4%	83.2%
NRCML [15]	72.5%	67.2%	67.2%	73.0%	79.8%	76.1%	79.8%	80.0%
K-BDPCA [25]	77.9%	78.0%	81.4%	87.9%	**88.8%**	81.8%	86.8%	87.2%
Faraki et al. [26]	71.3%	79.8%	73.7%	69.4%	75.6%	78.4%	72.6%	73.2%
Patel et al. [16]	83.4%	77.3%	75.8%	78.4%	84.0%	79.0%	79.2%	80.0%
DDMML [17]	86.4%	79.1%	81.4%	87.0%	87.4%	83.8%	83.2%	83.0%
MHDL3-L [27]	72.5%	71.7%	73.7%	87.0%	88.4%	84.0%	86.4%	89.2%
Zhang et al. [18]	75.7%	70.8%	73.4%	79.4%	84.9%	79.6%	**88.3%**	88.5%
DTL [28]	86.45%	**89.62%**	**89.57%**	**88.80%**	83.40	**89.40**	87.00	87.00
Our approach	73.4%	70.2%	72.4%	81.1%	84.0%	82.0%	83.2%	**89.6%**

Table 4. Comparison between the accuracy achieved by training the model on all subsets (A-SCNN) and the accuracy achieved by training on each subsets separately (SCNN).

Approach	KinfaceW-I				KinfaceW-II			
	FS	FD	MS	MD	FS	FD	MS	MD
SCNN	66.2%	67.4%	64.4%	72.8%	79.4%	74.2%	78.0%	83.6%
A-SCNN	72.8%	71.7%	70.7%	79.9%	83.4%	81.6%	83.8%	89.0%

5 Discussion

The findings from this study suggests that training on all facial images regardless of the gender of the subjects and their type of kinship relation improves the accuracy of the resulting model in the case of deep learning as shown in Table 4.

This is backed up in theory by the fact that adding more data reduces the degree of overfitting, which explains the higher performance of the model on test data. Although combining the training samples from all subsets removed the information about the gender of subjects, having more data fed to the deep learning model probably made the model able to infer it from raw image data or learn how to bypass it in the process of kinship verification, in fact, the whole idea behind deep learning is to be able to arrive to the final solution from a huge amount of raw data.

Table 5. Comparison between the total time needed for training the model on all subsets (A-SCNN) versus the total time needed for training on each subsets separately (SCNN).

Approach	Total time (s)
SCNN	396.56
A-SCNN	297.54

Compared to the other approaches in the literature, our approach achieves comparable results on all of the KinFaceW-II subsets especially on the mother-daughter subset where our approach performed better than all of the listed approaches (Table 3), On the other hand, the results of our approach on the KinFaceW-I subsets are still lacking (with the exception of the mother-daughter subset) and suggests further improvements, we think that this can be explained by the proportion of KinFaceW-I images to the total number of images which is less than **35%** (see Table 2) and the fact that the images of each pair from the KinFaceW-I dataset were taken from different photos which distinguishes them from KinfaceW-II images.

Training on all subsets required less execution time than training on each subset separately when using the same architecture, the same number of epochs and the same batch size. Theoretically both approaches should complete these 30 epochs in the same execution size if we train on all subsets, because the total number of image used for training is the same, which lead us to hypothesize that the difference in time should be due to the implementation details of Keras.

In summary, the method of training presented in this paper achieved better accuracies than the simple training on each subset separately when using the same siamese CNN architecture, which is the pusrpose of this study. On the other hand, comparison with state of the art approaches suggests that further improvements to this solution are needed.

6 Conclusions

In this paper we proposed a new training method for siamese convolutional neural network on kinship verification datasets. Experimental results demonstrated that better performance can be achieved in less execution time by training the deep learning model on all samples from every type of kinship relation. The proposed approach was able to achieve accuracies competitive with recent state of the art approaches.

In future studies we will investigate more sophisticated solutions based on other data augmentation techniques and transfer learning hoping to achieve better results in the field of kinship verification.

References

1. Fang, R., Tang, K.D., Snavely, N., Chen, T.: Towards computational models of kinship verification. In: 2010 17th IEEE International Conference on Image Processing (ICIP), pp. 1577–1580. IEEE (2010)
2. Zhou, X., Hu, J., Lu, J., Shang, Y., Guan, Y.: Kinship verification from facial images under uncontrolled conditions. In: Proceedings of the 19th ACM International Conference on Multimedia, pp. 953–956. ACM (2011)
3. Lu, J., Zhou, X., Tan, Y.P., Shang, Y., Zhou, J.: Neighborhood repulsed metric learning for kinship verification. IEEE Trans. Pattern Anal. Mach. Intell. **36**(2), 331–345 (2014)
4. Hu, J., Lu, J., Tan, Y.P.: Sharable and individual multi-view metric learning. IEEE Trans. Pattern Anal. Mach. Intell. (2017)
5. Hu, J., Lu, J., Tan, Y.P., Yuan, J., Zhou, J.: Local large-margin multi-metric learning for face and kinship verification. IEEE Trans. Circuits Syst. Video Technol. (2017)
6. Xu, M., Shang, Y.: Kinship measurement on face images by structured similarity fusion. IEEE Access **4**, 10280–10287 (2016)
7. Yan, H., Lu, J., Deng, W., Zhou, X.: Discriminative multimetric learning for kinship verification. IEEE Trans. Inf. Forensics Secur. **9**(7), 1169–1178 (2014)
8. Zhou, X., Shang, Y., Yan, H., Guo, G.: Ensemble similarity learning for kinship verification from facial images in the wild. Inf. Fusion **32**, 40–48 (2016)
9. Yan, H., Lu, J., Zhou, X.: Prototype-based discriminative feature learning for kinship verification. IEEE Trans. Cybern. **45**(11), 2535–2545 (2015)
10. Bottino, A., Vieira, T.F., Ul Islam, I.: Geometric and textural cues for automatic kinship verification. Int. J. Pattern Recognit. Artif. Intell. **29**(03), 1556001 (2015)
11. Zhou, X., Yan, H., Shang, Y.: Kinship verification from facial images by scalable similarity fusion. Neurocomputing **197**, 136–142 (2016)
12. Qin, X., Tan, X., Chen, S.: Mixed bi-subject kinship verification via multi-view multi-task learning. Neurocomputing **214**, 350–357 (2016)
13. Lan, R., Zhou, Y.: Quaternion-michelson descriptor for color image classification. IEEE Trans. Image Process. **25**(11), 5281–5292 (2016)
14. Lan, R., Zhou, Y., Tang, Y.Y.: Quaternionic weber local descriptor of color images. IEEE Trans. Circuits Syst. Video Technol. **27**(2), 261–274 (2017)
15. Yan, H.: Kinship verification using neighborhood repulsed correlation metric learning. Image Vis. Comput. **60**, 91–97 (2017)
16. Patel, B., Maheshwari, R., Raman, B.: Evaluation of periocular features for kinship verification in the wild. Comput. Vis. Image Underst. **160**, 24–35 (2017)
17. Lu, J., Hu, J., Tan, Y.P.: Discriminative deep metric learning for face and kinship verification. IEEE Trans. Image Process. **26**(9), 4269–4282 (2017)
18. Zhang, K., Huang, Y., Song, C., Wu, H., Wang, L.: Kinship verification with deep convolutional neural networks. In: Proceedings of the British Machine Vision Conference (BMVC), pp. 148.1–148.12. BMVA Press (September 2015). https://doi.org/10.5244/C.29.148
19. Bromley, J., Guyon, I., LeCun, Y., Säckinger, E., Shah, R.: Signature verification using a "siamese" time delay neural network. In: Advances in Neural Information Processing Systems, pp. 737–744 (1994)
20. Taigman, Y., Yang, M., Ranzato, M., Wolf, L.: Deepface: closing the gap to human-level performance in face verification. In: Proceedings of the IEEE Conference on Computer Vision and Pattern Recognition, pp. 1701–1708 (2014)

21. Koch, G., Zemel, R., Salakhutdinov, R.: Siamese neural networks for one-shot image recognition. In: ICML Deep Learning Workshop. vol. 2 (2015)
22. Sellam, A., Azzoune, H.: All-subsets siamese convolutional neural network for kinship verification (2018). https://github.com/asellam/ASCNN
23. Lu, J., Hu, J., Zhou, X., Shang, Y., Tan, Y.P., Wang, G.: Neighborhood repulsed metric learning for kinship verification. In: 2012 IEEE Conference on Computer Vision and Pattern Recognition (CVPR), pp. 2594–2601. IEEE (2012)
24. Chen, X., An, L., Yang, S., Wu, W.: Kinship verification in multi-linear coherent spaces. Multimed. Tools Appl. **76**(3), 4105–4122 (2017)
25. Dehshibi, M.M., Shanbehzadeh, J.: Cubic norm and kernel-based bi-directional PCA: toward age-aware facial kinship verification. Vis. Comput., 1–18 (2017)
26. Faraki, M., Harandi, M.T., Porikli, F.: No fuss metric learning, a hilbert space scenario. Pattern Recognit. Lett. **98**, 83–89 (2017)
27. Mahpod, S., Keller, Y.: Kinship verification using multiview hybrid distance learning. Comput. Vis. Image Underst. **167**, 28–36 (2018)
28. Yang, Y., Wu, Q.: A novel kinship verification method based on deep transfer learning and feature nonlinear mapping. DEStech Transactions on Computer Science and Engineering (aiea) (2017)

A Context-Aware Distributed Protocol for Updating BDI Agents Abilities

Hichem Baitiche(✉), Mourad Bouzenada, Djamel Eddine Saidouni, Youcef Berkane, and Hichem Chama

MISC Laboratory, University of Abdelhamid Mehri Constantine 2, Ali Mendjeli Campus, 25000 Constantine, Algeria
{hichem.baitiche,mourad.bouzenada,djamel.saidouni}@univ-constantine2.dz

Abstract. In this paper, we propose a context-aware distributed protocol, that improves agents awareness of their abilities. Based on the agent's context, the protocol allows the exploration of agent's neighborhood in order to detect the new available actions, and validate the existing ones. When a change in agent's abilities is detected, the new available actions will be added to the agent's set of available actions, and the invalid ones will be removed. The protocol is implemented in Jason, and tested in a smart laboratory scenario.

Keywords: AmI systems · MAS · BDI agents · Context awareness

1 Introduction

Ambient Intelligence (AmI) is an emerging discipline that aims to bring intelligence to our lives. It uses all the available technologies to build an environment, which is capable of understanding and anticipating our needs to provide us with the services we need [1,2]. Multi-agent systems (MAS) is one of the most promising approaches for the design and development of AmI systems [3]. In particular, Belief-Desire-Intention (BDI) model [4] offers a good alternative for reactive systems whenever planning is required.

BDI architectures have generally preferred the use of predefined plans library over the use of first principle planning approaches, due to the high computational cost of generating new plans. In fact, a wide variety of interpreters based on BDI model have been developed using this style (e.g., [5,6]). Though it allows the development of efficient agent reasoning cycles, the use of predefined plans affects the performance of the resulting agent, as it limits its autonomy and flexibility.

Despite that generating new plans is computationally expensive, the inclusion of planning capability substantially increases the autonomy and flexibility of agents. Planning is useful when [7]: (i) synthesize a new plan to achieve a goal for which no predefined plan worked (or found); (ii) make viable the precondition of a relevant, but not applicable, plan to achieve a given goal. Consequently, many

S. Chikhi et al. (Eds.): MISC 2018, LNNS 64, pp. 243–256, 2019.
https://doi.org/10.1007/978-3-030-05481-6_19

researches have been done to extend the standard BDI architectures with the ability to use planners in the aforementioned scenarios (e.g., [8–10]).

In the literature, there are many approaches to develop automated planners and integrate it in BDI agents. In [7], a survey has been conducted on the techniques and systems aimed at integrating planning algorithms and BDI agent reasoning (e.g., [11,12]). These approaches can be classified in two principles classes: (i) domain independent planning (first principle planning); (ii) domain dependent planning. Both independent and dependent domain planning require as input the set of available actions (SAA) from which plans can be generated. SAA is a static set defined at development time by developers (same as the predefined plan library). By using a predefined static SAA, agent won't be able to perceive the changes in its abilities (available actions) as shown in the following scenarios:

- Scalable Environments: if we scale up/down the environment by adding/ removing actuators, agent gains/loses access to the actions performed by these actuators;
- Mobile Agents: if the agent moves in its environment, and changes its context; it may gains/loses access to some actions based on its location.

In this paper, we develop a distributed protocol that resolves the agent's lack of awareness of its abilities. The protocol's main goal is to maintain the SAA consistent and updated. By continuously exploring the agent neighborhood, the protocol detects the new available actions, and validates the exiting ones. When a change in agent's abilities is detected, the new available actions will be added to the agent's SAA, and the invalid ones will be removed. Using a dynamic up-to-date SAA improves the performance and outcomes of the planners, as it contains all the available valid actions and doesn't incorporate invalid ones.

We implement the proposed protocol in Jason [13], a well-know multi-agent platform for the development of multi-agent systems. We test the protocol in a smart laboratory scenario where each researcher has an assistant BDI agent in its smart phone. The assistant agent is designed to assist researchers in their daily routines within the laboratory.

The rest of this paper is structured as follows. In Sect. 2, we define the problem related to our work, and detail our approach for keeping agent's SAA consistent and updated. In Sect. 3, we present our distributed protocol as a solution for the defined problem. In Sect. 4, we implement our approach, and illustrate the changes occur in agents abilities. In Sect. 5, we conclude our work and discuss future directions.

2 Approach Overview

In this section, we detail our approach for updating BDI agents abilities. First, we define the problem we aim to solve in this paper. Next, we present our solution for this problem.

2.1 Problem Definition

In an AmI system, agents have a repertory of *possible actions* (SAA) that they can perform on its environment using *actuators.* Agents have a partial knowledge and control over the environment [13]. Based on its location, an agent may or may not has access to some actuators. Changes in agents' abilities, can be occur in the following ways:

- A mobile agent may gain access to some actuators, and lose access to others.
- An Agent gains access to new installed actuators.
- Agents lose access to an actuator if it has been removed form the environment.
- Agents lose access to an actuator if it fails.

The problems that arise are: How to update agents capabilities in a dynamic environment? and how to organize the negotiation between agents and actuators?

2.2 Required Properties

In this subsection, we present the required properties for every approach that aims to solve the defined problem. The validation of the required properties ensures that the proposed approach presents a good solution to the problems that the agent may face in a dynamic environment. the required properties are defined as follows:

Safety (nothing bad happens). Each solution for the problem must ensure the availability of actions to be added to agent's SAA.

The property in (1), states that if at a moment t, an agent Ag adds an action A to its SAA; Then, there is an actuator that performs the action A ($actuator(A)$), that belongs to the neighborhood[1] of the agent Ag ($actuator(A) \in neighborhood(Ag)$).

$$\begin{aligned} &if\, at\, a\, moment\, t,\; Ag\, adds\, A\, to\, its\, SAA \\ &then: \exists\, actuator(A) \,/\, actuator(A) \in neighborhood(Ag) \end{aligned} \tag{1}$$

Liveness (something good eventually happens). Each solution for the problem must ensure the following liveness properties:
(1) The detection of new available actions:

The property in (2), states that if the actuator Act that performs the action A belongs to the neighborhood of an agent Ag ($actuator(A) \in neighborhood(Ag)$), and A doesn't belong to the SAA of Ag ($A \notin abilities(Ag)$); Then, A eventually belongs to the SAA of Ag.

$$\begin{aligned} &if\, actuator(A) \in neighborhood(Ag) \wedge A \notin abilities(Ag) \\ &then: A\, eventually\, belongs\, to\, abilities(Ag) \end{aligned} \tag{2}$$

[1] An actuator that is not available for an agent Ag, is considered not to belong to its neighborhood.

(2) The deletion of unavailable actions:

The property in (3), states that if the action A belongs to the SAA of an agent Ag ($A \in abilities(Ag)$), and the actuator Act that performs A becomes unavailable; Then, A eventually deleted from the SAA of Ag.

$$\begin{aligned} & if\, A \in abilities(Ag) \wedge actuator(A)\, becomes\, unavailable \\ & then : A\, eventually\, removed\, from\, abilities(Ag) \end{aligned} \tag{3}$$

2.3 Proposed Solution

An intuitive way to solve the problem presented in this section is to develop a distributed protocol that updates agents abilities. In this section, we present an overview of the distributed protocol we developed in this paper. First, we redefine the set of available actions to meet the needs of our approach. Next, we show how it can be managed and updated using the proposed protocol.

Set of Available Actions (SAA). SAA, is a set that contains all the actions from which the planner attempts to generate the resulting plans. As we discussed earlier, both dependent and independent domain planning consider a static pre-built set of action. Actions are generally **STRIPS-like** [11] operators, which are composed of three parts: (i) the *name of operator* and its *parameters*; (ii) the *precondition* that contains the conditions under which the operator can be used; and (iii) the *effects* of applying the operators.

In this paper we consider two types of actions, *persistent* and *dynamic* actions. *Persistent actions* are performed by the agent directly (e.g., $distance(A, B)$: calculate the distance between two points A and B). This type of actions is available to the agent all the time, regardless of its *context. Dynamic actions* are performed using actuators (e.g., $open(door)$: open the door if we can access the actuator). This type of actions is available to the agent based on its spatial context, i.e. whether or not the agent can reach it from its current location.

In [13], authors present six types of *formulas* (actions) that can be used in plans' body. In Table 1 we classify these formulas into *Persistent* and *Dynamic* actions. *External Actions* are the only actions that belong to the dynamic action set; they are used by the agent to act within its environment using *actuators. Actuators* are installed in agent's environment, and its state evolves during time (the actuator can be unavilable) [14].

The classification of actions yields two subsets: (i) the Set of Persistent Actions (*SPA*); and (ii) the Set of Dynamic Actions (*SDA*). In the rest of this paper, we think of SAA as the union of the two subsets *SPA* and *SDA*. Consequently, we define the SAA as follows:

$$SAA = SPA \cup SDA \tag{4}$$

$$\begin{aligned} where : & \\ & SPA = \{action \mid action\, isn't\, an\, external\, action\} \\ & SDA = \{action \mid action\, is\, an\, external\, action\} \end{aligned}$$

Table 1. Persistent vs Dynamic actions

Formula	Persistent actions	Dynamic actions
External actions		X
Achievement goals	X	
Test goals	X	
Mental notes	X	
Internal actions	X	
Expressions	X	

Protocol Overview. The protocol's main goal is to maintain the SAA consistent and updated. It updates the SAA by adding the new detected actions, and removing the invalid ones (inaccessible actions). In the previous subsection, the SAA is defined as the union of the two subsets SPA and $SDA(4)$. The SPA contains the static part of the SAA. It's assumed to be manually constructed, in advance, by the agent's developer. However, the SDA contains the dynamic part of the SAA, on which we'll focus in the the rest of this paper.

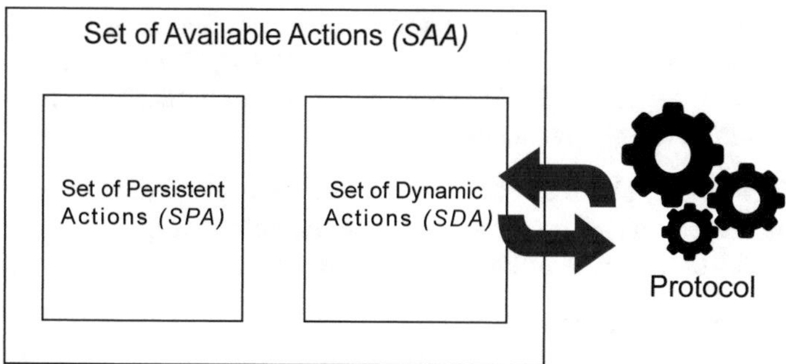

Fig. 1. Protocol Overview

In the introduction, we stated that, when the protocol detects changes in agents' abilities, the new available actions will be added to the SAA, and the invalid ones will be removed. More precisely, as shown in Fig. 1, the protocol focuses on the SDA as it's the dynamic part of the SAA. Initially, the SDA is empty and it will be initialized for the first time at system run-time, when the environment exploration for detecting abilities is started.

3 Distributed Protocol for Updating Agents Abilities

In this section, we present a distributed protocol for updating agents' abilities. First, we define the entities involved in our protocol. Next, we define the assumptions under which the proposed protocol works properly. Then, we present the notations that will be used to describe the involved entities. Finally, we detailed the behavior of both agents and actuators.

3.1 Entities

In a particular ambient intelligence environment, we have two types of BDI agents; *regular* and *actuator* agents. *Regular* agents[2] are normal BDI agents that can be mobile or static (fix). They have a repertoey of possible actions (SAA) that they can perform in order to modify their environment. We adopt agent definition in [7] for regular agents:

Definition 1. *(Agent) An agent is a tuple* $< Ag, Ev, Bel, Plib >$ *where* Ag *is the agent identifier;* Ev *is a queue of events;* Bel *is a belief base; and* $Plib$ *- the plan library - is a set of plan rules.*

In other hand, *actuator* agents[3] are a particular BDI agents that represent actuators in the environment. Actuators respond to agents' requests, and perform the actions they offer. As specified in Definition 2, an *actuator* considers a unique type of events (*+requestReceived*). It has as beliefs the information about the offered action, and a set of plans that will be used to respond to agents' requests.

Definition 2. *(Actuator) An actuator is a tuple* $< Act, actEv, ActBel, Plib >$ *where Act is the actuator identifier; actEvt is an event of type +requestReceived; ActBel is a static set of beliefs about the offered action; and Plib is the set of plans used to respond to agents' requests.*

Actuator has a basic abstract class, that must be implemented by all the agents of this type as they have the same behavior. As shown in Fig. 2, the belief base of these agent contains the following information:

- *state(S)*: The state of the actuator.
- *actionId(ID)*: The unique id of the action.
- *preConditions(P)*: The set of preconditions (it can be empty).
- *effects(E)*: The the set of effects (it can't be empty).

The library of plans contains two abstract plans that deal with the *+requestReceived* event, which is the only type of events considered by the *actuator*. The second plan is optional, as it deals with the actuator failure.

3.2 Assumptions

In designing our protocol, we assume that:

$$\varepsilon : time\, for\, message\, transmission.$$

$$d : the\, minimum\, distance\, travled\, by\, the\, agent,\ that\ causes\ change\, in\, its\, context.$$

$$v : movement\, speed.$$

[2] Refer to as agent in the rest of this paper.

[3] Refer to as actuator in the rest of this paper.

```
state(available|unavailable).
actionId(id).
preConditions(preCondition*).
effects(effect+).
---------------------------------------------
+requestReceived(cmd, source, data)
                : state(A) & A = 'available'
                <- ... /* treat the request */.

\\Optional plan
+requestReceived(cmd, source, data)
                : state(A) & A = 'unavailable'
                <- ... /* respond the unavilable state */.
```

Fig. 2. Actuator Abstract Class

- (A1) $\varepsilon \ll \frac{d}{v}$ if a mobile agent Ag sends a request, then it will receive a response before it changes its spatial context.
- (A2) Finite time for message transmission.
- (A3) Connection is reliable.

3.3 Distributed Protocol

In this subsection, we present the formal description that will be used later to specify the behavior of both agents and actuators.

Formalization. To simplify the behavior description of agents and actuators, we use the following notation:

- $A = A1 \cup A2$: The set of all agents.
- $A1 \subset A$: The set of all *regular* agents.
- $A2 \subset A$: The set of all *actuator* agents.
- L: The finite set of all the spatial localities of the system.

Agent Behavior. In this subsection, we present the behavior of an agent $Ag \in A1$, which is located in a location $l \in L$ at the moment $t \in T$.

Data sets and used variables: to implement the agent behavior, we use the following data sets and variables:

- SDA: the set of *dynamic* actions, initial($\emptyset$);
- $okReplayCount$: the number of OK messages still expected.
- $actionState[1:N]$: $actionState[action]$ is TRUE when the actuator Act that performs the $action$ responded with OK message (N is the size of SDA).

- (P1) When Ag invokes the *DISCOVER* routine:

```
broadcast(DISCOVER);
```

Agent *Ag* starts exploring its environment by broadcasting a *DISCOVER* request to detect all the available actions.

- (P2) When *Ag* invokes the *GET_STATE* routine:

```
1  okReplayCount = sizeOf(SDA);
2  foreach (action a of SDA) do
3  |   actionState[a] = false;
4  |   send(GET_STATE) to the actuator Act that executes the action a;
5  end
6  wait(t);
7  if (okReplayCount > 0) then
8  |   foreach (action a of SDA) do
9  |   |   if (not actionState[a]) then
10 |   |   |   remove a from SDA;
11 |   |   end
12 |   end
13 |   invoke the routine DISCOVER;
14 end
```

Agent *Ag* invokes this routine to check the validity of the available actions in *SDA*. It starts by initializing the counter of expected *OK* replays to the size of the *SDA* (line 1). Then, a *GET_STATE* request is sent to the actuators that execute the available actions in *SDA*, and their states is initialized to *FALSE* (lines 2–5). *Ag* waits t time ($t \gg 2 \times$ *messages transfer time*) before it resumes its execution (line 6). If *Ag* doesn't receive all the expected *OK* replays, then the unavailable actions are deleted from *SDA* and the *DISCOVER* routine is invoked (lines 7–14).

- (P3) When *Ag* receives a *RESPONSE(Data)* from an actuator *Act*:

```
1  switch (RESPONSE(Data)) do
2  |   case (EXIST(null))
3  |   |   if (Act ∉ SDA) then
4  |   |   |   send(INTRODUCE) to Act;
5  |   |   end
6  |   end
7  |   case (ABILITY(ability))
8  |   |   SDA = SDA ∪ ability;
9  |   end
10 |   case (AVAILABLE(null))
11 |   |   okReplayCount = okReplayCount − 1;
12 |   |   actionState[a] = true;
13 |   end
14 end
```

Based on the received *RESPONSE*, *Ag* behaves as follows:

- if $RESPONSE = EXIST$, then *Ag* checks the *none* existence of *Act* in its *SDA*, in order to send an *INTRODUCE* request (lines 2–5).

- if $RESPONSE = ABILITY(ability)$, then Ag adds the new *ability* to its SDA (lines 7–8).
- if $RESPONSE = AVAILABLE$, then The counter of expected OK replays is decremented by 1, and the state of the available action a executed by the actuator *Act* is assigned to $TRUE$ (lines 10–13).

Actuator Behavior. In this subsection, we present the behavior of an *actuator* $Act \in A2$, which is located in a location $l \in L$ at the moment $t \in T$.

Data sets and used variables: to implement the actuator behavior, we use the following data sets and variables:

- $state \in \{available, unavailable\}$: the state of the actuator.
- *ability*: the action performed by this actuator (as defined in Sect. 3.1).

  - (P6) When *Act* receives a $REQUEST$ from an agent *Ag*:

```
if (state == available) then
    switch REQUEST do
        case DISCOVER
            send(EXIST) to Ag;
        end
        case INTRODUCE
            send(ABILITY, ability) to Ag;
        end
        case GET_STATE
            send(AVAILABLE) to Ag;
        end
    end
end
```

Based on the $REQUEST$ received, *Ag* responds as follows:

- *Act* responds to the *DISCOVER* request with a response *EXIST*. The response *EXIST* states that the actuator is withing the agent's coverage area.
- *Act* responds to the *INTRODUCE* request with a response *ABILITY*. The response contains the action performed by *Act* as a second parameter *ability*.
- *Act* responds to the *GET_STATE* request with a response *AVAILABLE*. The response *AVAILABLE* states that the actuator is available.

Discussion: The frequency of using *DISCOVER* and *GET_STATE* routines must be defined when implementing the protocol, taking into account the use of *GET_STATE* instead of *DISCOVER* routine whenever it's possible to avoid *broadcasting*.

For space reason, we omit the validation of the required properties as its trivial under the assumptions we made in Sect. 3.2.

4 Implementation

Thus far, we have detailed our approach for updating agents' abilities. We now focus on the implementation of our approach to illustrate the behavior of agents and actuators in a smart laboratory scenario. First, we briefly present the *Smart Laboratory* scenario; and then, using our approach, we depict the changes occur in the agent's SAA.

4.1 Scenario (Smart Laboratory)

In a smart laboratory, each researcher has an assistant BDI agent in its smart phone. The assistant agent is designed to assist a busy researcher in its daily routines within the laboratory. Our goal in this scenario is to simulate the process of updating agents abilities using the proposed protocol.

As illustrated in Fig. 3, the laboratory consists of a corridor that connects three rooms in which researchers work. For the sake of simplicity, we are going to focus on one agent and five actuators, as the goal of this work is updating agents' abilities. The blue circle represents the agent; the red square represents the coverage area of the agent. Green squares represent the actuators installed in the environment. The label on the actuator indicates if the actuator works properly "On" (Off if the actuator fails). The agent moves within its environment and based on its location actuators may or may not be within its range.

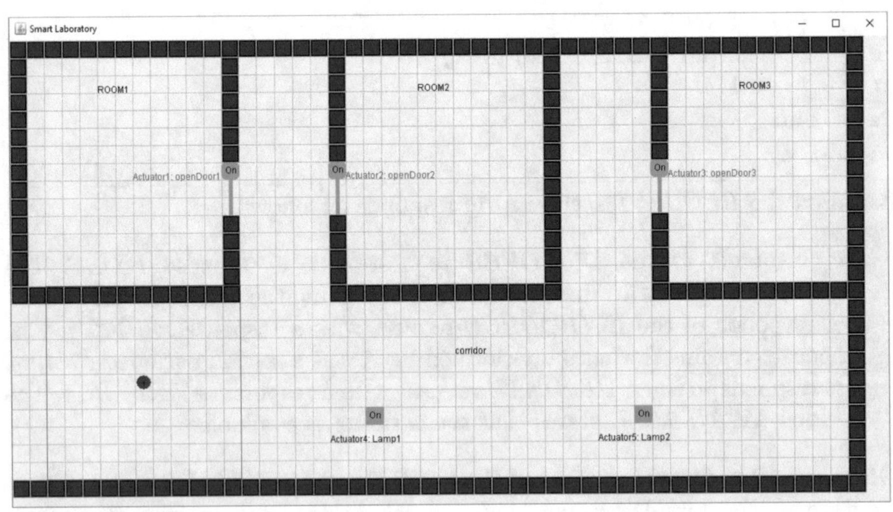

Fig. 3. Smart Laboratory

4.2 Illustration

In this subsection, we illustrate the changes occur in the agent's SAA based on the agent location. Initially the agent is located in the *start* location as shown in Fig. 3. Figure 4, illustrate that the agent's SAA in this case. Agent has no actuator within its coverage area, and its SAA is empty.

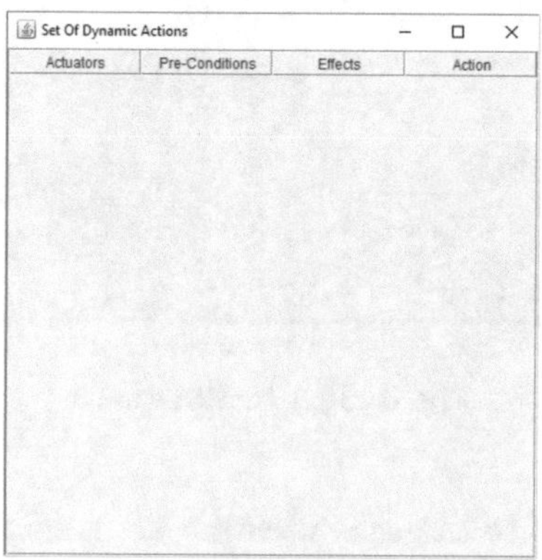

Fig. 4. Initial SAA

When the agent moves to the location shown in Fig. 5, two actuators are located within its coverage area *actuator1* and *actuator2*. The protocol allows the detection of these two actuators, and the actions, *openDoor1* and *openDoor2*, are added to the agent's SAA as shown in the Fig. 6.

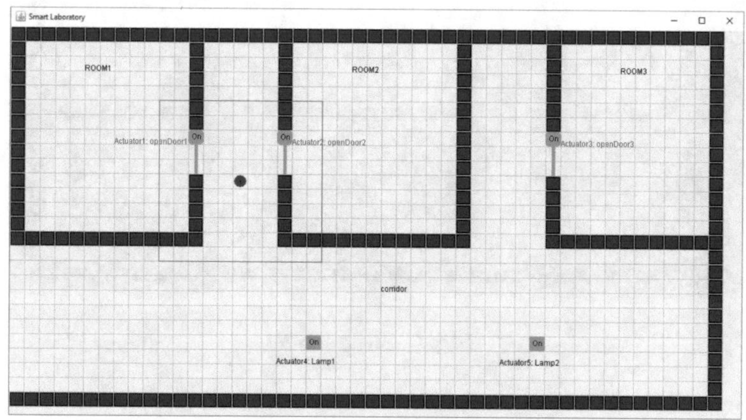

Fig. 5. Situation 1

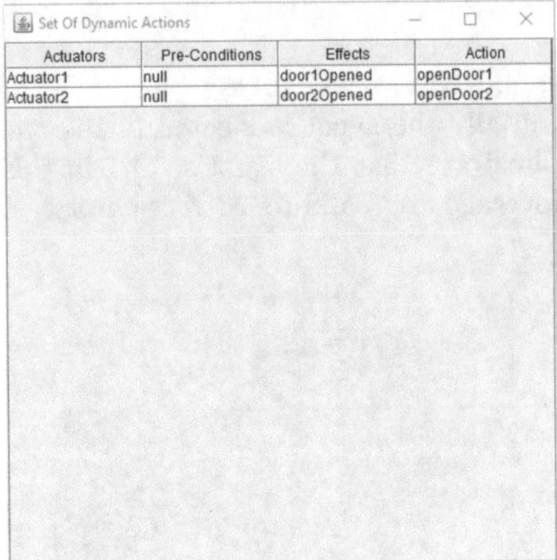

Set Of Dynamic Actions

Actuators	Pre-Conditions	Effects	Action
Actuator1	null	door1Opened	openDoor1
Actuator2	null	door2Opened	openDoor2

Fig. 6. SAA in Situation 1

In Fig. 7, we illustrate the case when an actuator fails, by turning off the *actuator2* (the label on the actuator becomes Off). The protocol allows the detection of the unavailable actuators; thus, the detection of the invalid actions (*openDoor2* in this case). The deletion of these actions is successfully performed as shown in Fig. 8.

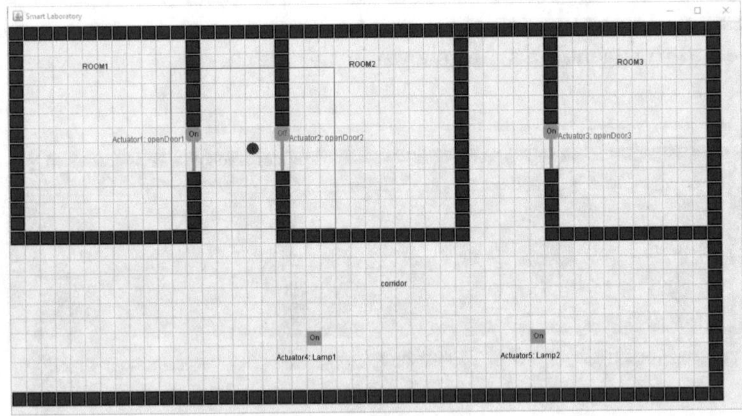

Fig. 7. Situation 2

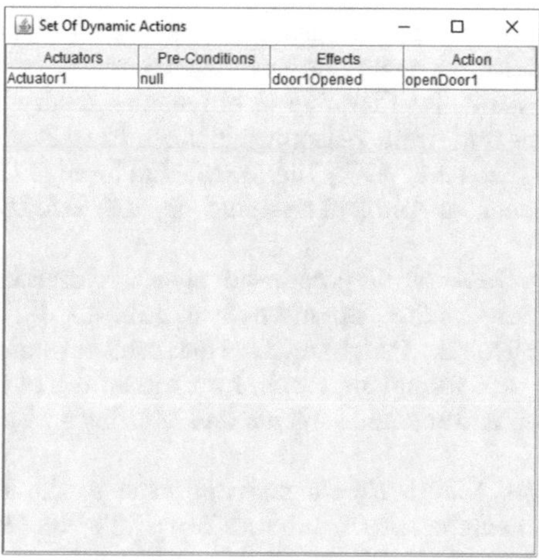

Fig. 8. SAA in Situation 2

5 Conclusion

In this paper, we presented the problem of updating agents abilities, and the required properties for each approach that aims to solve this problem. We also proposed a novel approach for updating BDI agents abilities. Our approach consists of a context-aware distributed protocol that resolves the agent's lack of awareness of its abilities. Based on the context of the agent, the protocol allows the exploration of agent's neighborhood in order to detect its abilities. Finally, we implemented the proposed protocol and proved its efficiency in a Smart Laboratory scenario.

As a future work, we intend to improve the proposed protocol by alleviate the assumptions to facilitate its implementation. Moreover, we plan to integrate the protocol with working planners, by using the updated SAA as the input for these planners. Then, we aim to study the impact of using the protocol on both performance and outcomes of planners.

References

1. Ilié, J.-M., Chaouche, A.-C.: Toward an efficient ambient guidance for transport applications. Procedia Comput. Sci. **110**, 190–198 (2017)
2. Roda, C., Rodríguez, A.C., López-Jaquero, V., Navarro, E., González, P.: A multi-agent system for acquired brain injury rehabilitation in ambient intelligence environments. Neurocomputing **231**, 11–18 (2017)
3. Tapia, D.I., Abraham, A., Corchado, J.M., Alonso, R.S.: Agents and ambient intelligence: case studies. J. Ambient Intell. Humaniz. Comput. **1**(2), 85–93 (2010)
4. Bratman, M.: Intention, plans, and practical reason (1987)
5. Bordini, R.H., Hübner, J.F.: A java-based interpreter for an extended version of agentspeak. University of Durham, Universidade Regional de Blumenau (2007)

6. Busetta, P., Rönnquist, R., Hodgson, A., Lucas, A.: Jack intelligent agents-components for intelligent agents in java. AgentLink News Lett. **2**(1), 2–5 (1999)
7. Meneguzzi, F., De Silva, L.: Planning in bdi agents: a survey of the integration of planning algorithms and agent reasoning. Knowl. Eng. Rev. **30**(1), 1–44 (2015)
8. Boukharrou, R., Chaouche, A.C., Ilié, J.-M., Saïdouni, D.E.: Contextual-Timed Planning Management for Ambient Systems, pp. 107–114. IEEE Computer Society (2014)
9. Rens, G., Moodley, D.: A hybrid pomdp-bdi agent architecture with online stochastic planning and plan caching. Cognitive Syst. Res. **43**, 1–20 (2017)
10. Sebastian, S., de Silva, L., Padgham, L.: Hierarchical planning in bdi agent programming languages: a formal approach. In: Proceedings of the Fifth International Joint Conference on Autonomous Agents and Multiagent Systems, pp. 1001–1008. ACM (2006)
11. Fikes, R.E., Nilsson, N.J.: Strips: a new approach to the application of theorem proving to problem solving. Artif. Intell. **2**(3–4), 189–208 (1971)
12. Nau, D., Cao, Y., Lotem, A., Munoz-Avila, H.: Shop: Simple Hierarchical Ordered Planner, vol. 2, pp. 968–973. Morgan Kaufmann Publishers Inc. (1999)
13. Bordini, R.H., Hübner, J.F., Wooldridge, M.: Programming Multi-agent Systems in AgentSpeak using Jason. Wiley (2007)
14. Baitiche, H., Bouzenada, M.: and Djamel Eddine Saïdouni. Towards a generic predictive-based plan selection approach for bdi agents. Procedia Comput. Sci. **113**, 41–48 (2017)

Software Engineering and Formal Methods

Effective Bridging Between Ecore and Coq: Case of a Type-Checker with Proof-Carrying Code

Jérémy Buisson[1(✉)] and Seidali Rehab[2]

[1] IRISA, Écoles de Saint-Cyr Coëtquidan, Guer, France
jeremy.buisson@irisa.fr

[2] MISC, University of Constantine 2 - Abdelhamid Mehri, Constantine, Algeria
seidali.rehab@misc-umc.org

Abstract. The work presented in this paper lies in the context of implementing supporting tools for a domain-specific language named SosADL, targeted at the description and analysis of architecture for systems of systems. While the language has formal definition rooted in the Cc-pi calculus, we have adopted the Eclipse ecosystem, including EMF, Ecore and Xtext for the convenience they provide in implementation tasks. Proof-carrying code is a well-known approach to ensure such an implementation involving non-formal technologies conforms to its formal definition, by making the implementation generate proof in addition to usual output artifacts. In this paper, we therefore investigate for an infrastructure that eases the development of proof-carrying code for an Eclipse/EMF/Ecore/Xtext-based tool in relation with the Coq proof assistant. At the core of our approach, we combine an automatic transformation of a metamodel into a set of inductive types, in conjunction with a second transformation of model elements into terms. The first one, reused from our previous work, provides necessary abstract syntax definitions such that the formal definition of the language can be mechanized using Coq. The second transformation is part of the proof generator.

Keywords: Ecore · Coq · Proof-carrying code · Model transformation

1 Introduction

In our previous work [1], we have presented a transformation that maps an Ecore metamodel [2] to a collection of inductive types, more specifically targeting Gallina and Vernacular, the language of the Coq proof assistant [3]. Thanks to this previous work, we are able not only to define instances of the metamodel within the proof assistant, but also to, e.g., quantify over objects of given classes in order to prove properties or provide specifications involving this metamodel. The latter is useful, for instance, to formally mechanize the semantics or the type system of the language whose abstract syntax is given by the metamodel.

S. Chikhi et al. (Eds.): MISC 2018, LNNS 64, pp. 259–273, 2019.
https://doi.org/10.1007/978-3-030-05481-6_20

Still, we think that this transformation, alone, is not sufficient to effectively bridge between the two technical spaces. Indeed, transforming model elements might be useful as well, especially when the application relying on the metamodel has to send parts of models to the proof assistant. This is for instance the case when this application generates proofs, e.g., in the context of proof-carrying code [4], to increase confidence in the implementation of the tools supporting the language.

This paper presents our preliminary work in this specific area. We extend our previous transformation [1] in order to generate a secondary transformation, which translates models to terms, consistently with inductive types produced accordingly to our previous work [1]. *Consistent* means here: when an EMF object is translated into a Gallina term, the object is an instance of an Ecore class (let name it C), and the term is of a inductive type; this inductive type is the result of the transformation of that Ecore class C. Then we combine the two transformations, yielding to an overall infrastructure for proof-carrying code.

In Sect. 2, we first present the context that motivates our work, here the development of supporting tools for SosADL [5], a domain-specific language for describing and analyzing architecture of systems of systems, and following the proof-carrying code approach. Then Sect. 3 gives some background on how an Ecore metamodel can be turned into a collection of inductive types, mechanization of the type system, and typical approach to the implementation of the type checker. Section 4 depicts how the type checker can be extended in order to generate proofs. Section 5 addresses the transformation of a model element into a term. Section 6 presents the related works. Last, Sect. 7 concludes the paper and gives our agenda for future work.

2 Context

To motivate our work, we consider the case of developing tools supporting a domain-specific language. In our case, we consider the implementation of the tools supporting SosADL [5], an architecture description language for system of systems. The language is intended to let an architect describe systems, which can be flexibly assembled into a larger system of systems by means of a constraint-based description of the assembly, based on the Cc-π formal calculus [6]. The resulting system of systems can be analyzed and simulated, e.g., in order to discover emerging behavior or to ensure the expected behavior is achieved. Analysis and simulation are enabled by the formal definition of the language.

The supporting tools for SosADL are developed using model-driven engineering, and more specifically Ecore/EMF [2] and Xtext [7] technologies from the Eclipse ecosystem. The formal definition of the language is mechanized using the Coq proof assistant [3], which enables to verify that the language definition is sound. In order to ensure that the tools conform to the formal definition, we set up a proof-carrying code infrastructure [4]. That is, in addition to performing analysis or producing output artifacts, the SosADL supporting tools issue a proof that the produced outcomes are correct. By checking the proof, the user ensures that the tools performed in conformance to the formal definition.

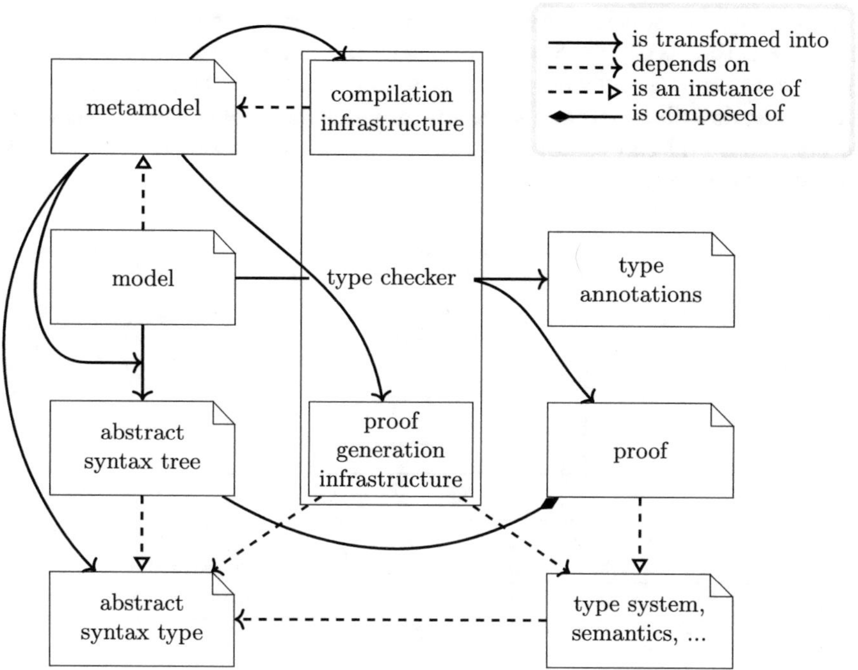

Fig. 1. Overview of applying the proof-carrying code approach to a type checker, by using model-driven engineering.

We have more specifically applied the proof-carrying code approach to the type checker in SosADL supporting tools. Figure 1 is an overview of it, described in subsequent sections. In the spirit of model-driven engineering, our challenge is to generate automatically (part of) the proof-carrying code infrastructure.

3 Background

In this section, we describe first in Sect. 3.1 how an Ecore metamodel can be turned into a collection of inductive types, following our previous work [1]. Then in Sect. 3.2, we depict a typical approach to mechanize a type system and to implement the corresponding type checker.

3.1 Transformation of a Metamodel to Inductive Types

On the one side, in the Ecore metamodel, each class defines one type of nodes of abstract syntax trees. The fields of the class describe the attributes of the nodes of that type, and the composition relationships between classes encode the parent-children relationships between the nodes in the abstract syntax tree. On the other side, an inductive type is a typical type definition in functional programming. An inductive type is made of a collection of constructors, such that any value of that type is the result of one of these constructors. Each inductive type defines one type of nodes of abstract syntax trees. The parameters of a constructor describe the data structure of the value, hence they encode both the attributes and the parent-children relationships between the nodes in the abstract syntax tree.

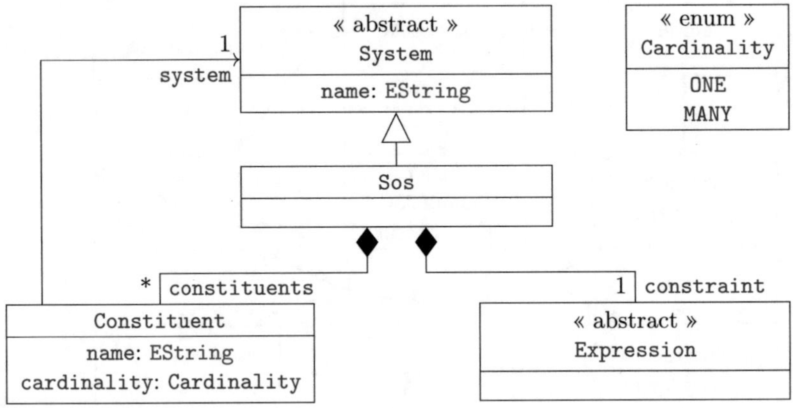

Fig. 2. An example metamodel, inspired by the SosADL case study.

Following the principle drawn by Djeddai et al. [8], each abstract class A is turned into an inductive type A, and any concrete class C that specializes A is transformed to a constructor C that belongs to inductive type A. In the example of Fig. 2 transformed into Fig. 3, abstract class System is mapped to inductive type sosadl_System, which declares constructor sosadl_System_sosadl_Sos, which is mapped from concrete class Sos that specializes System.

Members of concrete classes are mapped to parameters of the corresponding constructors. For example, members cardinality, name and system of class Constituent are mapped to parameters *cardinality*, *name* and *system* of constructor sosadl_Constituent_sosadl_Constituent.

Klint and van der Storm [9] propose preprocessing steps to support arbitrary source metamodels. The transformation first pulls class members down the specialization relationship, generalizes referenced classes then flattens the specialization relationship. Then the same principle as in [8] is applied to generate inductive types.

We have further improved the transformation in [1] in order to better support multiple inheritance. Constructors are duplicated in as many inductive types as classes in the specialization relationship. In the example, class Sos illustrates this approach. First, each class is mapped to an inductive type: class System is mapped to sosadl_System, and class Sos is mapped to sosadl_Sos_0. Then, concrete classes are mapped to constructors that belong to the types

```
Inductive sosadl_Expression_0: Type := .
Inductive sosadl_Cardinality_0: Type :=
| sosadl_Cardinality_MANY: sosadl_Cardinality_0
| sosadl_Cardinality_ONE: sosadl_Cardinality_0.
Definition ecore_EString_0: Type := string.
Definition _NoDupList_0: (Type → Type) := list.
Inductive sosadl_Constituent: Type :=
| sosadl_Constituent_sosadl_Constituent: (∀ (cardinality: sosadl_Cardinality_0),
   (∀ (name: ecore_EString_0), (∀ (system: (_URI_0 sosadl_System)),
     sosadl_Constituent)))
 with sosadl_System: Type :=
| sosadl_System_sosadl_Sos: (∀ (constituents: (_NoDupList_0 sosadl_Constituent)),
   (∀ (constraint: sosadl_Expression_0), (∀ (name_0: ecore_EString_0),
     sosadl_System))).
Inductive sosadl_Sos_0: Type :=
| sosadl_Sos_sosadl_Sos: (∀ (constituents_0: (_NoDupList_0 sosadl_Constituent)),
   (∀ (constraint_0: sosadl_Expression_0), (∀ (name_1: ecore_EString_0),
     sosadl_Sos_0))).
```

Fig. 3. Output of the transformation [1] for the metamodel of Fig. 2.

mapped from the super classes: class Sos is therefore mapped to two constructors, sosadl_System_sosadl_Sos (in type sosadl_System mapped from System) and sosadl_Sos_sosadl_Sos (in type sosadl_Sos mapped from Sos).

3.2 Mechanization of the Type System and Implementation of the Type Checker

By using inductive types generated like described in Sect. 3.1, we mechanize the type system of SosADL. Following the usual approach, the mechanized type system is itself a collection of inductive types, where each inductive type defines a judgment and its constructors are the axioms that encode rules for that judgment. Figure 4 illustrates the approach. Judgment system_is_well_typed asserts that a system declaration conforms to the type system. The inductive type encoding this judgment accepts two parameters: an environment and the system under consideration. In the figure, we provide only one rule for this judgment. This rule is encoded by constructor sos_is_well_typed. The first four parameters of the constructor (Γ, *constituents*, *constraint* and *name*) are the variables that have to be bound in order to apply the rule. Additional parameters (*constituents_exist* and *constraint_is_well_typed*) are the two premises of the rule. The return type of the constructor, where the inductive type has effective parameters, is the conclusion of the rule.

When the type system is syntax directed, a typical approach to implement a type checker is to follow the principle of Milner's algorithm $\mathcal{W}$. Like illustrated in Fig. 5, for each judgment, a function is implemented such that it attempts to

```
Inductive system_is_well_typed: environment → sosadl_System → Prop :=
| sos_is_well_typed: ∀ (Γ: environment)
        (constituents: _NoDupList_0 sosadl_Constituent)
        (constraint: sosadl_Expression_0) (name: ecore_EString_0),
    ∀ constituents_exist: (∀ c, c ∈ constituents → constituent_exists Γ c),
    ∀ constraint_is_well_typed: expression_has_type Γ constraint type_boolean,
  system_is_well_typed Γ (sosadl_System_sosadl_Sos constituents constraint name).
```

Fig. 4. Example of a mechanized rule.

```
public boolean proveSystemIsWellTyped(Environment g,
      System s) throws Unprovable {
  if(s instanceof Sos
    && ((Sos)s).constituents.stream()
         .allMatch(c -> proveConstituentExists(g, c))
    && proveExpressionHasType(g,
         ((Sos)s).constraint, BOOLEAN)){
    // proved by rule sos_is_well_typed
    return true;
  } else {
    throw new NoMatchingRule();
  }
}
```

Fig. 5. Typical code pattern for the type checker.

prove a goal by selecting the adequate typing rule according to the content of the abstract syntax tree, then recursively calling itself (and other functions mapped from other judgments) in order to try to prove the premises of the chosen typing rule. In Fig. 5, Java function proveSystemIsWellTyped is the function that aims at proving system_is_well_typed judgments. Depending on the syntactical type of the node s of the abstract syntax tree, it selects the rule it attempts. In the example, it attempts to prove the judgment by using rule sos_is_well_typed when s is an instance of Sos. If premises can in addition be proved, here by successfully calling proveConstituentExists and proveExpressionHasType, the function concludes that the judgment is successfully proved. Otherwise, if no rule applies, the function reports typing error, e.g., by throwing an exception. In addition to answering whether the source model is correctly typed or not, the type checker may annotate the source model with type information.

4 Generation of Proofs

By instrumenting the type checker, we extend it to generate a well-typed proof as well. Like illustrated by the example of Fig. 6, each function that proves a judgment is changed such that it returns a proof object, that is, an instance of a class that corresponds to the constructor encoding the rule in the mechanized type system. In the given example, class SosIsWellTyped is the concrete class that corresponds to rule sos_is_well_typed; it specializes abstract class SystemIsWellTyped that corresponds to judgment system_is_well_typed.

We have not worked yet on how these classes could be generated, but we think that Coq's extraction mechanism may be used to address this issue.

```
public SystemIsWellTyped proveSystemIsWellTyped(
      Environment g, System s) throws Unprovable {
  if(s instanceof Sos) {
    return new SosIsWellTyped(g, ((Sos)s).constituents,
      ((Sos)s).constraint, ((Sos)s).name,
      proveForAll((Sos)s).constituents,
          c -> proveConstituentExists(g, c),
      proveExpressionHasType(g,
          ((Sos)s).constraint, BOOLEAN));
    // proved by rule sos_is_well_typed, the proof object is returned
  } else {
    throw new NoMatchingRule();
  }
}
```

Fig. 6. Code pattern for the instrumented proof-generating type checker.

```
Definition proof: system_is_well_typed [("foo", foo); ("bar", bar)]
  (sosadl_System_sosadl_Sos
    [ sosadl_Constituent_sosadl_Constituent sosadl_Cardinality_ONE "a" ref_foo;
      sosadl_Constituent_sosadl_Constituent sosadl_Cardinality_MANY "b" ref_bar ]
    c "world") :=
  sos_is_well_typed [("foo", foo); ("bar", bar)]
    [ sosadl_Constituent_sosadl_Constituent sosadl_Cardinality_ONE "a" ref_foo;
      sosadl_Constituent_sosadl_Constituent sosadl_Cardinality_MANY "b" ref_bar ]
    c "world" P1 P2.
```

Fig. 7. Proof term, after generation of the Vernacular definition.

The proof object is then serialized into a Vernacular definition like the example of Fig. 7, such that Coq's compiler can check whether the proof is correct. In this excerpt, proof is defined to be a proof of system_is_well_typed with the given parameters (environment and system definition). Its value is the proof term, here built by applying suitable parameters to constructor sos_is_well_typed. Section 5.2 explains the principles behind the generation of this term.

5 Transformation of Model Elements Into Terms

Like seen in the previous section, the generated proof contains terms that encode some elements from the model being type checked. In this section, we first extend

```
// x: model element that is going to be transformed to a term
// c: class that denotes the type of the generated term
// trace: correspondence information between the metamodel and inductive types
element_to_term(x, c, trace) {
  inductive ← trace.inductives[c]
  ctor ← filter(trace.constructors[x.eClass],
      f ↦ f.inductive = inductive)
  return apply(ctor, map(ctor.parameters,
      p ↦ feature_to_term(t.features[p], x, trace)))
}

// x: value that is going to be transformed to a term
// t: data type of the value
attribute_to_term(x, t) { // ad-hoc code that deals with primitive types
  if (EINTEGER == t) { // EInteger may be mapped to nat
    return x.toString
  } else if (ESTRING == t) { // EString may be mapped to string;
    return '"' + x + '"'
  } else ... // and so on
}

// f: structural feature (either attribute or reference) to be transformed to a term
// x: model element to which the structural feature belongs
// trace: correspondence information between the metamodel and inductive types
feature_to_term(f, x, trace) {
  if (f.isMany) { // generate a list if the structural feature is a collection one
    return list(map(x.eGet(f), o ↦ value_to_term(o, f, x, trace)))
  } else {
    return value_to_term(x.eGet(f), f, x, trace)
  }
}

// o: value to be transformed to a term
// f: structural feature from which o comes from
// x: model element to which the structural feature belons
// trace: correspondence information between the metamodel and inductive types
value_to_term(o, f, x, trace) {
  if (EREFERENCE.isSuperTypeOf(f) ∧ f.isContainment) {
    return element_to_term(x.eGet(f), f.eType, trace)
  } else if (EREFERENCE.isSuperTypeOf(f) ∧ ¬f.isContainment) {
    return uri(x.eGet(f)) // non-containment references are mapped to URIs
  } else if (EDATA_TYPE.isSuperTypeOf(f)) {
    return attribute_to_term(x.eGet(f), f.eType)
  }
}
```

Fig. 8. Generic algorithm that transfoms any model element into a Gallina term.

in Sect. 5.1 the transformation of Sect. 3.1 with correspondence information. Correspondence information is used in 5.2 in order to first present a generic algorithm that transforms any model element into a term, whose type is the inductive type mapped from the class of the model element (mapped from by the transformation of Sect. 3.1). Then, to avoid runtime introspection of correspondence information, we present in Sect. 5.3 how we can generate an ad hoc transformation, which transforms any element whose class belongs to the given metamodel into a term of the inductive type mapped the element's class.

5.1 Correspondence Information

In addition to the inductive types generated like described in Sect. 3.1, correspondence information that maps between Ecore classes of the source metamodel and generated inductive types have to be produced. This information is going to be used like described in Sect. 5.2 in order to generate Gallina terms for model elements.

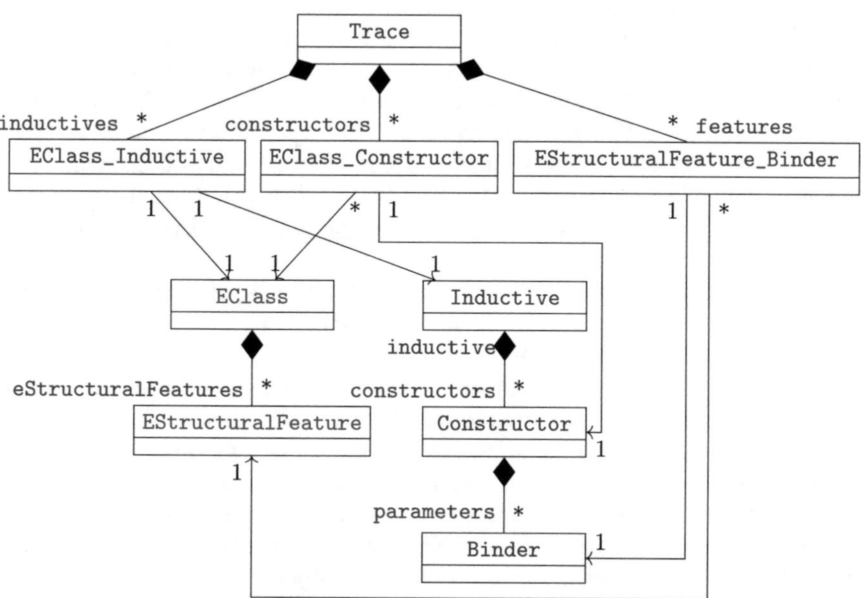

Fig. 9. Structure of correspondence information between a source Ecore metamodel and generated inductive types.

Figure 9 presents the structure of correspondence information. A Trace contains three reified associations. The first one named inductives contains one-to-one mapping between classes and inductive types. The second one named constructors contains many-to-one mapping between concrete classes and constructors. The last one named features contains many-to-one correspondence information between structural features and constructor parameters.

In this class diagram, EClass and EStructuralFeature are imported from the Ecore metametamodel, and Inductive, Constructor and Binder are imported from our own metamodel for Gallina and Vernacular.

5.2 Generic Transformation

The Vernacular definition of Fig. 7 is composed of two Gallina terms that have to be generated: the value (after the := symbol – here sos_is_well_typed ... *P2*) and the type (between : and := – here system_is_well_typed ... "world").

To begin with, we consider the value bound by the definition. It is the translation of the objects lying in the Ecore technological space. Figure 8 provides the pseudo-code of a generic algorithm that transforms a model element into such a term. It proceeds by introspecting the type of object by means of EMF reflection facilities. For each object x (function element_to_term) the constructor corresponding to its class is applied in order to build the subterm. The constructor is found thanks to the trace parameter, which is the object that records correspondence information described in Sect. 5.1 and obtained after the transformation of the metamodel into inductive types. Because the transformation may generate several constructors for each concrete class of the metamodel, function element_to_term has to select the right one. The suitable constructor depends on the inductive type the generated term is expected to have. This is the reason motivating the c parameter of element_to_term: this parameter c is the static type of the reference from which x has been got. Like shown in function feature_to_term, the effective parameter for c is indeed the static type of the reference in the parent object.

Once the constructor has been found, a function application term is generated by calling function apply. To generate effective parameters, each formal parameter of the constructor is first mapped back to the Ecore structural feature it comes from. Then function feature_to_term is called to generate the term for the effective parameter. This function deals with collections by issuing a list if necessary, and it calls value_to_term to convert individual objects. This latter function discriminates between containment references, non-containment references and attributes. The first ones, i.e., containment references, are converted by a recursive call to element_to_term. Non-containment references are translated into an URI, i.e., an identifier of the referenced object. Attributes are plain-old Java objects, which are translated to corresponding Coq terms by hard-coded rules.

The second term, the type of the generated definition, is the translation of the class of the object translated by the algorithm of Fig. 8. Because of how the type system is mechanized, the type has parameters that are values, like in the example of Fig. 7. However, none of Java nor EMF supports using an object as a type parameter in static types, and both Java and EMF erase type parameters from dynamic types. Therefore, we rely instead on type inference in Coq's compiler in order to suitably generate the type of the definition. In SosADL supporting tools, we have not faced any case when type inference fails.

5.3 Generation of the Transformation

The algorithm presented in the previous subsection is a generic one that uses Java and EMF reflection at runtime, in conjunction with correspondence information issued at the same time as inductive types generated by the transformation of Sect. 3.1. Instead of interpreting these data structures at runtime, a model element transformation can be statically generated specifically for the metamodel.

```
sosadl_Constituent(Constituent c) {
  return apply(sosadl_Constituent_sosadl_Constituent,
      sosadl_Cardinality(c.cardinality), string(c.name),
      uri(c.system));
}

sosadl_System(System s) {
   if (SOS.isSuperTypeOf(s.eClass)) {
    return apply(sosadl_System_sosadl_Sos,
        map(s.constituents, c ↦ sosadl_Constituent(c)),
        sosadl_Expression(s.constraint), string(s.name))
   } else {
      raise error
   }
}

sosadl_Sos(Sos s) {
  return apply(sosadl_Sos_sosadl_Sos,
      sosadl_Constituent(s.constituents),
      sosadl_Expression(s.constraint), string(s.name))
}

// and so on
```

Fig. 10. Generated algorithm that transfoms a SosADL model element into a term.

Figure 10 illustrates the generated code. For each inductive type, that is, for each class that may be used as a static type in the EMF technological space, a function is generated. When the class is concrete and when it has no known specializing class, the corresponding inductive type owns a single constructor. The generated function, e.g., sosadl_Constituent and sosadl_Sos in Fig. 10, applies that constructor to effective parameters got by (possibly recursively calling) other generated functions. When the class is abstract or when it has specializing classes, the generated function, e.g., sosadl_System in Fig. 10, uses Java or EMF reflection to find out the concrete class of the object and select the constructor accordingly.

Figure 11 outlines an algorithm to automatically generate such functions from correspondence information depicted at Sect. 5.1. To generate the ad-hoc transformation, function generate_transformation generates a function for each class

```
// trace: correspondence information from which the transformation is derived
generate_transformation(trace) {
 return map(trace.inductives,
    (c,i) ↦ generate_fun(trace, c, i))
}

// class: Ecore class (abstract or concrete) that corresponds to the inductive type
// inductive: inductive type generated that corresponds to the class
generate_fun(trace, class, inductive) {
  if (class.isAbstract ∨ ∃ c, class.isSuperTypeOf(c)) {
    cases ← map( filter (trace.constructors,
        (cl, ctor) ↦ ctor.inductive == inductive),
      (cl, ctor) ↦ generate_case(trace, cl, ctor, «x»))
   return function(inductive.name, [(class, «x»)], cases)
  } else {
    ctor ← filter (trace.constructors[class],
      c ↦ c.inductive == inductive)
   return function(inductive.name, [(class, «x»)],
      generate_generate(trace, ctor, «x»)
  }
}

// class: Ecore concrete class of the object that corresponds to the constructor
// constructor: constructor mapped from the class
// object: name of the parameter in the generated function
generate_case(trace, class, constructor, object) {
 return « if (» object «instanceof» class.name «) {»
    generate_generate(trace, constructor, object) «}»
}

// constructor: constructor that is going to be issued to transform the object
generate_generate(trace, constructor, object) {
 return «return apply(» constructor.name «(»
    map(constructors.parameters,
      p ↦ generate_call(trace, p, object)) «)»
}

// parameter: binder (in Gallina) that declares the parameter of the constructor
generate_call(trace, parameter, object) {
  feature ← trace.features[parameter]
  if (¬feature.isMany) {
    inductive ← trace.inductives[feature.eType]
   return inductive.name «(» object «.» feature.name «)»
  } else // deal with lists, and so on
}
```

Fig. 11. Automatic generation of the ad-hoc transformation of Fig. 10.

(or inductive type) of the metamodel for which the transformation is generated. This is done by invoking generate_function on each class. Like its name tells, this function generates one generator function, for one inductive type. Regarding the body of the generated function, generate_function first checks whether the class under consideration is abstract or has any specializing class. If so, it generates tests for each specializing class (by calling generate_case). If not, it directly invokes generate_generate, which generates instructions to issue a call to the constructor. Function generate_generate uses generate_call for each formal parameter of the constructor in order to generate function calls that issue terms for effective parameters of the constructor.

In this paper, we omit details to deal with collections and attributes.

6 Related Works

To the best of our knowledge, no previous work has studied automatic generation of an infrastructure for proof-carrying code for a language whose abstract syntax is described by a metamodel. Though, the approach is appealing since several popular language workbenches such as the Eclipse/EMF/Ecore/Xtext combination or MPS hardly integrate formal tools that may help in the verification of language definitions and implementations. In this regard, even if our work is still preliminary, it provides a novel step towards bridging semi-formal metamodels and formal approaches.

Like stated in Sect. 3.1, our work is based on and improves previous work on transforming a metamodel into a group of inductive types. In comparison to [8], our improved transformation does not suffer from any restriction on the source metamodel. In comparison to [9], we further improve support for multiple inheritance as we need not assume existence of a unique most-general super class for any class. As a consequence, generated inductive types are stricter. In comparison to our own previous work [1], improvements cover the handling of correspondence information and of model elements. In addition of transforming the metamodel, in this paper, we propose two approaches to consistently transform instances of that metamodel, i.e., model elements into terms: a generic algorithm that introspects the metamodel at runtime, and a algorithm that automatically generates ad-hoc code, hence avoiding the need for runtime introspection. In this paper, we also propose the combination of the transformations in order to build an infrastructure for proof-carrying code in the context of Eclipse and related DSL technologies (EMF/Ecore/Xtext).

Several previous work such as [10–13] have proposed approaches to transform OMT or UML class diagrams into terms or values in various formal calculus, hence enabling formal verification of these class diagrams. If we consider the abstract syntax for OMT or UML class diagrams (or even the Ecore metametamodel) as the metamodel, we think that our work may be able to generate automatically one such transformation, instead of hard-coding the transformation. Additional work is required in order to better evaluate how our own work might be usable in such a context.

7 Conclusion

In this paper, we have proposed to automatically generate an infrastructure for proof-carrying code given a metamodel produced in the context of the Eclipse/EMF/Ecore/Xtext [2, 7] language workbench. The work presented in this paper is motivated by our effort on providing supporting tools for the SosADL domain-specific language [5]. Our proposal is the combination of transforming the Xtext-generated metamodel into a collection of inductive types suitable for the Coq proof assistant. Then, from the same metamodel, we automatically derive a transformation that, consistently with the generated inductive types, transforms any model element into a term that can be successfully compiled by Coq.

Even if the infrastructure has been fully implemented in the type checker of SosADL supporting tools, we think that the area needs further investigation. This work allows us to define our agenda for future work in the area.

First, we plan to further study the transformation of model elements into terms. In addition to using this transformation in the context of our proof-carrying code infrastructure, we plan to assess how this automatic transformation could be used to verify properties of some models like done with various hard-coded UML-to-B transformations proposed in previous work.

Second, we have left open the question of defining classes that implement in the Java or Ecore the inductive types encoding the mechanized type system. These classes are indeed required in order to instrument the type checker such that it produces proofs. Existing Coq's extraction mechanism translates Gallina and Vernacular definitions into other languages. While this mechanism is a basis, it is designed to skip any proof-related item from the translation, which are precisely the items we want to translate to Java or Ecore. Changing the mechanism in this regard would need to study how it must be adapted in order to conform to restrictions and constraints imposed by Java and Ecore type systems.

Last, one may ask how much confidence can be put in our proposed approach. To address this issue, we consider applying our proof-carrying code infrastructure to itself. Namely, we consider instrumenting the transformations involved in our approach in order to generate conformance proofs that could be checked by the Coq proof assistant.

References

1. Buisson, J., Rehab, S.: Automatic transformation from ecore metamodels towards gallina inductive types. In: Hammoudi, S., Pires, L.F., Selic, B. (eds.) Proceedings of the 6th International Conference on Model-Driven Engineering and Software Development, MODELSWARD 2018, Funchal, Madeira, Portugal, 22–24 Jan 2018. pp. 488–495. SciTePress (2018). https://doi.org/10.5220/0006608604880495
2. Steinberg, D., Budinsky, F., Paternostro, M., Merks, E.: EMF: Eclipse Modeling Framework 2.0, 2nd edn. Addison-Wesley Professional, New York (2009)

3. Bertot, Y., Castéran, P.: Interactive Theorem Proving and Program Development—Coq'Art: The Calculus of Inductive Constructions. Texts in Theoretical Computer Science. An EATCS Series. Springer, Berlin (2004). https://doi.org/10.1007/978-3-662-07964-5
4. Necula, G.C.: Proof-carrying code. In: Lee, P., Henglein, F., Jones, N.D. (eds.) Conference Record of POPL'97: The 24th ACM SIGPLAN-SIGACT Symposium on Principles of Programming Languages. Papers Presented at the Symposium, Paris, France, 15–17 Jan 1997, pp. 106–119. ACM Press (1997). https://doi.org/10.1145/263699.263712
5. Oquendo, F., Buisson, J., Leroux, E., Moguérou, G.: A formal approach for architecting software-intensive systems-of-systems with guarantees. In: 13th Annual Conference on System of Systems Engineering, SoSE 2018, Paris, France, 19–22 June 2018, pp. 14–21. IEEE (2018). https://doi.org/10.1109/SYSOSE.2018.8428726
6. Buscemi, M.G., Montanari, U.: Cc-pi: a constraint language for service negotiation and composition. In: Wirsing, M., Hölzl, M.M. (eds.) Rigorous Software Engineering for Service-Oriented Systems—Results of the SENSORIA Project on Software Engineering for Service-Oriented Computing. Lecture Notes in Computer Science, vol. 6582, pp. 262–281. Springer (2011). https://doi.org/10.1007/978-3-642-20401-2_12
7. Bettini, L.: Implementing Domain-Specific Languages with Xtext and Xtend. Packt Publishing, Birmingham (2013)
8. Djeddai, S., Strecker, M., Mezghiche, M.: Integrating a formal development for DSLs into meta-modeling. J. Data Semant. **3**(3), 143–155 (2014). https://doi.org/10.1007/s13740-013-0030-4
9. Klint, P., van der Storm, T.: Model transformation with immutable data. In: Gorp, P.V., Engels, G. (eds.) Theory and Practice of Model Transformations—9th International Conference, ICMT 2016, Held as Part of STAF 2016, Vienna, Austria, 4–5 July 2016, Proceedings. Lecture Notes in Computer Science, vol. 9765, pp. 19–35. Springer (2016). https://doi.org/10.1007/978-3-319-42064-6_2
10. Meyer, E., Souquières, J.: A systematic approach to transform OMT diagrams to a B specification. In: Wing, J.M., Woodcock, J., Davies, J. (eds.) FM'99—Formal Methods, World Congress on Formal Methods in the Development of Computing Systems, Toulouse, France, 20–24 Sept 1999, Proceedings, Volume I. Lecture Notes in Computer Science, vol. 1708, pp. 875–895. Springer (1999). https://doi.org/10.1007/3-540-48119-2_48
11. Lano, K., Clark, D., Androutsopoulos, K.: UML to B: formal verification of object-oriented models. In: Boiten, E.A., Derrick, J., Smith, G. (eds.) Integrated Formal Methods, 4th International Conference, IFM 2004, Canterbury, UK, 4–7 April 2004, Proceedings. Lecture Notes in Computer Science, vol. 2999, pp. 187–206. Springer (2004). https://doi.org/10.1007/978-3-540-24756-2_11
12. Barbier, F., Cariou, E.: Inductive UML. In: Abelló, A., Bellatreche, L., Benatallah, B. (eds.) Model and Data Engineering—2nd International Conference, MEDI 2012, Poitiers, France, 3–5 Oct 2012. Proceedings. Lecture Notes in Computer Science, vol. 7602, pp. 153–161. Springer (2012). https://doi.org/10.1007/978-3-642-33609-6_15
13. Cabot, J., Clarisó, R., Riera, D.: On the verification of UML/OCL class diagrams using constraint programming. J. Syst. Softw. **93**, 1–23 (2014). https://doi.org/10.1016/j.jss.2014.03.023

Optimization of Component-Based Systems Run Time Verification

Lina Aliouat[1] and Makhlouf Aliouat[2(✉)]

[1] FEMTO-ST Institute, University Bourgogne Franche-Comt, CNRS, Montbliard, France
lina.aliouat@femto-st.fr

[2] LRSD Laboratory, University Ferhat Abbes Setif1, Sétif, Algeria
malioua@univ-setif.dz

Abstract. As technology evolves, software systems become more and more voluminous and complex. Being currently unable to produce programs free of errors, and in order to ensure that program behaviors comply with their specifications, formal verification of their essential properties is paramount. To this end, model-checking verification approach has been widely used and continues to be. However, if these properties have been verified on a system model, would they still true during any real system execution? So, modeled behavior of a system would be exactly the same in real executions when interacting with its environment? That is why verification during current system execution is essential stage even as a complementary way to other verification approaches. In this paper, we are concerned by runtime verification optimization of component-based systems in Behavior Interactions Priority (BIP) framework in order to significantly reduce time overhead. Our contribution in this paper is to consider only component states involved in the property being verified. The required states imply their associated components to be activated and those useless are disabled. Also, when a steady state is reached during monitoring process, the monitor is stopped which reduces system consumption resources. Our experiment results showed that a non negligible amount of space-time overhead was avoided.

Keywords: Runtime verification
Time and space overhead optimization
Component based systems · BIP

1 Introduction

Component-Based Systems (CBS) Approach (CBSA) has been showed to significantly enhance software-based system development by coping with design complexity. Therefore, complex systems are built from smaller components with attainable complexity using the concept of component-based which provides a family of operators for building composite components from simpler ones [1]. The CBS approach could potentially overcome many difficulties associated with

S. Chikhi et al. (Eds.): MISC 2018, LNNS 64, pp. 274–288, 2019.
https://doi.org/10.1007/978-3-030-05481-6_21

developing and maintaining monolithic software applications. In particular, the approach should result in better quality products, more rapid development, and increased capability to reuse of system subsets and accommodate change which enables efficiency and flexibility in maintaining these systems updated or complying with prospective new requirements [2].

Building systems from components is essential in any engineering discipline. Components are abstract building blocks encapsulating behavior. Their composition should be rigorously defined so that it is possible to infer the behavior of composite components from the behavior of their constituents as well as global properties from the properties of individual components. The CBS verification, we consider in this paper, is dedicated to the context of Behavior Interaction Priority (BIP) framework, a toolkit developed in VERIMAG laboratory [3]. BIP is a general framework supporting rigorous design allowing building complex systems by coordinating the behavior of a set of atomic components. The combination of interactions and priorities characterizes the overall architecture of a system and confers strong expressiveness that cannot be matched by other existing formalism.

Although CBSA offers many advantages, it is not immunized from designer faults or bugs. These faults must be detected before they can produce disasters especially in critical systems. So, it is necessary to ensure correct system behavior even if the latter is being to deviate from its initial specification. Runtime Verification (RV) is a promising method allowing checking good behaving or misbehaving of a CBS. RV is concerned with monitoring and analysis of software and hardware system executions. RV techniques are crucial for system correctness, reliability, and robustness. They are significantly more powerful and versatile than conventional testing, and more practical than exhaustive formal verification. RV can be used prior to deployment, for testing, verifying, and debugging purposes, and after deployment for ensuring reliability, safety, and security and for providing fault containment and as well as online system recovery.

In RV, a run of the system under scrutiny is analyzed incrementally using a decision procedure namely: a monitor. This monitor may be generated from a user-provided high level specification. This monitor aims to detect violation or satisfaction with regard to a given specification. RV avoids the complexity of traditional formal verification techniques, such as model checking and theorem proving, by analyzing only one or a few execution traces and by working directly with the actual system. However, observing an executing system typically incurs some runtime overhead. It is important to reduce the overhead of runtime verification tools as much as possible, particularly when the generated monitors are deployed with the system and the parts of that system under monitoring are considered as whole without possibility of targeting just what is needed.

Tuning RV process may be of great importance in order to reduce the generated overhead, knowing that one is facing the limitation of CBS verification induced by the possible state space explosion related to the large number of different components in interaction within a heterogeneous environment. Thus, a process of RV could not achieve the expected objective unless the temporal overhead incurred remains within acceptable limits. This is our main objective in this paper. The paper is organized as follows. After a brief related work, Sect. 3 gives a background of monitoring and runtime verification. Section 4 presents runtime verification in BIP framework. Section 5 is dedicated to the improvement of BIP optimization process. Section 6 concludes the paper by a conclusion and prospective future work.

2 Related Work

In [4], authors proposed a method to optimize runtime verification process by using monitor patterns and taking usage of arrays and functions which reduce the amount of monitor codes of the targeted system. This may enhance the employ of RV methods about scarce resource and safety critical systems. The improvement concerns lessening the impact on the application programs and speeding up the process by which monitors being generated from properties being verified. The authors in [5] proposed a sampling based program monitoring approach where predictable monitors are added to observe the behavior of existing program. They used a novel time-triggered approach, where the monitor frequently takes samples from the system in order to analyze the system health instead of using the commonly used event trigged approach. The new approach significantly reduces the overhead incurred in the system. In [6], the authors proposed an approach to control the overhead of software monitoring using control theory for discrete event systems. In their paper, overhead control is achieved by means of temporarily disabling involved monitor which contribute to avoid the overhead to exceed a threshold defined for user. In [3], the authors included RV in BIP framework by means of monitors checking specifications at runtime in order to get a consistent verdict about properties being verified. This valuable attempt did not take into account the overhead in time-space generated by needless runtime checks performed by monitors and issued from useless connectors.

3 Monitoring and Runtime Verification

Continuously monitoring of a running system is a complementary approach to increase the assurance of correct system execution. When running a program, only executions can be observed while in verification process, we concentrate attention to determine whether a run of a system respects given correctness properties. Executions are the primary object analyzed in the setting of Runtime Verification. Checking whether an execution meets a correctness property is realized using a monitor. A monitor decides whether the current system execution satisfies a given correctness property by giving a verdict of type either yes/true or no/false. According to frequent use, verification techniques such as theorem proving, model checking, and testing [2] have dominated the formal verification area in specialized literature. However, each of them exhibits drawbacks in their utilization. Therefore, model checking techniques suffer from state explosion which limits the size of systems that can be verified. Moreover, model checkers operate on models and thus introduce additional proof obligations on the correctness of abstraction or model creation. As far as, verification based theorem provers usually involves significant amount of manual effort that limits seriously the size of the system amenable to be verified. Testing covers a wide field of diverse, often ad hoc, and incomplete methods for proving correctness, or, more precisely, for detecting bugs. Runtime verification focusing on the current execution of a system, allows getting automatic analysis which is less dependent on the size of the system and, at the same time, does not require as much abstraction. The fact of being in action at runtime confers RV the possibility to intervene whenever incorrect behavior of a software system is detected. This enables RV to be eligible as a good mechanism of fault detection and recovery for software systems [3]. The real advantage of RV is the ease of checking an actual system execution to ensure that the implementation effectively meets its correctness properties. Moreover, there are many situations where some information items are only available at runtime or are suitably being checked at this point of time, thus, in such cases, runtime verification is an alternative to others verification methods. The behavior of a system may greatly depend on the environment of its real execution, but the environment of its experiment when using model checking or theorem proving is different. Then it is not possible to obtain the necessary information to test the system in an adequate manner. Furthermore, formal correctness proofs by model checking or theorem proving may only be achievable by taking certain assumptions on the behavior of the environment which should be checked at runtime. In this case, runtime verification outperforms classical testing and adds on formal correctness proofs by model checking and theorem proving. In the case of safety-critical systems, as a fault may lead to a catastrophe, it would be necessary to monitor properties that have been already statically proved or tested in order to put more confidence in the system under verification. Once more, runtime verification acts as a partner of theorem proving, model checking, and testing. The behavior of highly dynamic systems [7] depends heavily on the environment and changes over time, which makes their evolving hard to predict and hard to analyze prior to execution. To assure certain correctness properties of such systems, it is expected that runtime verification will become a major verification technique, and runtime verification based component to be part of the architecture of such dynamic systems. Ultimately, what especially characterizes runtime verification from other methods of verification is its online monitoring and analysis carried out during program execution which can lead to the capabilities of immediately responding to specified correctness properties violation. This way of treating system behavior deviations lacks in other verification techniques. Since RV is a technique mainly focusing on event-triggered solutions with monitor as essential active part, most of time and space overhead is generated by frequent invocations of that monitor. Indeed, every change in system state, the monitor is triggered to verify system properties. Besides the difficulty of accessing a global system state (Fully distributed system), the major disadvantages are: overuse of system resources for performing checking (overhead) and the lack of predictability.

4 Runtime Verification in Framework BIP

The approach of building complex systems from basic components is essential in software engineering domain. More precisely, the methodology from which a complete system is generated from a set of predefined components and associated requirements is called component-based design. According to this design principle, a composite component can be built from basic components which are considered in BIP as abstract building blocks including behavior. The components composition should be rigorously defined so that it is possible to infer the behavior of composite components from the behavior of their constituents as well as global properties from the properties of individual components. This design principle is faced with difficulty of verifying system properties using only static verification techniques. This obstacle is due to the fact that certain proprieties can only be verified at system execution when some information is available (this is not allowed in static verification methods). The state-explosion problem, created by a huge number of interacting components, is also a handicap for verifying system properties.

These limitations impose then runtime verification as a complementary verification technique which offers ability of dynamic analysis and scalability.

Runtime Verification [8](monitoring) is an effective technique to ensure, at runtime, that a system respects a desirable behavior. It can be used in numerous application domains, more particularly when integrating together unreliable software components.

In monitoring [9], a run of the system under scrutiny is analyzed incrementally using a decision procedure named: a monitor. A monitor may be generated from a user-provided high-level specification (e.g., a temporal formula, an automaton) and is dedicated to provide a verdict represented by a violation or satisfaction of a given specification.

Monitoring [10] has several advantages when compared with static validation techniques. Compared with static analysis, monitoring allows to check more expressive behavioral specifications. Moreover, monitoring does not rely on abstracting or over-approximating the state space, and thus does not produce false positives. Compared with model-checking, monitoring is less sensitive to the state-explosion problem which is rapidly occurring when composing the behavior of several components. Compared with compositional verification techniques, monitoring remains applicable for BIP component-based systems (where external functions can be called).

4.1 Runtime Verification Functions

Runtime Verification is a dynamic analysis method aiming at checking whether a run of the system under scrutiny satisfies a given correctness property. The inputs to an RV system are: (1) a system to be checked, and (2) a set of properties to be checked against the system execution. The properties can be expressed in a formal specification language (e.g., a temporal formula or automata-based formalism), or even as a program in a general-purpose programming language. A runtime verification process typically consists of the following three stages. First, from a property is generated a monitor, i.e., a decision procedure for the property. This step is often referred as to monitor synthesis. The monitor is capable of consuming events produced by a running system and emits verdicts according to the current satisfaction of the property based on the history of received events. Second, the system under scrutiny is instrumented. The purpose of this stage is to be able to generate the relevant events to be fed to the monitor. This step is often referred as to system instrumentation. Third, the systems execution is analyzed by the monitor. This analysis can occur either during the execution in a lock-step manner, or after the execution has finished assuming that events have been written to a log. This step is often referred as to execution analysis.

Runtime Verification Phases Before giving formal definitions of runtime verification concepts, we give a brief overview of the stages involved in monitoring a system, captured in Fig. 1:

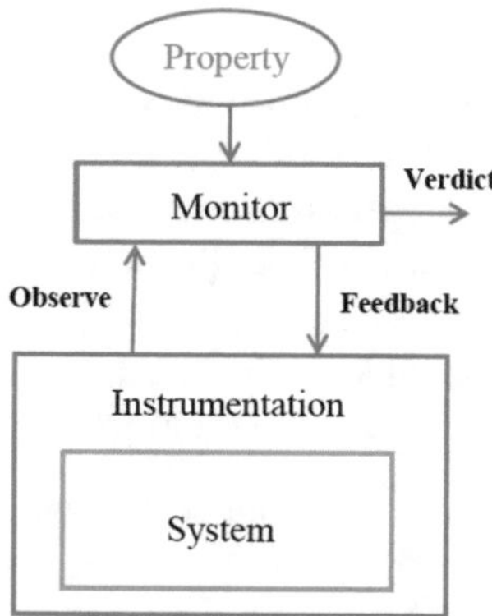

Fig. 1. An overview of the runtime verification process

1. Monitor creation: A monitor is created, potentially from a formal property.
2. Instrumentation: The system is instrumented to generate events for the monitor.
3. Execution: The system is executed, generating events for the monitor.
4. Responses:

- The monitor produces for each consumed event a verdict indicating the status of the property depending on the event sequence seen so far.
- The monitor sends feedback to the system - This may give further information to the system, so that more specific corrective actions can be taken.

Events, Traces and Properties The behaviour of a system for a given execution can be captured as a finite sequence of events describing selected actions taken by the system, or selected states of the system. In general, an event consists of a name and sequence of data values.

Different runtime verification systems use different formalisms or logics, and thus can express different sets of properties. We say that one system is more expressive than another if the properties the first can express properly contain those the second can express.

Verdicts and Feedback One advantage of verifying a system at runtime is that the system can take corrective action if a property is violated, using the results of verification to steer itself towards more desirable behaviors. To achieve this, monitors communicate with the system through verdicts and feedback. Verdicts give the status of the monitored system with respect to a property and feedback provides additional information to the monitored system.

4.2 Verification Monitors

A monitor is a procedure that consumes events fed by a BIP system and produces an appraisal on the sequence of events read so far. We follow a general approach in which verification monitors are deterministic finite-state machines that produce a sequence of truth-values (a sequence of verdicts) in an expressive 4-valued truth-domain $\mathbb{B}_4 \stackrel{\text{def}}{=} \{\bot, \bot_c, \top_c, \top\}$, as introduced in [8] and used in [11]. $\mathbb{B}_4$ consists of the possible evaluations of a sequence of events and its possible futures relatively to the specification used to generate the monitor:

- The truth-value $\top_c$ (resp. $\bot_c$) denotes "currently true" (resp. "currently false") and expresses the satisfaction (resp. violation) of the specification "if the system execution stops here".
- The truth-value $\top$ (resp. $\bot$) is a definitive verdict denoting the satisfaction (resp. violation) of the specification: the monitor can be stopped.

The notion of monitor defined relatively to a set of events Σ expressed on a composite component. Monitors are deterministic Moore (finite-state) machines emitting a verdict on each state.

Definition 1 (Monitor). A *monitor* $\mathcal{A}$ is a tuple $(\Theta^{\mathcal{A}}, \theta^{\mathcal{A}}_{\text{init}}, \Sigma, \longrightarrow_{\mathcal{A}}, \mathbb{B}_4, ver^{\mathcal{A}})$. The finite set $\Theta^{\mathcal{A}}$ denotes the control states and $\theta^{\mathcal{A}}_{\text{init}} \in \Theta^{\mathcal{A}}$ is the initial state. The complete function $\longrightarrow_{\mathcal{A}}: \Theta^{\mathcal{A}} \times \Sigma \rightarrow \Theta^{\mathcal{A}}$ is the transition function. In the following we abbreviate $\longrightarrow_{\mathcal{A}} (\theta, a) = \theta'$ by $\theta \xrightarrow{a}_{\mathcal{A}} \theta'$. The function $ver^{\mathcal{A}} : \Theta^{\mathcal{A}} \rightarrow \mathbb{B}_4$ is an output function, producing verdicts (i.e., truth-values) in $\mathbb{B}_4$ from control states.

Such monitors are independent from any specification formalism used to generate them and are able to check any specification expressing a linear temporal specification [11]. Intuitively, runtime verification of a specification with such monitors works as follows. An execution sequence is processed in a lock-step manner. On each received event, the monitor produces an appraisal on the sequence read so far. For a formal presentation of the semantics of the monitor and a formal definition of sequence checking, we refer to [11].

In the remainder, we consider a monitor $\mathcal{A} = (\Theta^{\mathcal{A}}, \theta^{\mathcal{A}}_{\text{init}}, \Sigma, \longrightarrow_{\mathcal{A}}, \mathbb{B}_4, ver^{\mathcal{A}})$.

4.3 Example

A task system, called Task, illustrated in this section, as example, will be used in our approach in Sect. 5. The system consists of a task generator (Generator) along with 3 task executors (Workers). Each newly generated task is handled whenever two cooperating workers are available. An atomic component is endowed with a finite set of local variables X taking values in a domain Data. Atomic components synchronize and exchange data with other components through ports.

Figure 2 shows the atomic components of system Task.

Figure 2a depicts a model of component *Generator* defined as $Generator.P = \{deliver[\emptyset], newtask[\emptyset]\}$, $Generator.L = \{ready\ ,\ delivered\}$, $Generator.T = \{(ready, deliver,\ true, [\ ],\ delivered),\ (delivered\ , newtask, true,\ [\ ], ready)\}$, $Generator.X = \emptyset$.

Figure 2b depicts a model of worker. Component *Worker* is defined as $Worker.P = \{exec[\emptyset],\ finish[\emptyset], reset[\emptyset]\}$, $Worker.L = \{free,\ done\}$, $Worker.T = \{(free,\ exec, true, [x := x + 1], done),\ (done, finish, (x \leqslant 10), [\ ], free),\ (done, reset, (x > 10), [x := 0], free)\}$, $Worker.X = \{x\}$.

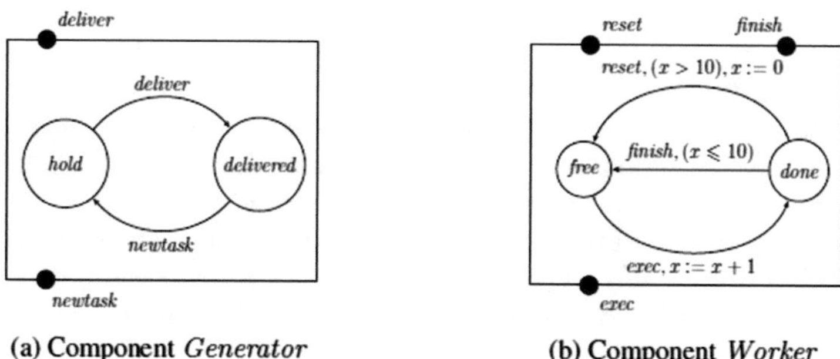

Fig. 2. Atomic components of system task

Figure 3 depicts the composite component $\gamma(Worker_1, Worker_2, Worker_3)$ of system Task, where each $Worker_i$ is identical to the component in Fig. 2(b) and *Generator* is the component depicted in Fig. 2a. The set of interactions is $\gamma = \{ex_{12}, ex_{13}, ex_{23}, r_1, r_2, r_3, f_1, f_2, f_3, n_t\}$. We have $ex_{12} = (\{deliver, exec_1, exec_2\}, [\ ])$, $ex_{23} = (\{deliver, exec_2, exec_3\}, [\ ])$, $ex_{13} = (\{deliver, exec_1, exec_3\}, [\ ])$, $r_1 = (\{reset_1\}, [\ ])$, $r_2 = (\{reset_2\}, [\ ])$, $r_3 = (\{reset_3\}, [\ ])$,

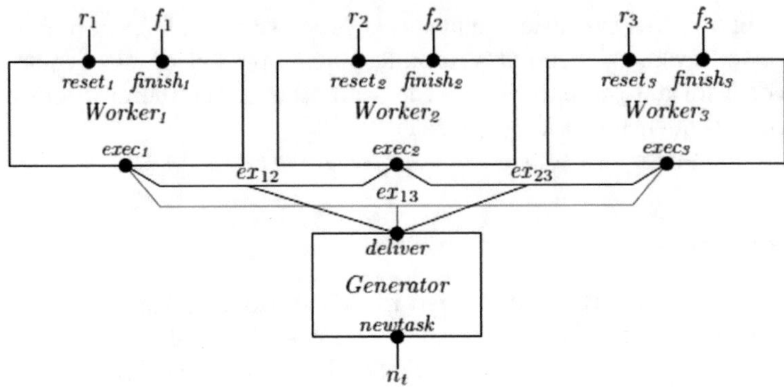

Fig. 3. Composite component of system task

$f_1 = (\{finish_1\}, [\,])$, $f_2 = (\{finish_2\}, [\,])$, $f_3 = (\{finish_3\}, [\,])$, and $n_t = (\{newtask\}, [\,])$. One of the possible traces of system Task is: $(free, free, free, hold) \cdot \mathtt{ex}_{12} \cdot (done, done, free, delivered) \cdot n_t \cdot (done, done, free, hold)$ such that from the initial state $(free, free, free, hold)$, where workers are at location *free* and task generator is ready to deliver a task, interaction ex_{12} is fired and $worker_1$ and $worker_2$ move to location *done* and *Generator* moves to location *delivered*. Then, a new task is generated by the execution of interaction n_t so that *Generator* moves to location *hold*.

5 Optimization of Runtime Verification in BIP

In the precedent work, the process of runtime verification in BIP framework includes all components of a system under monitoring. That is, the part of components related to property correctness being verified and those are not involved in. This situation of no components distinction is made leads to incur an overhead. This overhead may be important particularly when the number of components under monitoring and analysis is important. Our main objective is then to make engaged only the components just needed by the property to verify without impacting other components of the system. To this end, we follow two arguments:

1. From the monitor analysis, to verify a property we only need a few component states but. Thus, we have to disable unused components and activate the required ones at every property station.
2. If the property becomes false at a certain point in time, it will always remain false, this means that a stable system state is reached. So, the monitor stops checking the property up to the end of system execution (Monitoring definition). The monitor stops working and the system will proceed.

The two previous proposals can improve the performance of the system by reducing the overhead generated by the useless monitoring activity. To do this, we are going to make changes at the level of components and the monitor, changes which would result in appropriate transformations.

5.1 Atomic Components Transformations

Let $B^{\perp} = (P^{\perp}, L^{\perp}, T^{\perp}, X^{\perp})$ be a component at runtime verification with initial location $l_0 \in L$ (According to atomic component instrumentation).

We define a new component $B^o = (P^o, L^o, T^o, X^o)$ where:

- $X^o = X^{\perp} \cup \{d\}$; is set of variables where d is Boolean variable denoting the path being followed by the component, if d = false. the monitor will disable this component and it takes the initial BIP path, otherwise, it takes the BIP runtime verification path.
- $P^o = P^{\perp} \cup \{act_i, dact_i\}$; Set of ports where act_i and $deact_i$ are ports through which the component is activated or deactivated
- $L^o = L \cup L^{\perp}$; Set of locations, i.e., the union of the BIP initial locations (when the component is deactivated or disabled) and the locations of RV BIP (When the component is activated).
- We have $T^{\perp} = T_1 \cup T_2$ Where $T_1 = \{a \mid (l, p, g_{\tau}, [\ ], l_{\tau}^{\perp})\}$, $T_2 = \{b \mid (l_{\tau}^{\perp}, \beta, true, f_{\tau}, l')\}$
  - $\forall a \in T^{\perp}$, we transform a to a' such that: $a' = (l, p, g_{\tau}; [d = true], [\ ], l_{\tau}^{\perp}) \in T^o$
  - $\forall a', b \in T^o$, where $a' = (l, p, g_{\tau}; [d = true], [\ ], l_{\tau}^{\perp}) \in T^o$, $b = (l_{\tau}^{\perp}, \beta, true, f_{\tau}, l')\}$ we add a new transition c such that $c = (l, p, g_{\tau}; [d = false], f_{\tau}, l')$
  - $\forall l \in L^o$ we add two new transitions $d_l, a_l \in T^o$ such that: $d_l = (l, deact, [\,], [d = false], l)$ and $a_l = (l, act, [\,], [d = true], l)$

Thus $T^o = \tau_1 \cup \tau_2 \cup \tau_3 \cup \tau_4$ where:
$\tau_1 = \{a' \mid (l, p, g_{\tau}; [d = true], [\,], l_{\tau}^{\perp})\}$, $\tau_2 = \{b \mid (l_{\tau}^{\perp}, \beta, true, f_{\tau}, l')\}$, $\tau_3 = \{c \mid (l, p, g_{\tau}; [d = false], f_{\tau}, l')\}$, $\tau_4 = \{a_l \mid (l, act, [\ ], [d = true], l)\}$, $\tau_5 = \{d_l \mid (l, deact, [\,], [d = false], l)\}$

Example Figure 4 presents the new version of components according to BIP optimization in task system (depicted in Fig. 2)

- Figure 4a represents component task generator, where $\texttt{Generator}^o.P^o = \{\texttt{deliver}[\emptyset], \texttt{newtask}[\emptyset], \beta[\{loc\}]\}$, $\texttt{Generator}^o.T^o = \{(\texttt{hold}, \texttt{deliver}, \texttt{true}, [\ \ ], \perp), (\perp, \beta, \texttt{true}, [loc := \texttt{delivered}], \texttt{delivered}), (\texttt{delivered}, \texttt{newtask}, \texttt{true}, [\ \ ], \perp), (\perp, \beta, \texttt{true}, [loc := \texttt{hold}], \texttt{hold})\}$, $\texttt{Generator}^o.X^o = \{loc\}$.
- Figure 4b depicts a worker component, where $\texttt{Worker}^o.P^o = \{\texttt{exec}[\emptyset], \texttt{finish}[\emptyset], \texttt{reset}[\emptyset], \beta[\{x, loc, d\}], act[\{d\}], dact[\{d\}]\}$, $\texttt{Worker}^o.T^o = \{ (\texttt{free}, \texttt{exec}, (d = true), [\,], \perp), (\perp, \beta, \texttt{true}, [x := x+1; loc := \texttt{done}], \texttt{done}), (\texttt{done}, \texttt{finish}, (x \leqslant 10 \wedge d = true), [\,], \perp), (\perp, \beta, \texttt{true}, [loc := \texttt{free}], \texttt{free}), (\texttt{done}, \texttt{reset}, (x > 10 \wedge d = true), [\,], \perp), (\perp, \beta, \texttt{true}, [x := 0; loc := \texttt{free}], \texttt{free}), (\texttt{free}, \texttt{exec}, (d = false), [x := x + 1; loc := \texttt{done}], \texttt{done}),$

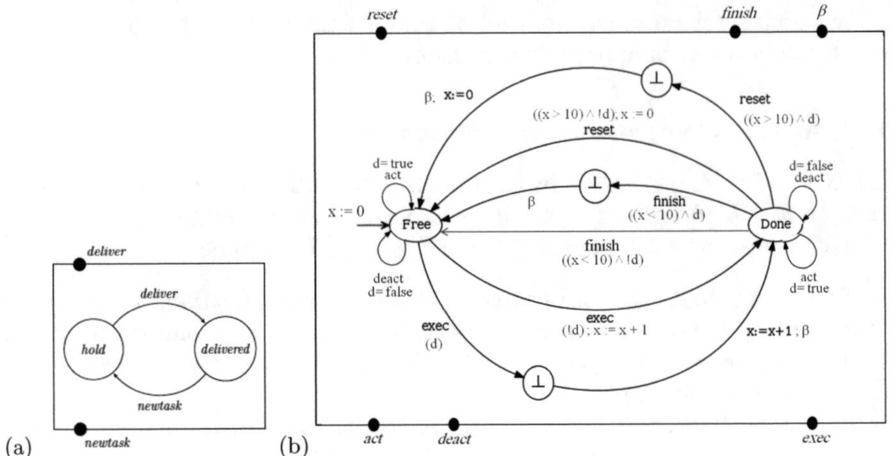

Fig. 4. Atomic component in RV optimization. **a** Generator and **b** worker.

(done, reset, $(x > 10 \wedge d = false)$, $[x := 0;\ loc :=$ free], free), (done, finish, $(x \leqslant 10 \wedge d = false)$, $[loc :=$ free], free), (free, act, true, $[d = true;\ loc :=$ free], free), (free, $dact$, true, $[d = false;\ loc :=$ free], free), (done, act, true, $[d = true;\ loc :=$ done], done), (done, $dact$, true, $[d = false;\ loc :=$ done], done)}, $\texttt{Worker}^o.X^o = \{x, loc, d\}$.

5.2 Transformed Monitor

Transformation of BIP Monitor into Optimized BIP Monitor. We define a new monitor with new constraints in such a way we can activate or deactivate components if needed and at every station (location) within a property. The monitor will be deactivated when a stable state is encountered because no change will occur after that state.

Definition (Monitor Atom): Let $M^A = (P, L, T, X)$ be a BIP monitor; we define a new Monitor component such that: $M^o = (P^o, L^o, T^o, X^o)$ where (P^o, L^o, T^o, X^o) are defined as follow:

- $X^o = X$: Set of variables defined in X
- $P^o = P \cup \{act_i, dact_i \mid i \in [1, n]\}$: Set of ports of BIP monitor. Furthermore, we add ports act_i and $deact_i$ after every monitor station according its needs.
- $L^o = L \cup \{Ld(l_z) \cup La(l_z) \mid z \in [1, n]\}$: Set of locations of BIP monitor plus set of locations of deactivated components Ld, where Ld is initially empty.
  - $Ld(l) = \emptyset$
  - If there exists $l_i \longrightarrow_A l_z, j \in \{ind_i - (ind_i \cap ind_z)\} : Ld(l_z)^i = Ld(l_z)^{i-1}.l_{deact_{zj}}$

  If there exists a transition between l_i and l_z then we have for every index j belonging to the sequence of indexes of deactivated components, a location of that index ldeactj, thus we add this location to the sequence Ld of the deactivated components.

And set of locations of activated components La, where La is initially empty.

- $La(l) = \emptyset$
- If there exists $l_i \longrightarrow_A l_z, j \in \{ind_z - (ind_i \cap ind_z)\} : La(l_z)^i = La(l_z)^{i-1}.l_{act_{zj}}$

If there exists a transition between l_z and l_i , then we have for every index j belonging to the sequence of indexes of activated components $\{ind_z - (ind_i \cap ind_z)\}$, a location $lact_j$ of that index, we add this location to the sequence of activated components.

- $\forall (l, e, l') \in T$, there exists T^o for which 7 cases of transitions are possible:
  1. Case where we have a single component to deactivate and no components to activate: length $(Ld(l_z)) = 1$ and length$(La(l_z)) = 0$ $\{(l, deact_{Ld(l)_1}, l_{deact_{Ld(l)_1}}).(l_{deact_{Ld(l)_1}}, e, l')\}$
  2. Case where several components have to be deactivated and no component to activate: length $(Ld(l_z)) > 1$ and length$(La(l_z)) = 0$ $\{(l, deact_{Ld(l)_1}, l_{deact_{Ld(l)_1}})...(l_{deact_{Ld(l)_{k-1}}}, deact_{Ld(l)_k}, l_{deact_{Ld(l)_k}})$ $.(l_{deact_{Ld(l)_k}}, e, l')\}$
  3. We have one component to activate et no components to deactivate: length $(Ld(l_z)) = 0$ and length$(La(l_z)) = 1$ $\{(l, act_{La(l)_1}, l_{act_{La(l)_1}}).(l_{act_{La(l)_1}}, e, l')\}$
  4. We have no component to deactivate and many components to activate: length $(Ld(l_z)) = 0$ and length$(La(l_z)) > 1$ $\{(l, act_{La(l)_1}, l_{act_{La(l)_1}})...(l_{act_{La(l)_{k-1}}}, act_{La(l)_k}, l_{act_{La(l)_k}}).(l_{act_{La(l)_k}}, e, l')\}$
  5. There is only one component to deactivate and only one component to activate: length $(Ld(l_z)) = 1$ and length$(La(l_z)) = 1$ $\{(l, deact_{Ld(l)_1}, l_{deact_{Ld(l)_1}}).(l_{deact_{Ld(l)_1}}, act_{La(l)_1}, l_{act_{La(l)_1}}).(l_{act_{La(l)_1}}, e, l')\}$
  6. There are many components to deactivate and many components to activate: length $(Ld(l_z)) > 1$ and length$(La(l_z)) > 1$ $\{(l, deact_{Ld(l)_1}, l_{deact_{Ld(l)_1}})...(l_{deact_{Ld(l)_{k-1}}}, deact_{Ld(l)_k}, l_{deact_{Ld(l)_k}}).$ $.(l_{deact_{Ld(l)_k}}, act_{La(l)_1}, l_{act_{La(l)_1}})...(l_{act_{La(l)_{k-1}}}, act_{La(l)_k}, l_{act_{La(l)_k}})$ $.(l_{act_{La(l)_k}}, e, l')\}$
  7. There is no components to deactivate and activate: length $(Ld(l_z)) = 0$ and length$(La(l_z)) = 0$ $\{(l, e, l')\}$

Example of transformed monitor For tasks system, we have the following property (Fig. 5).

For each station in the monitor there is a need for a number of specific components: At the station S1, we need for comp1 and comp2, so we have to activate the two components. After that, at the state S2, we activate the component 3 and deactivate the component 1 as indicated in Table 1. In the state S4, we stop the monitor because we reach the state bad of the property; we then deactivate all the components. In order to do all that, the monitor will be as described in Fig. 6.

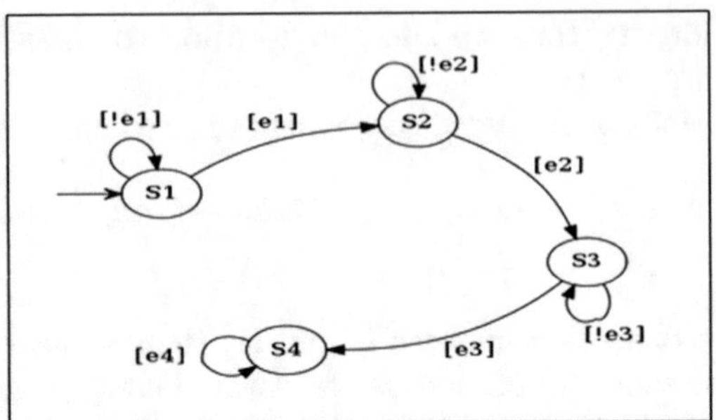

Fig. 5. Property.

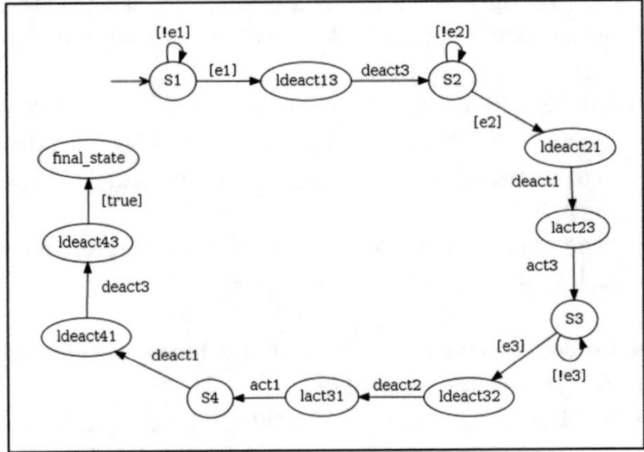

Fig. 6. Property of OP BIP.

Table 1. Needed components

Monitor state	Involved components
S1	comp1, comp2
S2	comp2, comp3
S3	comp1, comp3
S4	–

5.3 Experimental Result

Following the work flow presented in Sect. 4, for each system and its properties, we modified its components and BIP monitor to obtain an optimized RV BIP. In Table 2, the columns have the following meanings:

- Column interaction: shows the number of functional step of system
- Column event: the number of the snapshot of the system send to the monitor
- Column time: shows the execution time of the system

Initially, we tested our example of Task using Runtime Verification BIP without any optimization. The execution time of the initial monitored system is 13.1 s with the overhead of 13% and the number of executed interactions is 50,018. Then we tested our example using optimized RV BIP level 1 where components are activated or deactivated by the monitor depending of each state of the property. The execution time of the monitored system is 12.94 s with the overhead of 11.4% and the number of executed interactions is 50,330.

Table 2. Result table

		Interactions	Events	Time (s)	Extra interactions	Over-head(%)
	Initial	39,999	–	11,800	–	–
Q1	RV_BIP	50,018	10,019	13,100	10,019	13
	OP_BIP 1	50,330	10,018	12,940	10,331	11.4
	OP_BIP 2	45,770	4083	12,337	5771	5.37

Our optimization technique introduced in optimized RV BIP level 1 guarantees 2% reduction in the overhead, and the number of executed interactions is bigger than without optimization because activating and deactivating components.

Afterwards we tested the example using optimized RV BIP level 2 where in addition to the optimization level 1, the monitor will be stopped if the property takes a stable station. The execution time of the monitored system is 12.33 s with the overhead of 5.37% and the number of executed interactions is 45770.

Our optimization technique introduced in optimized RV BIP level 1 caused 8% reduction in the overhead, and the number of executed interactions is less by 4248 interactions. As expected the induced overhead in optimization level 1 is higher than the optimization level 2. Overall, the overhead reduction is half from the Runtime verification BIP (for this specific example).

6 Conclusion

Developing framework and toolset for meaningfully enhancing CBS development in such a way as to master software complexity by rapidly producing quality engineering products and increasing capability to accommodate change is the main aim of BIP framework. However, as complexity increases, it is hard to ensure system correctness. So, verification formal methods have been designed and used to verify the main system properties. Nevertheless, these classical formal methods while intensively actually used have some drawbacks and need to be supplemented by runtime verification technique. The primary advantage provided by the latter is its action and reaction to properties violation during system execution. However, this advantage can be countered by the cost of its use except if measures of optimization are used. The intended cost is mainly represented by time and space overhead to be incurred. The paper contribution in this context is to improve the existing Runtime Verification in BIP (RV BIP) framework by an optimization process. Indeed, in the actual BIP runtime verification, verifying a system property involves the use of all component states by the monitoring process. Our contribution is to include only those components directly related to the property being verified. Consequently, the time and space overhead incurred has been significantly reduced. Another improvement is when a steady state is reached during the monitoring process, the monitor is stopped, which has not been the case previously in RV BIP. Thus, unnecessary consumption of system resources is lessened as well. As prospective future work, we envision to tackle the hard issue of runtime verification in a context of decentralized monitors evolving in a distributed context.

References

1. Basu, A., Bensalem, S., Bozga, M., Combaz, J., Jaber, M., Nguyen, T.H., Sifakis, J.: Rigorous component-based system design using the bip framework. IEEE Softw. **28**, 41–48 (2011)
2. Leucker, M., Schallhart, C.: A brief account of runtime verification. J. Logic Algebr. Programm. **78**, 293–303 (2009)
3. Falcone, Y., Jaber, M., Nguyen, T.H., Bozga, M., Bensalem, S.: Runtime verification of component-based systems in the bip framework with formally-proved sound and complete instrumentation. Softw. Syst. Model. **14**, 173–199 (2015)
4. Zhou, G., Dong, W., Liu, W., Shi, H., Hu, C., Yin, L.: Optimizing monitor code based on patterns in runtime verification. In: 2017 IEEE International Conference on Software Quality, Reliability and Security Companion (QRS-C), pp. 348–354. IEEE (2017)
5. Bonakdarpour, B., Navabpour, S., Fischmeister, S.: Sampling-based runtime verification. In: International Symposium on Formal Methods, pp. 88–102. Springer (2011)
6. Huang, X., Seyster, J., Callanan, S., Dixit, K., Grosu, R., Smolka, S.A., Stoller, S.D., Zadok, E.: Software monitoring with controllable overhead. Int. J. Softw. Tools Technol. Transf. **14**, 327–347 (2012)
7. Nazarpour, H., Falcone, Y., Bensalem, S., Bozga, M., Combaz, J.: Monitoring multi-threaded component-based systems. Technical Report TR-2015-5, Verimag Research Report (2015)
8. Bauer, A., Leucker, M., Schallhart, C.: Comparing LTL semantics for runtime verification. J. Logic Comput. **20**, 651–674 (2010)
9. Falcone, Y., Fernandez, J.C., Mounier, L.: What Can You Verify and Enforce at Runtime?, vol. 14, pp. 349–382. Springer, Berlin (2012)
10. Bauer, A., Leucker, M., Schallhart, C.: The Good, the Bad, and the Ugly, But How Ugly is Ugly?, pp. 126–138. Springer, Berlin (2007)
11. Falcone, Y., Fernandez, J.C., Mounier, L.: Runtime verification of safety-progress properties. International Workshop on Runtime Verification, pp. 40–59. Springer, Berlin (2009)

Meta-ECATNets for Modelling and Analyzing Clinical Pathways

Abdelkader Moudjari[1(✉)], Fateh Latreche[2], and Hichem Talbi[1]

[1] MISC Laboratory Constantine 2 University, Constantine, Algeria
moudjariabdelkader@gmail.com, hichem.talbi@univ-constatntine2.dz
[2] LIRE Laboratory Constantine 2 University, Constantine, Algeria
fateh.latreche@univ-constatntine2.dz

Abstract. Recently, the paradigm of clinical pathway has taken more attention in the field of healthcare system. The clinical pathway is complex, dynamic and flexible. The use of basic Petri nets for modelling clinical pathways generates a complex and extremely large nets (Chincholkar and Chetty in Int J Advan Manuf Technol 12(5):339–348 (1996), [1]). Meta-ECATNets are a kind of high-level Petri nets with two levels, in which meta places control elements of the lower level. In this paper, Meta-ECATNets are used to model the clinical pathway of the Chronic Obstructive Pulmonary Disease (CODP), the correctness of this process is done by TCTL model-checker of Real-Time Maude.

Keywords: Clinical pathway · Meta-ECATNets · Flexibility Dynamic

1 Introduction

In the last two decades, the workflow paradigm occupied an important place inside the enterprise world. Therefore, the way of achieving goals is completely evolved. By its definition, the workflow aims to automate tasks in order to gain time and reduce cost. The hospital is considered as an enterprise and it engaged to adopt the workflow paradigm to cure patients. However, the nature of patient care is more complex and involves many people and resources at the same time. In addition, the patient's health state may worsen during treatment and then requires a particular procedure.

The clinical processes are defined by the World Health Organization (WHO) and are adopted by governments and healthcare ministries among world. In general, the processes are applied independently. They may or not use computer tools. Hospitals have implemented clinical processes using workflow management systems in order to give a better treatment for patients and to preserve money and effort. However, problems may occur during the process implementation and cause disasters in the clinical organisation. Thus, before deploying the clinical

Supported by organization x.

S. Chikhi et al. (Eds.): MISC 2018, LNNS 64, pp. 289–298, 2019.
https://doi.org/10.1007/978-3-030-05481-6_22

processes some properties should be verified such as: a clinical process should always terminate correctly, never administer medicines when contraindicated [2].

To resolve this problem, formal methods have been used because they provide a solid mathematical foundation and rigorous checking techniques. Some works focus on modelling the clinical process and checking its properties. However, despite the deployed effort, many problems remain without solution. Among those problems, one could cite the problem of flexibility control that allows modification of pre-established healthcare processes.

In the field of modelling and analysis of healthcare processes, several research works have been done. In [2], authors have proposed the use of recursive and algebraic Petri nets to consider exceptions and deviation that may occur, but there is no clear separation between control and functional levels.

To make clinical pathway more controllable, we propose Meta-ECATNets [3] to model and analyse the clinical pathway of the COPD. A Meta-ECATNet is a layered Petri net model for flexible systems that takes time and data type aspects into account. In the present work, meta transitions of the higher level control transitions that represent treatment tasks. In addition, time is associated with lower level transitions in the form of firing duration. To execute and check Meta-ECATNet COPD models, we use the Real-time Maude system and its TCTL model checker.

This paper is organised as follow. First, we theoretically analyse some related works and give their limits. Then, we give basic concepts allowing a good comprehension of the proposed model. In the third section, the proposed approach is detailed. To show the usefulness of our approach, we apply it on COPD clinical pathway. Finally, a conclusion and some future works are given.

2 Related Work

Several works tackling challenging aspects of clinical pathways have been proposed. However, few of them use formal methods. In this section, we present a review of some important works dealing with healthcare processes assessment while adopting the Petri net formalism.

Authors of [2] adopt the Recursive and algebraic Workflow Nets (*Rec WF-Nets*) to tackle exceptions and deviations occurring during medical treatment process. At the modelling step, two types of transition: *elementary* and *abstract* are used to model medical tasks. Firing of an abstract transition generates a new plan of actions in the workflow process (the lower level), this plan terminates when it reaches the terminated state or when the plan is interrupted by occurring of an exception of the higher level. At the analysis step of the health care processes model, authors have simulated and verified *Rec WF-Nets* models using the Maude system and its LTL model checker.

This work is promising, it shares with our work the use of rewriting logic and the Maude system as checking and execution platform. However no clear separation between the control and functional levels is done, also, the time aspect is not treated.

In [4] authors have proposed a formal method based time recursive ECATNets for modelling and analyzing home care plans. Authors believe that time recursive ECATNets are well-suited for describing duration and delays of activities and cancellation delays on sub-processes care plans. Furthermore, the model transformation approach is applied to generate the equivalent Maude representation. At the checking step, the Maude TCTL model checker is used to evaluate some temporal properties.

This formal approach is useful, it is another way to control functioning of care plans. But it merges the basic treatment actions with abstract transitions that involve repetitive activities.

Authors in [5] proposed a method based on *Modular Temporised Coloured Petri Net* with changeable structure. This work adopted the timed arc expression functions to specify the time delay. In addition, the clinical pathway workflow model can be dynamically updated and deal with deviations that occur during the patient care process. Two kinds of mechanism are used to carrying out structure change, by modification and composition. In this work, authors have used a modular formalism that considers both time aspect, and dynamic evolution of patient state. But, like the two previous works, no clear separation between the two levels control and functional is done. Besides, it used the concept of port, which does not belong to Petri net formalism.

In [6], authors have proposed a translation of Little-JIL [7] healthcare workflow model into Component Timed-Arc Petri nets (CTAPNs), a component-based version of Petri nets with timing information attached to tokens. The translation aims at tackling the aspects : time, modular (composition) modelling and formal verification. The model-checker TAPAAL [8] has been used to verify some properties for the blood transfusion case study. The main drawback of this work is that the translation to CTAPNs model was performed manually, also untimed TCTL formulae are used to express properties.

3 Basic Concepts

ECATNets are high level algebraic Petri nets initially proposed by [9]. They combine both expression power of Petri nets and abstract data types. The use of *ECATNets* provides a highly condensed and concurrent model, with a natural formal and semantics definition in terms of rewriting logic.

Definition 1 (ECATNet). An *ECATNet* is a high level Petri net having the structure $(P, T, sort, IC, DT, CT, TC, M_i, M_f)$, where

- P is a set of places having a well-defined sort;
- T is the set of transitions, with $P \cap T = \varnothing$
- $sort : P \rightarrow S$ is a function that associates to each place an algebraic sort s belonging to Σ;
- IC (*Input Condition*) : $P \times T \rightarrow MT_{\Sigma/EUA}(X)$, is a function that specifies partial conditions on input place markings;

- DT (*Destroyed Tokens*) : $P \times T \rightarrow MT_{\Sigma/EUA}(X)$, is a function that associates to each input arc $(p \times t)$ of a transition t, a multi-set of algebraic terms to be consumed from input place;
- CT (*Created Tokens*) : $P \times T \rightarrow MT_{\Sigma/EUA}(X)$ associates to each output place of P, a multiset of algebraic terms which may be added when a transition is fired;
- TC is an additional condition, its default value is the term true;
- M_i and M_f are respectively the initial and the final markings.

Definition 2 (Meta-ECATNet). *Meta-ECATNets* are two levels meta transitional nets in which meta places control lower level transitions [3,10]. Each transition of the controlled level is presented if the controlling meta place contains a token. *Meta-ECATNets* have timing constraints: lower level transitions have firing duration, i.e tokens are available for fixed durations.

4 The Proposed Approach

In this paper, we use the Petri nets formalism to model clinical pathway [11]. Their main characteristics of these nets are: formal semantics, graphical notion, expressiveness, analysis methods, and vendor independent. Because the basic Petri nets formalism generates a complex nets, we have adopted high level algebraic Petri nets which give concise nets.

In our modelling approach of clinical pathway, two levels are considered. The meta level, or the control level, manages the lower, or functional, level. The transitions of the functional level represent the treatment tasks. As described in the Sect. 4.1, the clinical pathway of COPD has four (4) stages (from stage 1 to stage 4) according to the Forced Expiratory Volume at 1 second FEV_1 test value, and each stage has a fixed duration. The stages are exclusive, i.e at a given time, a patient could not be at more than one stage. Also, the stages are ordered from one to four and a patient cannot go back to a previously reached stage.

4.1 Case Study: Chronic Obstructive Pulmonary Disease (COPD)

COPD is a lung disease that makes it hard to breathe. Most people with COPD have both emphysema and chronic bronchitis. Emphysema damages the walls between the lungs air sacs and causes the air sacs to loose elasticity. This leads to a respiratory failure. The Chronic bronchitis inflames the lungs airways and causes them to become clogged with mucus. This makes breathing hard, and causes a chronic productive cough (with mucus ejection) [12].

According to the World Health Organisation (WHO) [13], the major factor of this disease is tobacco smoke (including second-hand or passive exposure). Other risk factors may include:

- indoor air pollution (such as solid fuel used for cooking and heating)
- outdoor air pollution

- occupational dusts and chemicals (such as vapours, irritants, and fumes)
- frequent lower respiratory infections during childhood.

The disease can affect both men and women around the world. More than 90% of COPD deaths occur in low and middle income countries where effective strategies for prevention and control are not always implemented or accessible. The COPD is now the fourth leading cause of death worldwide and it will become the third leading cause of death by 2020 [13,14]. In addition, the patient treatment procedure is expensive and often involves a whole medical staff for a long period.

Furthermore, during the patient's treatment phase, its health state can change and some complications or an aggravation may appear. These events must be considered and a solution should be planned to avoid any further deterioration of patient's health. In other words, this is called a dynamic monitoring of the patient's health state [15].

In our proposed approach, this aspect is taken into account by separating the treatment action applied on the patient and the control level. The latter is represented by the higher level (meta transition), and the functional level is dealt with through the lower level (normal transition) of our Petri nets model.

The severity of COPD is determined by a test from pulmonary function called Forced Expiratory Volume at 1 second (FEV_1). It shows the amount of air a person can forcefully exhale in one second. According to the value of FEV_1, four stages are drawn and the airflow becomes more limited with each stage [16]:

- stage 1 (Mild) : The airflow is somewhat limited, but we do not notice it much. The patient coughs and ejects mucus sometimes. FEV_1 is about 80 percent or more of normal.
- stage 2 (Moderate): The airflow is worse. The patient is often short of breath after doing something active. This is the point where most people notice symptoms and require help. FEV_1 is between 50 and 80 percent of normal.
- stage 3 (Severe): The airflow and shortness of breath are worse. One cannot do normal exercises any more. And the symptoms flare up frequently, also called an "exacerbation". FEV_1 is between 30 and 50 percent of normal.
- stage 4 (Very severe): The airflow is limited, the flares are more regular and intense, and the patient's quality of life worsens. FEV_1 is lower than that in stage 3.

4.2 Description

Figure 1 contains the proposed Meta-ECATNet of COPD pathway. It involves two levels. The figure shows only a sub net of the whole clinical process. In fact, we started from the point when the COPD is confirmed (transition *t12*), then the meta transitions *mt1, mt2, mt3, and mt4* activate the lower transitions *t16, t17, t18, and t19* that represent the four stages of COPD treatment.

The order between treatment stages is guaranteed by the higher-level net component. For example, after being at the stage one (meta place *mp1* is marked), only two behaviours are possible, the state either remains at stage one (firing the meta transition *mt9*) or evolves to stage 2 (fire of the meta transition *mt10*). In the lower level component, transitions *t16, t17, t18, and t19* are endowed with firing durations expressed in days (corresponding to 6 months, 6 months, 3 months, and 1 month respectively). The proposed Meta-ECATNet is implemented using Real-Time Maude in order to execute and analyse the deduced model. Figure 2 shows the Real-Time Maude modules implementing Meta-ECATNets.

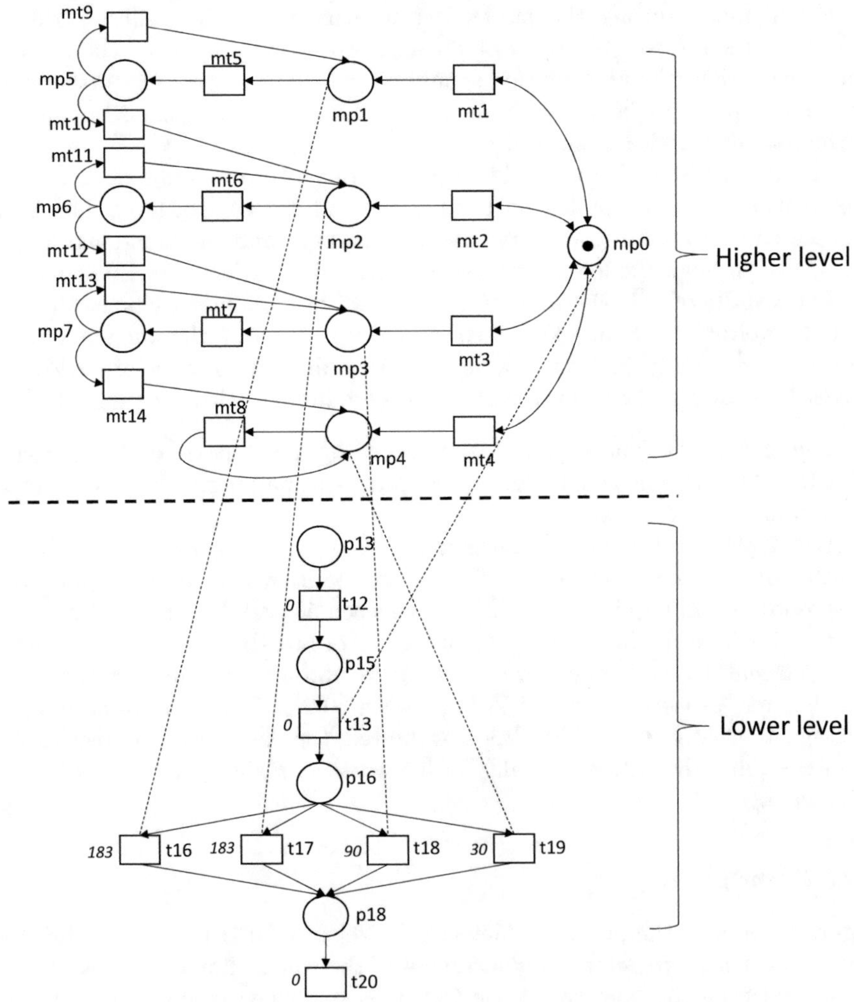

Fig. 1. The Meta-ECATNet model for COPD.

The Listing 1.1 includes the timed module *COPD* that implements COPD pathway Meta-ECATNet. At first, the module imports the module

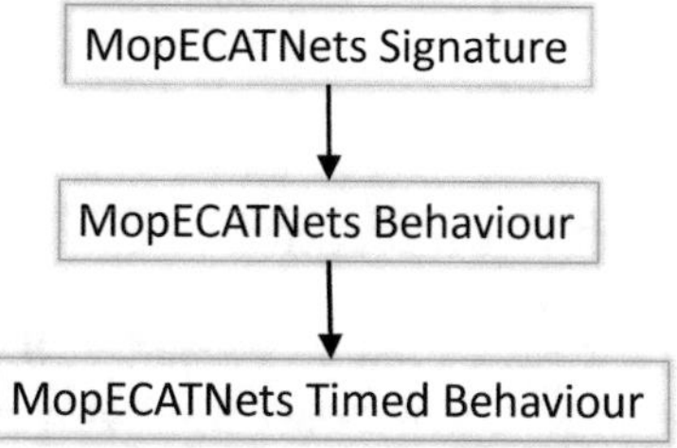

Fig. 2. Real-Time Maude modules implementing Meta-ECATNets

Listing 1.1. Real-Time Maude code of the COPD Meta-ECATNet

```
1 (tmod COPD is
2 including MetaECATNetTimedBehavior .
3 ops mp0 mp1 mp2 mp3 mp4 mp5 mp6 mp7 p0 p1 p2 p3 p4 p5 p6 p7 p8 p9 p10 p11 p12
    p13 p14 p15 p16 p17 p18 : -> PlaceId [ctor] .
4 ops mt1 mt2 mt3 mt4 mt5 mt6 mt7 mt8 mt9 mt10 mt11 mt12 mt13 mt14 t0 t1 t2 t3 t4
    t5 t6 t7 t8 t9 t10 t11 t12 t13 t14 t15 t16 t17 t18 t19 t20 : -> TransitionId [ctor] .
5 op copd-process : -> MetaECATNet [ctor] .
6 var p : PlaceId .   var T : TimeInf .
7 var tks : Tokens .
8 eq copd-process = < (mp0 , t13 , 0 : tk(INF)) (mp1 , t16 , 0 : noneTk ) (mp2 , t17 , 0
    : noneTk ) (mp3 , t18 , 0 : noneTk ) (mp4 , t19 , 0 : noneTk ) (mp5 , noneTi , 0
    : noneTk ) (mp6 , noneTi , 0 : noneTk ) (mp7 , noneTi , 0 : noneTk ) , mp0 ->[mt1
    ]-> mp0 mp1 ; mp0 ->[mt2]-> mp0 mp2 ; mp0 ->[mt3]-> mp0 mp3 ; mp0 ->[
    mt4]-> mp0 mp4 ; mp1 ->[mt5]-> mp5 ; mp2 ->[mt6]-> mp6 ; mp3 ->[mt7]->
    mp7 ; mp4 ->[mt8]-> mp4 ; mp5 ->[mt9]-> mp1 ; mp5 ->[mt10]-> mp2 ; mp6
    ->[mt10]-> mp2 ; mp6 ->[mt11]-> mp3 ; mp7 ->[mt12]-> mp3 ; mp7 ->[mt13
    ]-> mp4 > < (p0 , tk(0)) ( p1 , noneTk ) (p2 , noneTk ) (p3 , noneTk ) (p4 , noneTk
    ) (p5 , noneTk ) (p6 , noneTk )(p7 , noneTk ) (p8 , noneTk ) (p9 , noneTk )(p10 ,
    noneTk ) (p11 , noneTk ) (p12 , noneTk ) (p13 , noneTk ) (p14 , noneTk ) (p15 ,
    noneTk ) (p16 , noneTk ) (p17 , noneTk ) (p18 , noneTk ) , p0 ->[ t0 , 0 ]-> p1 ; p1
    ->[ t1 , 0 ]-> p2 p3 ; p2 ->[ t2 , 0 ]-> p4 ; p3 p4 ->[ t3 , 0 ]-> p5 ; p5 ->[ t4 ,
    0 ]-> p6 ; p5 ->[ t5 , 0 ]-> p7 ; p7 ->[ t6 , 0 ]-> p8 ; p8 ->[ t7 , 0 ]-> p9 ; p9
    ->[ t8 , 0 ]-> p10 p11 ; p10 ->[ t9 , 0 ]-> p12 ; p11 p12 ->[ t10 , 0 ]-> p13 ; p13
    ->[ t11 , 0 ]-> p14 ; p13 ->[ t12 , 0 ]-> p15 ; p15 ->[ t13 , 0 ]-> p16 ; p16 ->[
    t16 , 183 ]-> p16 p18 ; p16 ->[ t17 , 183 ]-> p16 p18 ; p16 ->[ t18 , 90 ]-> p16
    p18 ; p16 ->[ t19 , 30 ]-> p16 p18 ; p18 ->[ t20 , 0 ]-> nonePi > .
9 endtm)
```

MopECATNetTimedBehavior. Then, meta places, places, meta transitions and transitions identifiers are declared by means of algebraic constructor terms (see Lines 3 and 4). At last, The initial state of the COPD Meta-ECATNet is fixed in the timed module *COPD* using the the equation of Line 8.

Checking of the COPD Meta-ECATNet model is done using the TCTL Real-Time Maude model-checker [17]. This tool accepts as input Meta-ECATNet model of the COPD and the property to be analyzed expressed as a TCTL formula, then it checks if all behaviours satisfy this property.

Listing 1.2. Checking of the COPD Meta-ECATNet

```
(tmod COPD-MC is
protecting COPD .
including TCTL-MODEL-CHECKER .
vars ts ts' : Transitions .
vars M M' : Marking . var P : Prop .   var S : System . var T : Time .
ops StageFix atStage1 atStage2 atStage3 atStage4 : -> Prop [ctor] .
eq { < (mp0, t13, 0 : tk(0)) M , ts >< M' , ts' > } |= StageFix = true .
eq { < (mp1, t16, 0 : tk(183)) M , ts >< M' , ts' > } |= atStage1 = true .
eq { < (mp2, t17, 0 : tk(183)) M , ts >< M' , ts' > } |= atStage2 = true .
eq { < (mp3, t18, 0 : tk(90)) M , ts >< M' , ts' > } |= atStage3 = true .
eq { < (mp4, t19, 0 : tk(30)) M , ts >< M' , ts' > } |= atStage4 = true .
eq { < M , ts >< M' , ts' > } |= P = false [owise] .
endtm)
```

Listing 1.3. Analyzed properties

```
Full Maude 2.3 '(February 12th', 2007') Real-Time Maude 2.3 TCTL
Model Checker extension April 19, 2013
==================
mc-tctl { copd-process } |= EF[ <= than 30 ] atStage4
rewrites: 1990684 in 6374638180ms cpu (5534ms real) (0 rewrites/second)
Checking equivalent property:
mc-tctl_|=_. {copd-process} |= E tt U[c 0,30 c] atStage4 .
Property satisfied
==================
mc-tctl { copd-process } |= AG( atStage1 implies ( EF[<= than 184 ] atStage2 ) ) .
rewrites: 2214471 in 6374638180ms cpu (9580ms real) (0 rewrites/second)
Checking equivalent property:
mc-tctl_|=_. {copd-process} |= not (E tt U[c 0,INF o] atStage1 and not (E tt
U[c 0,184 c] atStage2)) .
Property satisfied
==================
mc-tctl { copd-process } |= AG( atStage2 implies ( EF[<= than 184 ] atStage1 ) ) .
ewrites: 2312076 in 6374638180ms cpu (12254ms real) (0 rewrites/second)
Checking equivalent property:
mc-tctl_|=_. {copd-process} |= not (E tt U[c 0,INF o] atStage2 and not (E tt
U[c 0,184 c] atStage1)) .
Property not satisfied
```

In the Listing 1.2, we declare some useful atomic propositions and give their semantics.

In this work, we are interested by checking the following properties (see Listing 1.3):

- In the first TCTL formula, we check if a patient having COPD can be at a severe stage after its first diagnostic. For this property the result is :"Property satisfied".

- The second TCTL formula makes certain that the order relation between the stage 1 and the stage 2 is respected, i.e always the stage 1 can be followed by the stage 2.
- The last formula tests if it is possible to execute stage 1 after having executed the stage 2. The obtained response for this formula is : "Property not satisfied", because the COPD is chronic cannot be cured or reversed by medication.

5 Conclusion

In this paper, the Meta-ECATNet formalism is applied to model the COPD patient's clinical pathway. Meta-ECATNets are based on separating the higher level (what the system do) and the lower level (how is it done). This allows to be dynamic and flexible during the patient's treatment. Further, our proposed model can deal with exception and aggravation easily. The higher level represented by meta transition controls the functional level and can monitor the evolution of the patient's health state efficiently.

As future extensions of this work, we aims at exploring in depth the data aspect of Meta-ECATNets as well as considering clinical pathways of other diseases.

References

1. Chincholkar, A., Chetty, O.K.: Stochastic coloured petri nets for modelling and evaluation, and heuristic rule base for scheduling of fms. Int. J. Advan. Manuf. Technol. **12**(5), 339–348 (1996)
2. Hicheur, A., Dhieb, A.B., Barkaoui, K.: Modelling and analysis of flexible healthcare processes based on algebraic and recursive petri nets. In: International Symposium on Foundations of Health Informatics Engineering and Systems, pp. 1–18. Springer (2012)
3. Latreche, F., Belala, F.: A layered petri net model to formally analyse time critical web service composition. IJCCBS **7**(2), 119–137 (2017)
4. Barkaoui, K., Hicheur, A., Kheldoun, A., Liu, D.: Modelling and analyzing home care plans using high-level petri nets. In: 13th International Workshop on Discrete Event Systems, WODES 2016, Xi'an, China, May 30–June 1, 2016, pp. 284–290 (2016)
5. Du, G., Jiang, Z., Diao, X., Yao, Y.: Workflow modelling of clinical pathway based on modular temporised coloured petri net with changeable structure. Int. J. Serv. Oper. Inf. **6**(3), 183–210 (2011)
6. Bertolini, C., Liu, Z., Srba, J.: Verification of timed healthcare workflows using component timed-arc petri nets. In: International Symposium on Foundations of Health Informatics Engineering and Systems, pp. 19–36. Springer (2012)
7. Group, L.P.W., et al.: Little-jil 1.5 language report. Laboratory for Advanced Software Engineering Research, Department of Computer Science, University of Massachusetts-Amherst, MA (2006)

8. David, A., Jacobsen, L., Jacobsen, M., Jørgensen, K.Y., Møller, M.H., Srba, J.: Tapaal 2.0: Integrated development environment for timed-arc petri nets. In: International Conference on Tools and Algorithms for the Construction and Analysis of Systems, pp. 492–497. Springer (2012)
9. Bettaz, M., Maouche, M.: How to specify non-determinism and true concurrency with algebraic term nets. In: Recent Trends in Data Type Specification, 8th Workshop on Specification of Abstract Data Types Joint with the 3rd COMPASS Workshop, Dourdan, France, 26–30 Aug 1991, Selected Papers, pp. 164–180 (1991)
10. Latreche, F., Belala, F.: Mop-ecatnets for formal modeling dynamic web services. In: Proceedings of the 1st International Conference on Advanced Aspects of Software Engineering, ICAASE 2014, Constantine, Algeria, 2–4 Nov 2014, pp. 27–34 (2014)
11. van der Aalst, W.M.P., Van Hee, K., Houben, G.: Modelling and analysing workflow using a petri-net based approach. In: Proceedings of the second Workshop on Computer-Supported Cooperative Work, Petri Nets and Related Formalisms, pp. 31–50 (1994)
12. Littner, M.R.: Chronic obstructive pulmonary disease. Ann. Intern. Med. **154**(7) (2011). (ITC4–1)
13. World-Health-Organization: Global burden of disease 2004 update: part 2, causes of death (2008). http://www.who.int/healthinfo/global_burden_disease/GBD_report_2004update_part2.pdf
14. Celli, B.R.: Predictors of mortality in copd. Respir. Med. **104**(6), 773–779 (2010)
15. Augusto, V., Xie, X.: A modeling and simulation framework for health care systems. IEEE Trans. Syst. Man Cybern. Syst. **44**(1), 30–46 (2014)
16. Pauwels, R.A., Buist, A.S., Calverley, P.M., Jenkins, C.R., Hurd, S.S.: Global strategy for the diagnosis, management, and prevention of chronic obstructive pulmonary disease: Nhlbi/who global initiative for chronic obstructive lung disease (gold) workshop summary. Am. J. Respir. Crit. Care Med. **163**(5), 1256–1276 (2001)
17. Lepri, D., Ábrahám, E., Ölveczky, P.C.: Timed CTL model checking in real-time maude. In: Rewriting Logic and Its Applications-9th International Workshop, WRLA 2012, Held as a Satellite Event of ETAPS, Tallinn, Estonia, 24–25 Mar 2012, Revised Selected Papers, pp. 182–200 (2012)

Ontologies-Based Process in Software Reengineering

Moussa Saker(✉) and Nora Bounour

LISCO Laboratory, Badji Mokhtar-Annaba University, Annaba 23000, Algeria
moussa.saker@hotmail.com, nora_bounour@yahoo.fr

Abstract. Since the emergence of term "Ontology" in the philosophical sphere, its application is widespread in multiplicity of research fields. In software engineering ontologies were adopted more than two decades ago, including requirements analysis, specification and design phases. Unfortunately, this vivacity and richness in software engineering is not the same in software reengineering where there is a lack of interest from the research community. In this paper, we present a survey about the most significant works regarding the use of ontologies in software reengineering process according to two perspectives: as software artefacts and as domain knowledge. Through our study, we highlight some of produced tools and frameworks in addition to open challenges in the field.

Keywords: Ontology · Software reengineering · Software understanding Software maintenance · Software quality · Software test

1 Introduction

The term ontology was first defined in philosophy as a part of metaphysics. After, it has been widely used in different research fields such as biology, medicine, law, computer science etc. The main reason behind this intensive use of ontology in these various areas backs to the increasing demand for a reliable and an unambiguous conceptual foundation and to the need of a common understanding knowledge base for reuse of knowledge. In computer science ontologies were adopted more than two decades ago, they had been used in knowledge engineering, requirement elicitation, artificial intelligence, software maintenance and evolution, etc.

In literature there is a diversity of works about the use of ontologies in software engineering, as information retrieval, object oriented analysis, agent based system design, software understanding. Regrettably, this vivacity and richness in software engineering is not the same in software reengineering where there is a lack of interest from the research community (except software understanding). That is what drove us to produce a survey about the most significant works regarding the use of the ontologies in software reengineering process and trying to highlight new opportunities in this domain, our selection criteria is based on three parts: program analysis (retrieve information from source code and establish appropriate models), information focusing (filter information) and program visualization (depiction of retrieved and filtered information). We organized different works cited in this paper according to two axes as

S. Chikhi et al. (Eds.): MISC 2018, LNNS 64, pp. 299–311, 2019.
https://doi.org/10.1007/978-3-030-05481-6_23

showed in Fig. 1: the first axe defines the use of ontologies as software artefacts and its use as domain knowledge in different software reengineering activities. The second axe give the percentage of works cited in this paper.

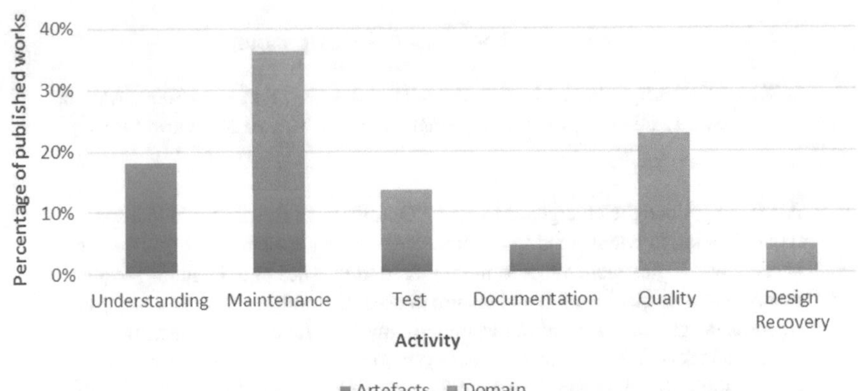

Fig. 1. Ontologies use in software reengineering.

This paper is organized as follow; the Sect. 2 will introduce a background about ontologies. We demonstrate the use of ontologies as software artefacts in Sect. 3. Section 4 will present ontologies as domain knowledge. While Sect. 5 will present a comparative study and Sect. 6 will introduce open challenges in the field; finally, the conclusion and future work are arranged in last section.

2 Background

While the term "Ontology" was limited to the philosophical sphere in the past, it was introduced in computer science for first time at the end of the XXth century by artificial intelligence researchers. At the beginning of XXIth century; there had been a multitude of research works that concerned the use of ontologies in software engineering.

Despite the lack of a unified definition of ontology in literature, Gruber's definition is the most suitable one "an explicit specification of a conceptualization" [1], we consider as well Guarino's definition which refines the precedent definition "An ontology is a logical theory accounting for the intended meaning of a formal vocabulary. The intended models of a logical language using such a vocabulary are constrained by its ontological commitment. Ontology indirectly reflects this commitment (and the underlying conceptualization) by approximating these intended models" [2].

While formally an ontology O is defined by 5 attributes O = {C, P, R, A, I}, where:

- C is a set of concepts, which represents different entities in some domain being modeled.
- P is a set of properties that defines a concept for example: for concept "Home", "have door" is a property.

- R is a set of relationships between concepts; the most used relations are "instance of" and "kind of".
- A is a set of axioms, usually formalized into some logic language.
- I is a set of instances.

To describe its structure and to perform queries; ontologies requires means, we highlight: OWL (Ontology Web Language) used in domain formalization defining classes and properties of this classes, individuals and statements of this individuals. Besides, reasoning about these classes and individuals according to formal semantics defined by the language [3], RDF (Resource Description Language), OIL (Ontology Inference Layer) and SPARQL a query language for RDF and can be used for retrieving individuals and properties in OWL model [4].

Now ontology is recognized and spread in almost all software engineering activities (as for software requirements, design, testing, maintenance, quality, software comprehension, etc.). Through our study, we are perceived that different types of ontologies are used. The ontology has been used as a conceptual model to represent software artefacts for example documentation and source code or as domain knowledge to represent a process or an activity for example ontology for software maintenance [5].

In what comes next, we will present the most notable works about the use of ontologies in software reengineering as software artefacts.

3 Ontology as Software Artefacts

Software reengineering is a multidimensional problem that creates challenges for both the research community and the tool developers. These challenges are due to the variety of software artefacts, knowledge resources, and existing relationships between them. Generally existing solutions that address these aspects are not integrated together.

In matter of fact, the research community focused on the use of ontologies in software reengineering activities.

Artefact ontology defines the concepts of multiple software artefacts generated during development; these concepts contain metadata information (source code, xml configuration file, log file, versions or releases). Next, we will introduce the most cited works in the reengineering process combined with the use of ontologies for representing a software artefact.

Devanbu et al. [6] proposed "LaSSIE" one of the first applications of description logic in software engineering it handles the problem of complexity and invisibility in large systems, with the use of explicit knowledge representation and reasoning. As shown in Fig. 2, the "LaSSIE" taxonomy is frame-based representation with frames represent classes of objects, starting from general classes to more specialized ones, with the most general class "Thing". "LaSSIE" system has the ability to provide semantic retrieval of instances from an information base. But, since the mapping between the instances, concepts and the source code is manual so the modification of this system or the addition of new features is considered very difficult; in addition, it is not applicable on object-oriented code.

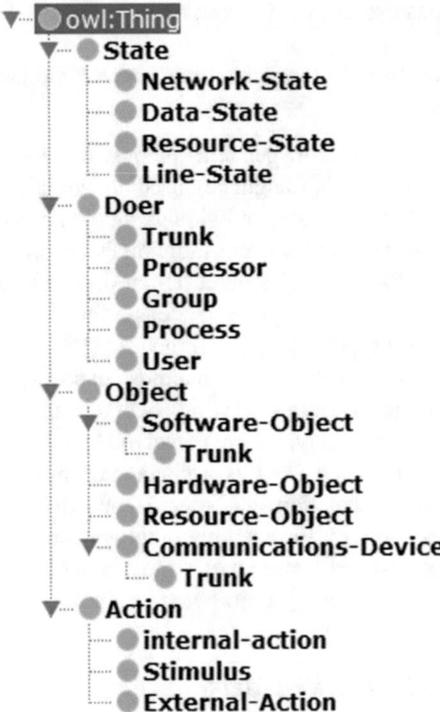

Fig. 2. The LaSSIE taxonomy [6].

A complementary system of LaSSIE is CODE-BASE introduced by Selfridge [7]; the system extracts the information automatically from the source code and stores them in a database. CODE-BASE designed so that the user can extend the knowledge base and build new concepts. In addition, it offers a graphical interface that allows users to make specific requests. Nevertheless, it inherits the limits of LaSSIE and it's difficult to add new features to it.

Another system is KITSS presented by Nonnenmann et al. [4]; an automated testing system "black box testing", it transforms natural language test cases specifications to test scripts as output, with the use of knowledge base to resolve reference in noun

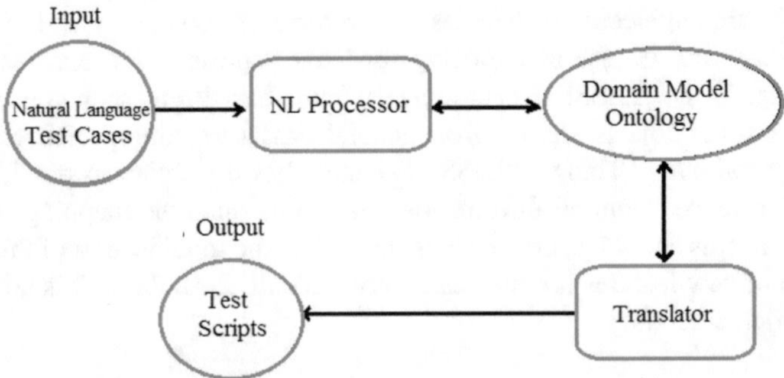

Fig. 3. KITSS architecture [4].

phrases. Figure 3 shows the tree modules of KITSS, natural language processor, interaction analyzer and the translator. The software test model is described via an ontology. However, KITSS is limited to just black box testing only.

Yang et al. [8] discussed a method for the ontology extraction from legacy systems written in COBOL language, the integrated information in these systems has been organized in three levels: level of requirements and environment, schema level, data and application code level. A tool covering reengineering process called Reengineering Assistant (RA) has been used to extract high-level specifications from software source code. However, this method focuses only on one programming language (COBOL).

Hwang et al. [9] proposed a formal approach to construct ontologies from source code by using Formal Concept Analysis (FCA). They proposed FCA as framework for identifying concepts and for constructing concepts hierarchies ("concept lattices"). They implemented a tool called "FCAWizard" to support the ontology construction, it creates hierarchy of concepts in several steps: first, it builds list of names of objects and attributes, and their relations; from this, it creates a list of formal concepts. There are then put into the algorithm to build the concept lattice, and displayed to a line diagram (Hasse diagram) format. Hence, it is computationally intensive and lacks of simplicity.

Zhang et al. [10] introduced an approach for traceability recovery that explicitly includes the integration of structural and semantic information, by providing an ontological representation for the two types of software artefacts (source code and documentation). Instances of concepts from source code and documentation are detected throw the analysis of concrete systems, then traceability links are created throw ontology alignment. They developed a system of Text Mining (TM) that semantically analyze software documents. The formal ontological representation also allows them to gain automated reasoning services provided by the ontology reasoners to infer the implicit relations (links) between the two types of artefacts and to define new concepts and relations. Software ontologies based on two sub-ontologies: source code ontology and documentation ontology to use the structural and semantic information in various software artefacts. Extraction system of concepts and instances of concepts for source code ontology is based on SOUND [11], while text extraction system for documentation ontology is based on GATE [12]. However, some inconsistencies between source code and documentation cannot be discovered using this approach.

Using ontologies and description language, Rilling et al. [11] formally modeled the major information resources used in software maintenance. The modeling process is based on two main components:

- Ontology manager: used to manage the infrastructure of the process model.
- Process manager: built on top of ontology manager and provide users with both the context and iterative guidance for the comprehension process.

López et al. [13] presented an approach for architecture rationale recovery called TREx (Toeska Rationale extraction) which allows the extraction, representation and the exploration of architecture rationale knowledge from text documents. To integrate this information into rationale repository, enable manipulation by other tools. The whole process is ontology-driven. Ontologies were used in all the rational recovery problem steps (identification, representation and manipulation). TREx combines four conceptual components: (1) Software architecture (of systems) ontology and software architecture rationale (of projects) ontology. (2) Ontology-based extraction of segments of plain-texts documents (knowledge units) that are relevant according to a given ontology. (3) Expert-supervised validation and inconsistency management of the connection of automatically extracted knowledge units. (4) Facet-based exploration of the knowledge store and the original project documents.

4 Ontology as Domain Knowledge

In this section, we will discuss the use of ontologies to represent particular domain knowledge of software reengineering and the relationships that exist between different concepts of this domain. We noticed a multitude of works in what comes we will present some of works that we considered as the most interesting.

In software test, Souza et al. [14] presented a system for software testing based on reference domain ontology called ROoST (Reference ontology for software testing). In order to develop the SABiO method (Systematic Approach for Building Ontologies), in this system they reused and extended SP-OPL a core ontology on software processes, that contains a set of interrelated ontology patterns. ROoST focused on software testing process and activities, artefacts and techniques for test case design.

Li et al. [15] proposed an ontology-based approach for GUI (Graphical User Interface) testing, using the reverse engineering techniques it captures elements from source code (concepts, database, file, variable) and then stores them in GUI ontology, based on the knowledge stored in this ontology, test case generation rules are extracted from the tester's experience.

In software maintenance, Kitchenham et al. [3] suggest an ontology for software maintenance process. They aimed to help researchers to report and understand the results of empirical studies. The ontology has an enhancement and correction entities; enhancement throw changes in implementation or in requirements where in correction, defects are recognized and corrected throw the comparison of observed application behavior and the required one. Two maintenance scenarios were presented: evolutionary development (optimization and prediction of release quality), independent maintenance scenario (whether the maintenance task is performed by the organization or by a third party). The maintenance ontology is divided to four sub-ontologies:

- Product ontology: defines concepts used in the ontology (size, age, maturity, etc.).
- Activity ontology: As shown in Fig. 4; maintenance activity ontology defines the different maintenance activities (investigation activity, modification activity, quality assurance activity, management activity).
- Process ontology: based on two maintenance process (software procedure ontology, maintenance organization process).
- Agent ontology: it involves stuff affecting maintenance activity (maintenance organization staff, customer and user).

Ruiz et al. [16] proposed a semi-formal ontology for managing software maintenance projects based on Kitchenham et al. [3], to represent static aspects; this ontology (Fig. 5) is based on several sub-ontologies: product, activities, process, agents, and

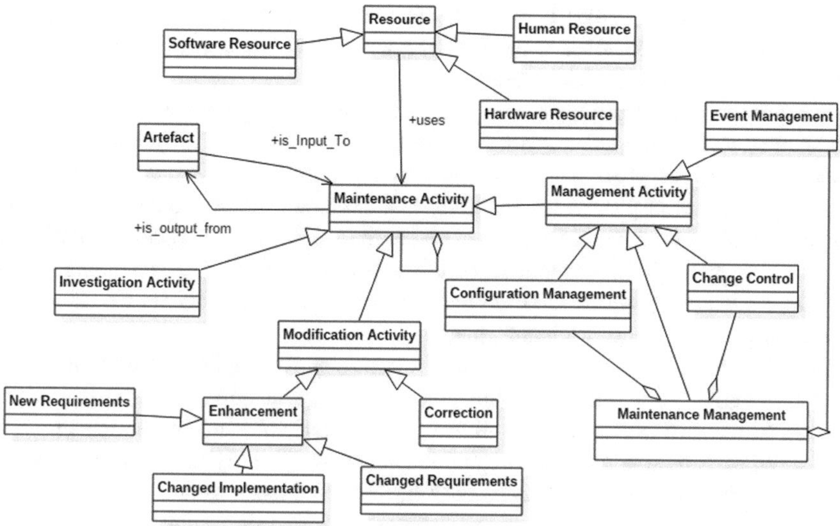

Fig. 4. The maintenance activity ontology. [3]

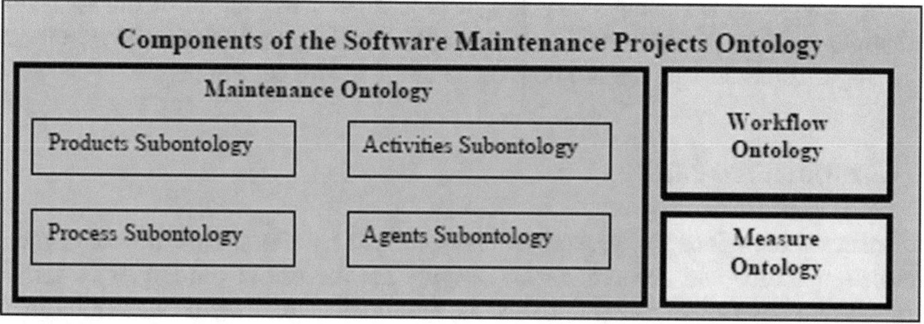

Fig. 5. Structure of the software maintenance projects ontology [16].

measures. The dynamic aspects are represented by ontology named "workflow ontology". The ontology called "measure ontology" represents both the static and the dynamic aspects related to the software measures. A system of knowledge management KM-MANTIS [17] has been developed to implement the proposed ontology.

Deridder [18] presented ontologies as means for concept-oriented approach to support activities of reuse and maintenance in software. This approach captures the implicit knowledge with the use of ontology. To link the concepts with the code/artefacts, the author has proposed the use of concept definition types of the intentional and extensional of SOUL (SmalTalk Open Unification Language) [19]. It should be mentioned that this approach captures only static aspects.

In software quality, Boehm and In [20] developed a knowledge base called "QARCC", it can alert different member involved in development of an information system (users, developers and stakeholders) about conflicts among their software quality requirements, besides it can help to identify additional quality requirements. However, this approach focuses only on requirement conflicts disregarding most frequent problems in reverse engineering.

In the work carried out by Kayed et al. [21], the ontology was used to achieve an understandable and unified semantic framework to measure the quality of web or desktop applications that their concepts and terminologies are inconsistent among studies and reports in progress. First, they had excluded the documents that included more than 80% of source code, then they had extracted the various concepts, definitions and terminology of different reports, then they had formalized these concepts in order to produce a coherent and well-structured terminologies and concepts common to the measures and attributes of software quality.

Da Silva et al. [22] proposed a domain ontology for quality software inspection called "OntoQAI", this ontology maps concepts, properties and restrictions related to three areas; the software process, software project and software inspection throw the use of OWL-DL as specification language, and SPARQL [23] as query language. To support quality assurance inspection, they developed a system named "QAISys" this system is based on agents and is able to produce an inspection checklist and allocate noncompliance issues, trying to minimize human interpretation and inspection effort. Both [21, 22] lacks of an integrated model for representing software evaluation.

Ruiz et al. [24] proposed a semi-formal ontology for software measurement; the ontology was designed using UML language and used adaptation of the REFSENO methodology. They presented tables of concepts, attributes and relationships in the aim of getting a common agreement about these terms in this discipline.

5 Comparative Study

The software reengineering process is characterized by the use of different types of information source and different tools. Despite that the use of ontologies in software reengineering have a lot of benefits, such as, eliminating conceptual and terminological mismatch and ambiguity, providing features such as the genericity and the extensibility. Ontology favorites the reuse of knowledge and common understanding and simplifying communication among different tools. Some methodologies focus on specific development activities rather than others as [11, 13, 22, 25].

Compared to the software engineering process; a very little work has been done in software reengineering, thus, there is some potential opportunities to use ontologies in reengineering to explore, reuse, organize knowledge in this process. We noticed that software maintenance and understanding are the center point of ontology researches in software reengineering process, while there is a lack of interest in domains that handles refactoring or versioning.

In the followed organization of this paper, we have classified different works by exploring the type of the ontology. This classification has given us the opportunity to see the ontologies in software reengineering on more understandable and specified pattern, in sort that every work based on domain or artefacts ontologies.

Table 1 summarize the comparison between research works and shows different ontologies types and their application tasks, objectives, contributions, limitations.

Table 1. A comparative table between different works in software engineering

Ontology type	Task	Propositions	Objective	Contribution	Limitation
Artefact	Software comprehension	Extracting ontologies from legacy systems	Understandingand re-engineering. [8]	Building ontology from COBOL Code	Focuses only on one programming language (COBOL)
		A knowledge-based software information system	provides semantic retrieval of instances from information base [6]	Retrieves instances from DB.	-difficult to modify and to maintain. -not applicable to object-oriented code
		Knowledge representation support for a software information system	automatic extraction of concepts from the source code [7]	Concepts extraction	-difficult to add new features
		A FCA-based ontology construction for the design of class hierarchy	identifying concepts and building concepts hierarchies [9]	Building ontologies	Computationally intensive and lacks simplicity
	Test	KITSS: A functional software testing system using a hybrid domain model	the use of knowledge base to resolve references in noun phrases [4]	-test case generation	-limited to black box testing
	Documentation	Ontological approach for the semantic recovery of traceability links between software artefacts	Analyzing documents and discover concepts to be aligned with source code and recover traceability links [10]	Establishing links between source code and documentation	-some inconsistencies between source code and documentation cannot be discovered
	Maintenance	Empowering software maintainers with semantic web technologies	Discover concepts and artefacts using code analysis and text mining. [26]	Establishing traceability links between software source code and documentation	/
		A unified ontology-based process model for software maintenance and comprehension	a software understanding model that analyze and restores traceability links between artefacts [11]	A software-understanding model to restore tractability links between different artefacts	/
		Ontology-based software analysis and reengineering tool integration: The OASIS service-sharing methodology	Integration methodology to help software maintainers to understand and analyze programs. [27]	Establish sematic connection among artefacts (.i.e. source code and documentation)	Focuses only on service integration among reverse engineering activities
Domain	Design recovery Process	Bridging the gap between software architecture rationale formalisms and actual architecture documents: An ontology-driven approach	Architectural rationale recovery from text documents. [13]	Extraction and exploration and representation of architectural rationale knowledge	/
	Test Process	Using Ontology Patterns for Building a Reference Software Testing Ontology	A system for software testing process based on domain ontology. [14]	Software test -techniques for test case design	

(*continued*)

Table 1. (*continued*)

Ontology type	Task	Propositions	Objective	Contribution	Limitation
		An ontology-based approach for GUI testing	GUI testing approach using reverse engineering techniques. [15]	GUI test	The test case generation is limited to the tester's experience
	Maintenance process	Towards an ontology of software maintenance	ontology that covers all maintenance aspects [3]	Identify factors affecting results of empirical studies	-no formal semantics were presented.
		An ontology for the management of software maintenance projects	semi-formal ontology to model both static and dynamic aspects software maintenance [16]	Represent both static and dynamic aspects (i.e. workflow) related to software maintenance	No implementation was presented
		A unified ontology-based process model for software maintenance and comprehension	Modeled formally the major information resources for software maintenance [11]	-Model and framework for software maintenance	/
		A concept-oriented approach to support software maintenance and reuse activities	Captures and links concepts with the use of ontologies [24]	-Link artefacts and help maintainers to find them	Representing just static aspects
		Environment for Managing Software Maintenance Projects	a maintenance ontology to share knowledge among maintainers. [17]	A maintenance environment to help maintainers sharing knowledge	No implementation was presented
	Estimation of Software Quality	Identifying Quality Requirements Conflicts	Alert the stakeholders about conflicts in their software quality requirements [20]	-A tool to help developers analyzing requirements and identifying conflicts among them	Focused on only one problem (requirements conflicts) and disregard most frequent problems in Reverse Engineering
		Towards an ontology for software, product quality attributes	A framework to measure the quality of desktop and web application. [21]	A common understanding for quality attributes	Lack of an integrated model for representing software evaluation
		OntoQAI: An Ontology to Support Quality Assurance Inspections	An ontology for quality inspection [22]	An ontology and a system for quality assurance inspection	
		A proposal of a Software Measurement Ontology	Semi-formal ontology for software measurements [24]	Tables of concepts, attributes and relationships to get a common agreement	/

6 Open Challenges

The following open challenges have been identified across literature and emerged as potential scope for future contributions

- Using software artefacts ontologies or domain ontologies such as (design ontology, version ontology, testing and quality ontologies) to enrich software maintenance process by representing information about system components and the relations among them.
- Supporting model evolution and achieving a traceability model between different types of artefacts to enhance software understanding.
- Preserving software quality is a main challenge for maintainers, so, using quality ontology and architecture ontology to detect errors and conflicts in feature models is crucial part to preserve software quality.
- In documentation, developing an automated documentation ontology emerges as a potential opportunity, this ontology should gender software domain, artefacts, architecture sub-ontologies and could be transformed into text file.
- Finally, using semantic web ontologies in software reverse engineering could be also explored, to extract hidden domain information.

7 Conclusion

Despite the immense number of research topics in software engineering, reengineering process lacks of interest from research community, in this paper we have discussed the most noteworthy researches that used ontology in software reengineering besides the open challenges in this domain.

Our work provides a clear view and classification of the presented works based on activity of software reengineering process and ontology types; where researches are organized according to two axes: ontologies used as software artefacts, ontologies used as domain knowledge. We noticed that the software maintenance and understanding are the center of interest of these works, while there is a lack of research in domains of refactoring, versioning, and architecture recovery.

References

1. Gruber, Thomas R.: Toward principles for the design of ontologies used for knowledge sharing. Int. J. Hum. Comput. Stud. **43**(5), 907–928 (1995)
2. Guarino, N. ed.: Formal ontology in information systems. In: Proceedings of the first international conference on formal ontology and information systems (FOIS'98). IOS press. Trento, Italy, vol. 46, no. 1998. pp. 81–97 (1998)
3. Kitchenham, B.A., Travassos, G.H., Von Mayrhauser, A., Niessink, F., Schneidewind, N.F., Singer, J., Takada, S., Vehvilainen, R., Yang, H.: Towards an ontology of software maintenance. J. Softw. Maint. **11**(6), 365–389 (1999)

4. Nonnenmann, U., Eddy, J.K.: KITSS-A functional software testing system using a hybrid domain model. In: Proceedings of the Eighth Conference on Artificial Intelligence for Applications, 1992, pp. 136–142. IEEE (1992)
5. Anquetil, N., de Oliveira, K.M., Dias, M.G.: Software Maintenance Ontology. In Ontologies for Software Engineering and Software Technology, pp. 153–173. Springer, Berlin (2006)
6. Devanbu, P.T., Selfridge, P.G., Ballard, B.W., Brachman, R.J.: A knowledge-based software information system. In: Proceeding of International Joint Conference on Artificial Intelligence (IJCAI), vol. 89. pp. 110–115 (1989)
7. Selfridge, P.G.: Knowledge representation support for a software information system. In: Proceedings, Seventh IEEE Conference on Artificial Intelligence Applications, 1991, Vol. 1, pp. 134–140. IEEE (1991)
8. Yang, H., Cui, Z., O'Brien, P.: Extracting ontologies from legacy systems for understanding and re-engineering. In: Proceedings of the Computer Software and Applications Conference, COMPSAC'99, pp. 21–26. The Twenty-Third Annual International, IEEE (1999)
9. Hwang, S.H., Kim, H.G., Yang, H.S.: A FCA-based ontology construction for the design of class hierarchy. In: Gervasi O. et al. (eds.) Computational Science and Its Applications (ICCSA). Lecture notes in computer science, vol. 3482, pp. 827–835. Springer Berlin (2005)
10. Zhang, Y., Witte, R., Rilling, J., Haarslev, V.: Ontological approach for the semantic recovery of traceability links between software artefacts. IET Softw. **2**(3), 185–203 (2008)
11. Rilling, J., Zhang, Y., Meng, W.J., Witte, R., Haarslev, V., Charland, P.: A unified ontology-based process model for software maintenance and comprehension.In: Dingel, J., Solberg A. (eds.): Models in Software Engineering. Lecture Notes in Computer Science, vol. 4364, pp. 56–65. Springer, Berlin (2007)
12. Cunningham, H., Maynard, D., Bontcheva, K., Tablan, V.: GATE: a framework and graphical development environment for robust NLP tools and applications. In: Proceeding of the 40th Anniversary Meeting of the Association for Computational Linguistics (ACL). Philadelphia, USA, pp. 168–175 (2002)
13. López, C., Codocedo, V., Astudillo, H., Cysneiros, L.M.: Bridging the gap between software architecture rationale formalisms and actual architecture documents: an ontology-driven approach. Sci. Comput. Program. **77**(1), 66–80 (2012)
14. Souza, E.F., Falbo, R.A., Vijaykumar, N.L.: Using ontology patterns for building a reference software testing ontology. In: 2013 17th IEEE International Enterprise Distributed Object Computing Conference Workshops, pp. 21–30. IEEE (2013)
15. Li, H., Chen, F., Yang, H., Guo, H., Chu, W.C.C., Yang, Y.: An ontology-based approach for GUI testing. In: 2009 33rd Annual International Computer Software IEEE and Applications Conference, vol. 1, pp. 632–633. IEEE (2009)
16. Ruiz, F., Vizcaíno, A., Piattini, M., García, F.: An ontology for the management of software maintenance projects. Int. J. Softw. Eng. Knowl. Eng. **14**(03), 323–349 (2004)
17. Ruiz, F., Garcia, F., Piattini, M., Polo, M.: Environment for managing software maintenance projects. In: Advances in Software Maintenance Management: Technologies and Solutions. Idea Group Publication, USA, pp. 255–290 (2002)
18. Deridder, D.A.: A concept-oriented approach to support software maintenance and reuse activities. In: Proceedings of the 5th Joint Conference on Knowledge Based Software Engineering. IOS press (2002)
19. Wuyts, R.: A logic meta-programming approach to support the co-evolution of object-oriented design and implementation. Doctoral dissertation, Vrije Universiteit Brussel (2001)
20. Boehm, B., In, H.: Identifying quality requirement conflicts. IEEE Softw. **13**(2), 25–35 (1996)

21. Kayed, A., Hirzalla, N., Samhan, A.A., Alfayoumi, M.: Towards an ontology for software product quality attributes. In: Fourth International Conference on Internet and Web Applications and Services, ICIW'09, pp. 200–204. IEEE (2009)
22. Da Silva, J.P.S., DallOglio, P., Pinto, S.C.C.D.S., Bittencourt, I.I., Mergen, S.L.S.: OntoQAI: an ontology to support quality assurance inspections. In: Brazilian Symposium on Software Engineering (SBES), pp. 11–20. IEEE (2015)
23. Prud, E., Seaborne, A.: Sparql query language for rdf. http://www.w3.org/TR/rdf-sparql-query/
24. Ruiz, F., Genero, M., García, F., Piattini, M., Calero, C.: A proposal of a software measurement ontology. Department of Computer Science University of Castilla-La Mancha (2003)
25. Terkaj, W., Urgo, M.: Ontology-based modeling of production systems for design and performance evaluation. In: 2014 12th IEEE International Conference on Industrial Informatics (INDIN), pp. 748–753. IEEE (2014)
26. Witte, R., Zhang, Y., Rilling, J.: Empowering software maintainers with semantic web technologies. In: The Semantic Web: Research and Applications. Lecture Notes in Computer Science, vol. 4519, pp. 37–52. Springer, Berlin (2007)
27. Jin, D., Cordy, J.R.: Ontology-based software analysis and reengineering tool integration: The OASIS service-sharing methodology. In: 21st IEEE International Conference on Software Maintenance (ICSM'05), pp. 613–616. IEEE (2005)

An Improved Fuzzy Analytical Hierarchy Process for K-Representative Skyline Web Services Selection

Abdelaziz Ouadah[1(✉)], Allel Hadjali[2], Fahima Nader[1], and Karim Benouaret[3]

[1] LMCS, Ecole Nationale Supérieure d'Informatique, Oued-Smar, Alger, Algérie
{a_ouadah, f_nader}@esi.dz
[2] LIAS/ENSMA, Poitier, France
allel.hadjali@ensma.fr
[3] LIRIS, Université Claude Bernard Lyon 1, Villeurbanne, France
karim.benouaret@liris.cnrs.fr

Abstract. Nowadays, the processes of selecting web services which give the same functionality with different quality of service (QoS) become an important issue. To deal with the large number of Web services candidates, K-representative Skyline is appeared as a Skyline variant to find the short list of the most relevant Web services that represent a summary about the full skyline result. However, it returns generally a conflicting result. The AHP (Analytical Hierarchic Processes) method and its variants as Fuzzy AHP are widely used in ranking incomparable alternatives. However, it requires a huge number of inputs for users to fulfill a multiple comparison matrix, which make it difficult to use in practice notably in Web services selection field. In this work, we propose an improved Fuzzy AHP called IFAHP which allows to: i) elicit the QoS importance level using linguistic terms based on natural language, asking fewer efforts to users, ii) group the QoS attributes according to their importance level, iii) reduce the number of inputs and generate automatically all pair-wise matrix with respect to each attribute, using a discretization algorithm. The experimental evaluation conducted on real world dataset illustrates the feasibility and the effectiveness of our approach.

Keywords: Web services selection · K-representative skyline
Fuzzy AHP · User preferences

1 Introduction

With the rapid development of Service Oriented Computing, the number of Web services published on the web has increased exponentially. To select the best Web services which provide the same functionality with different quality of service (QoS) is considered as a Multi-Criteria Decision Making (MCDM) problem [1].

S. Chikhi et al. (Eds.): MISC 2018, LNNS 64, pp. 312–328, 2019.
https://doi.org/10.1007/978-3-030-05481-6_24

To choose a Web service among several candidates, with similar functionality but with different qualities of services, the user cannot try all services by himself, but, he can leverage historical invocation experiences performed by other users [2].

In the last years, the Skyline operator is appeared as a new and popular method to select the best or the most relevant Web services [3, 4]. It is considered as a promising direction which allows reducing the space of decisions of the user by preselecting the best one and prunes others. However, dominance relationship used by Skyline presents some drawbacks: i) the number of Skyline Web services cannot be controlled, either a huge or a small number which can be more or less informative for the user, ii) all attributes have the same importance which doesn't allows to users to express their priorities about QoS attributes, for example in financial services, security attribute is very important than other attributes.

To deal with the large number of Skyline Web services candidates, the concept of K Representative Skyline [5, 6] has been proposed, which can return the k services that represent the full Skyline result [7]. However, K-Representative Skyline allows generally returning incomparable and conflicting results, or the user often encounters some difficulties to select the best services which answer most to his preferences.

The approach to use MCDM methods to resolve some selection problems is not new. Analytical Hierarchy Process (AHP) and its variants as Fuzzy AHP is the most significant method being implemented as a Decision Making tool, one of the major advantages of the AHP method is its ability to be combined with other MCDM methods [8]. Researchers have focused on developing hybrid decision methods by combining AHP with other MCDM methods like Promethee [9], Electre [10], Topsis [11], Fuzzy Topsis [12], Skyline and Prometee [13, 14], Skyline, Entropy and Prometee [15]. However, AHP and Fuzzy AHP methods requires to the user to compare between each two criteria, subcriteria and alternatives according to their importance using Saaty scale (1, 3, 5, 7 or 9), which needs to fulfill several comparison matrix which involves more efforts to users, making it difficult to use in practice, notably in Web services selection area. Also, the user is asked in many times to review his judgments about QoS attributes and alternatives if the consistency check is not satisfied, which represent a very difficult task.

To overcome drawbacks cited above, we propose a hybrid approach where:

- Skyline result is clustered in k-clusters which allow identifying a set of k-representative Skyline services that represent all trade-off several QoS parameters and giving users a summary about the full Skyline services result.
- An improved Fuzzy AHP method called IFAHP is proposed to: i) weight QoS properties based on elicitation of user preferences using natural language and minimizing the user efforts in the step of assignment of weights to the different QoS attributes, ii) grouping the QoS attribute according to their importance level in order to reduce the number of pairwise comparison between them, iii) compute the alternatives eigenvectors by generating automatically all comparison matrix from QoS data, using a discretization algorithm. To the best of our knowledge, there is no work in the literature that uses discretization technique to generate alternatives eigenvectors for AHP methods.

The rest of the paper is structured as follows: Sect. 2 provides a background on K-representative Skyline and Fuzzy AHP method. Section 3 presents the approach proposed to compute k-representative Skyline Web services, weight QoS criteria, discretize QoS data and rank-order the K-representative Skyline Web services. Section 4 presents the experimental evaluation. Section 5 concludes the paper.

2 Background

2.1 K-Representative Skyline Operator

The Skyline operator is a popular and relevant tool for modeling and processing preference database queries [16]. Let p_1, p_2 ...p_N be a group of objects, the Skyline operator returns all objects p_i, such that p_i is not dominated by any other object p_j. It relies on Pareto dominance relation which can be expressed as follows:

Given S_i and $S_j \in S$, we say that S_i dominate S_j, denoted by $S_i \prec S_j$, if S_i is better or equal to S_j on all attributes of QoS and strictly better at least on one attribute of QoS (without loss of generality, we assume that the smaller value, the better).

K-representative Skyline allows giving users a summary of the entire Skyline result [5, 6], it reduces considerably the decision space and returns only the most pertinent Skyline services with different trade-off QoS attributes.

Consider a request which illustrates the pertinence of the Skyline operator, concerning a set of web services payment which provides users for hotel and on-line reservation and payment; for each Web service, two QoS are taken into consideration: Security and Response time (Table 1). Assume that QoS data are normalized and the two QoS attributes are to be minimized. The user is looking for the best Web services w.r.t. to security and response time. The most interesting services are provided by the Skyline (see Fig. 1).

Table 1. QoS values

Service	QoS
S_1	(0.6, 0,5)
$\mathbf{S_2}$	**(0.1, 0.8)**
$\mathbf{S_3}$	**(0.7, 0.2)**
S_4	(0.3, 0.8)
$\mathbf{S_5}$	**(0.2, 0.6)**
S_6	(0.7, 0.9)
S_7	(0.3, 1.0)
S_8	(0.4, 0.4)
$\mathbf{S_9}$	**(0.8, 0.1)**
$\mathbf{S_{10}}$	**(0.3, 0.3)**
$\mathbf{S_{11}}$	**(0.25, 0.4)**

Table 2. Scale of pairwise comparisons using TFN.

Intensity of importance	Definition	Membership function
$\tilde{1}$	Equal importance	(1,1,2)
$\tilde{3}$	Moderate importance of one over another	(2,3,4)
$\tilde{5}$	Essential or strong importance	(4,5,6)
$\tilde{7}$	Very strong importance	(6,7,8)
$\tilde{9}$	Extreme importance	(8,9,10)

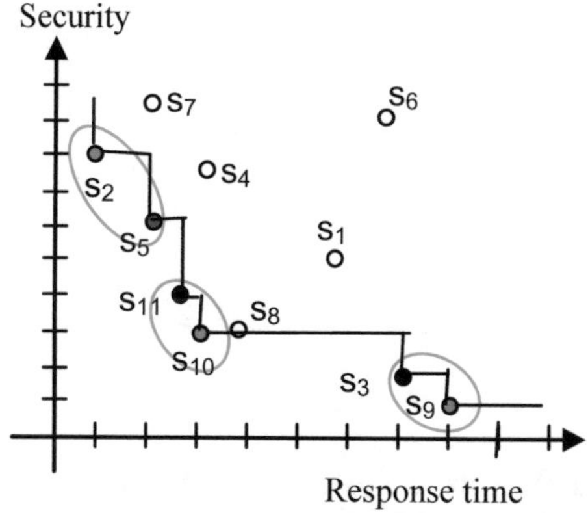

Fig. 1. K-representative Skyline Web services.

It's well known that Skyline assign the same importance to all attributes. Whenever, QoS attribute importance is very pertinent in many domains and change from domain to another, from context to another and from user to another; for example in financial domain, security attribute is prioritized than other attributes. In our example, the Skyline Service S_2 is more secured than others, so, the proposed approach should return the secured services at first. In other situation like information research, where user prefers Skyline services with the best response time, this attributes is to be prioritized than other attributes.

2.2 Fuzzy AHP Method

Due to the vagueness and the uncertainly of user judgments in the pair-wise comparison. AHP proposed by [17] seems inapt to capture user judgments with the better precision [18]. For this reason, fuzzy logic [19] was introduced to the pair-wise comparison of Saaty's AHP in order to overcome this limitation [20]. In the theory of fuzzy set, each element has a degree of membership in a fuzzy set which is defined by a membership function.

Different types of fuzzy membership function was proposed, three types are the most used: monotonic, triangular and trapezoidal. According to [18, 21, 22]: the Triangular Fuzzy Numbers (TFN) is the most suitable due to their computational simplicity and it has been successfully applied in many applications.

Fuzzy AHP steps [18]:

Step 1: In fuzzy AHP, crisp values are replaced by triangular fuzzy numbers ($\tilde{1}$, $\tilde{3}$, $\tilde{5}$, $\tilde{7}$, $\tilde{9}$), in order to indicate the importance scale of each element.

Step 2: constructing the fuzzy comparison, via pairwise comparison.

Via the pairwise comparison, the fuzzy judgment matrix $\tilde{A}$ is constructed as follows:

$$\tilde{A} = \begin{matrix} 1 & \tilde{a}_{12} & \ldots & \tilde{a}_{1n} \\ \tilde{a}_{21} & 1 & \ldots & \tilde{a}_{2n} \\ \\ \tilde{a}_{n1} & \tilde{a}_{n2} & \ldots & 1 \end{matrix} \quad (1)$$

Where

$$\tilde{a}_{ij} = \begin{cases} \tilde{1}, \tilde{3}, \tilde{5}, \tilde{7}, \tilde{9} \text{ or } \tilde{1}^{-1}, \tilde{3}^{-1}, \tilde{5}^{-1}, \tilde{7}^{-1}, \tilde{9}^{-1} & i \neq j \\ 1 & i = j \end{cases} \quad (2)$$

Step 3: solving fuzzy eigenvalues $\tilde{\lambda}$, which is a fuzzy number solution to:

$$\tilde{A}\tilde{x} = \tilde{\lambda}\tilde{x} \quad (3)$$

where $\tilde{A}$ is n-by-n fuzzy matrix, $\tilde{x}$ is a n-by-1 fuzzy eigenvalue containing fuzzy numbers, fuzzy arithmetic is used for the all operations.

The degree of satisfaction can be estimated from the judgment matrix $\tilde{A}$ by the index of optimism λ, the larger value of the index λ indicates the higher degree of optimism:

$$\widehat{a}_{ij}^{\alpha} = \lambda\tilde{a}_{iju}^{\alpha} + (1 - \lambda)\tilde{a}_{ijl}^{\alpha}, \forall\lambda \in [0, 1] \quad (4)$$

While α is fixed, the following matrix can be obtained after setting the index of optimism λ, in order to estimate the degree of satisfaction.

$$\tilde{A} = \begin{bmatrix} 1 & \widehat{a}_{12} & \ldots & \widehat{a}_{1n} \\ \widehat{a}_{21} & 1 & \ldots & \widehat{a}_{2n} \\ \\ \widehat{a}_{n1} & \widehat{a}_{n2} & \ldots & 1 \end{bmatrix} \quad (5)$$

The five TFNs are defined with corresponding intensity of importance as follows:

$$\begin{aligned} &\tilde{1}_\alpha = [1, 3 - 2\alpha] \\ &\tilde{3}_\alpha = [1 + 2\alpha, 5 - 2\alpha], \quad \tilde{3}_\alpha^{-1} = [1/(5 - 2\alpha),\ 1/(1 + 2\alpha)] \\ &\tilde{5}_\alpha = [3 + 2\alpha, 7 - 2\alpha], \quad \tilde{5}_\alpha^{-1} = [1/(7 - 2\alpha),\ 1/(3 + 2\alpha)] \\ &\tilde{7}_\alpha = [5 + 2\alpha, 9 - 2\alpha], \quad \tilde{7}_\alpha^{-1} = [1/(9 - 2\alpha),\ 1/(5 + 2\alpha)] \\ &\tilde{9}_\alpha = [7 + 2\alpha, 11 - 2\alpha], \quad \tilde{9}_\alpha^{-1} = [1/(11 - 2\alpha),\ 1/(7 + 2\alpha)] \end{aligned} \quad (6)$$

Step 4: checking the consistency.

Step 5: computing alternatives eigenvectors with respect each attribute and ranking alternatives.

3 An Improved Fuzzy AHP for Selecting K-Representative Skyline Web Services

The approach which we propose tries to answer the following objectives:

i) Reducing the space research and propose to the user only a the k-representative Skyline Web services that represent the full Skyline Web services
ii) Proposing an improved Fuzzy AHP, that must be able to be efficient in terms of:
   – expressing preference about the importance level of QoS attributes using linguistic terms;
   – grouping QoS attributes that represent the same importance level in order reduce the number of comparison in step of weighting of QoS attributes;
   – reducing the number of inputs and generating automatically all pair-wise comparison of Skyline Web services with respect to each attribute by the discretization of the performance matrix;
   – ranking the K-representative Skyline Web services.

The steps of our approach (see Fig. 2) are described as follows.

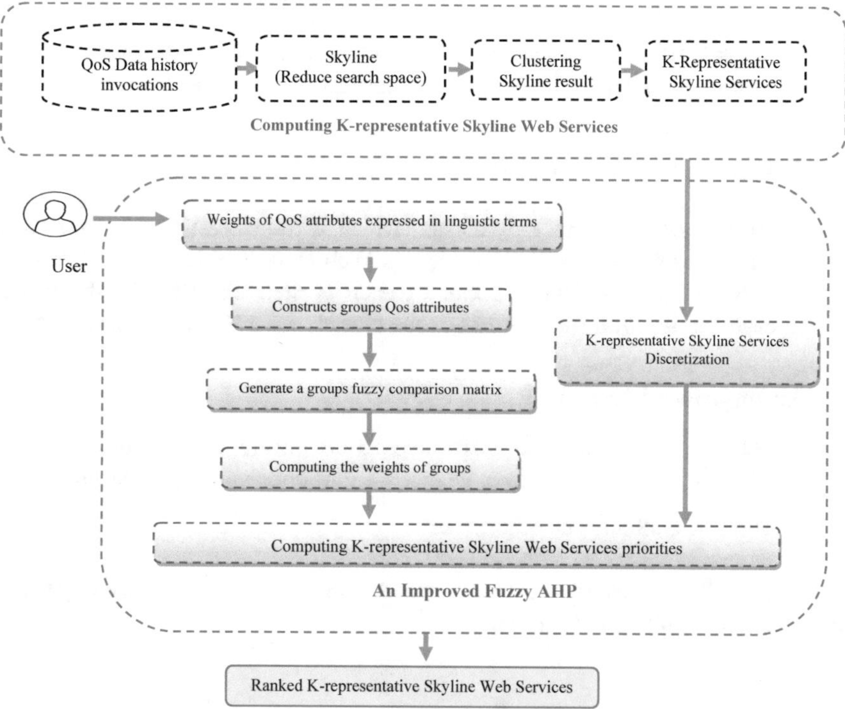

Fig. 2. An Improved Fuzzy AHP for Selecting K-representative Skyline Web Services.

In the first step of our approach, we use the Skyline operator for pruning the search space focusing only on interesting web services that are not dominated by any other service. The Skyline Web services dataset is clustered in K-clusters using K-means method. The best Skyline Web Service for each cluster is proposed to the user. At the end we will have K Skyline Web services which represent the K Representative Skyline Web Services.

In the second step, we propose an improved version of the well-known Fuzzy AHP method to compute the weights of QoS and rank-order the K-representative Skyline according to the user preferences.

3.1 Computing K-Representative Skyline Web Services

A) Computing Skyline

In the first step of our approach, we use the Skyline operator to reduce the search space. We use the well-known BNL (Block Nested Loops) algorithm to compute the Skyline Web services, due to its simplicity and large usability [16].

B) Clustering Skyline Web Services Result

The method K-means is used to cluster Skyline dataset in a k cluster, where every cluster contains all Skyline Web services which are similar between them within the same cluster and that are dissimilar in comparison with the Skyline Web services of other clusters.

K-means is still one of the most widely used algorithms for clustering [23]. Ease of implementation, simplicity, efficiency, and empirical success are the main reasons for its popularity.

C) Selecting the K-representative Skyline

We propose to take one service by cluster, which is the closest of the center of the cluster, that's allows to select the best Skyline Web service which have a good compromise between QoS attributes within of each cluster. To compute the distance between each service with the centroid of the cluster, we use the euclidean distance.

3.2 An Improved Fuzzy AHP

In this section we describe the steps needed to ensure the ranking of the K-representative Skyline Web Services, where, we introduce our Improved Fuzzy AHP.

A) Elicitation of QoS attributes Importance

In our Improved Fuzzy AHP, the importance of QoS attributes is expressed by the user using linguistic terms (see Table 3).

Table 3. Linguistic terms.

Label	Definition
VI	Very important
I	Important
N	Normal
NI	Not Important
NIA	Not Important at All

B) Grouping QoS attributes

In this step, each QoS attributes having the same importance level are affected in the same group. At the end, we consider that all QoS attributes which belongs to the same group will have the same weight.

QoS attributes are grouped in five groups with respect to their importance, noted as follows:

$G_i = (G_1, G_2, \ldots, G_5)$. Where:

G_1 contains Very Important (*VI*) QoS attributes;
G_2 contains Important (*I*) QoS attributes;
G_3 contains Normal (*N*) QoS attributes;
G_4 contains Not Important (*NI*) QoS attributes;
G_5 contains Not Important at All (*NIA*) QoS attributes.

In the example described in Sect. 2, if the importance levels given by user for the two QoS attributes security and response time are VI (Very Important) and NI (Not Important) respectively, so, the tow attributes will belong to two different groups G1 and G4. Where G1 contains the QoS attributes that have an importance level equal *VI*, and G4 contains the QoS attributes that have the importance level equal *NI*, So, G1 = {security} and G4 = {response time}.

C) Comparison matrix generation between groups

The QoS attributes groups comparison matrix can be constructed using the following linguistic terms dependence (*DEP*) which expresses the relationship between each tow linguistic terms using Triangular Fuzzy Numbers:

$$DEP = \begin{matrix} \\ VI \\ I \\ N \\ NI \\ NIA \end{matrix} \begin{matrix} VI \quad I \quad N \quad NI \quad NIA \\ \begin{bmatrix} 1 & \tilde{3} & \tilde{5} & \tilde{7} & \tilde{9} \\ \tilde{3}^{-1} & 1 & \tilde{3} & \tilde{5} & \tilde{7} \\ \tilde{5}^{-1} & \tilde{3}^{-1} & 1 & \tilde{3} & \tilde{5} \\ \tilde{7}^{-1} & \tilde{5}^{-1} & \tilde{3}^{-1} & 1 & \tilde{3} \\ \tilde{9}^{-1} & \tilde{7}^{-1} & \tilde{5}^{-1} & \tilde{3}^{-1} & 1 \end{bmatrix} \end{matrix} \tag{7}$$

The importance intensity between linguistic terms is generated using Table 2.

Very Important QoS attribute have the same importance with another *Very Important* QoS attribute. *Very Important* QoS attribute have a moderate importance with another *Important* QoS attribute. *Very important* QoS attribute have an extreme importance with another *not important at All* QoS attribute. …etc.

In our example, the comparison matrix between the two groups G1 (which contains the Very Important QoS attributes) and G2 (which contains the Not Important QoS attributes) is given as follows:

$$CM(G_1, G_4) = \begin{array}{c} \\ G_1 \\ G_4 \end{array} \begin{array}{c} \begin{array}{cc} G_1 & G_4 \end{array} \\ \begin{bmatrix} \tilde{1} & \tilde{7} \\ \tilde{7}^{-1} & \tilde{1} \end{bmatrix} \end{array} \tag{8}$$

D) Computing the weights *Wi*

Now, the comparison matrix between groups is generated, to compute the weights of QoS attributes groups we use the steps described in Sect. 2.1.

E) Performance Matrix Discretization

According to [24], the discretization transforms quantitative QoS attribute into qualitative attribute, that is, numeric attributes into discrete or nominal attributes with a finite number of intervals. In practice, discretization can be viewed as a data reduction method since it maps data from a huge spectrum of numeric values to a greatly reduced subset of discrete values.

In our approach, to discretize performance matrix, we follow this steps:

Step1: Data normalization: in this step all QoS attributes will be normalized that the higher value is preferred.

Step2: Discretization: in this step we transform continues values into discrete ones, using the well-known equal-width discretization method [25], which partition all continues attributes in k intervals using the following algorithm:

Algorithm 1. QoS data discretization

Input: *QoS data, the number of intervals K.*

Foreach *QoS attribute* ***do***

1- Compute the width of each interval, using the following equation :

W ← (min -max)/k ;

2- Define intervals using the following equation :

*[min + (k-1)*w, min+k*w]*

3- Labeling QoS data: using linguistic terms.

End.

Return Disc [i,j]: discretized QoS data

In our proposition we use k = 5, to compose 5 intervals; where, each label expresses the importance level of the QoS attribute value (see Table 4).

Table 4. Labeling QoS attributes values.

Intervals	QoS Label	Description
[min, min+w]	VI	Very important
[min+w, min+2*w]	I	Important
[min+2*w, min+3*w]	N	Normal
[min+3*w, min+4*w]	NI	Not important
[min+4*w, min+5*w]	NIA	Not important at all

Table 5. Example of QoS Skyline Web services discretization.

Skyline Web Service	Response time	Response time label	Security	Security label
S_2	0.1	VI	0.8	NIA
S_{10}	0.3	I	0.3	I
S_9	0.8	NIA	0.1	VI

The labeling step allows describing the importance level of each Skyline Web service with respect to each attribute. The importance in this case, describes the relevance of the Skyline Web service in this QoS attribute.

In our example described in Sect. 2, if we consider K = 5, then, w = (1 − 0)/ 5 = 0,2. The representative Skyline Web services can be discretized, as follows.

As we can see in Table 5, the Skyline service S_2 is very important with respect to the response time QoS attribute, but it's not important at all in security. S_{10} is important in both QoS attributes. S_9 is not important at all in response time but it's very important in security.

F) Generating comparison matrix with respect to each attribute

In this step, we leverage the discretized data to generate all comparison matrix with respect each attribute, we use the same procedure described in Sect. 3.2.C.

For example, to compare between S_2 and S_{10} in response time QoS attribute, S_2 is labeled as very important (VI), S_{10} is labeled as Important. So, to compare between the two linguistic terms VI and I (see Table 4), the comparison matrix is:

$$CM(S_2, S_{10}) = \begin{array}{c} \\ S_2 \\ S_{10} \end{array} \begin{array}{c} \begin{array}{cc} S_2 & S_{10} \end{array} \\ \begin{bmatrix} \tilde{1} & \tilde{3} \\ \tilde{3}^{-1} & \tilde{1} \end{bmatrix} \end{array} \quad (9)$$

After generating the pairwise comparison between each two Skyline services, the eigenvectors can be computed by the same way, as described in Sect. 3.2.C.

G) Ranking Skyline services

After computing the QoS weights and Skyline services eigenvectors, the Skyline services can be raked using the following score computed as follows:

$$\text{IFAHP Score}_i = \sum_j QoS\,Eigenvector_j \times W_i \tag{10}$$

4 Experiment

In our experiments, we use the publically available QWS DataSet[1] [26], which comprises of nine measurements of QoS attributes; it contains 2507 Real-World Web Services. We consider that all services for the QoS DataSet offer the same functionality. We are using eight QoS attributes (Q1: Response Time, Q2: Availability, Q3: Throughput, Q4: Successability, Q5: Reliability, Q6: Compliance, Q7: Best Practices, Q8: Latency) for the full QWS Dataset, skyline result return 117 Skyline Web services among 2507 proposed. As it can be seen, the search space is reduced; only pertinent Skyline Web services are returned which are not dominated by any other service are proposed.

4.1 Selecting K-Representative Skyline Web Services

In this step we use the clustering method K-means to segment Skyline services to 10 clusters. After computing the distance between each Skyline Web service with the centroid of the cluster, the K-representative Skyline is retrieved as shown in Table 6. We can see that the number of Skyline Web services is reduced considerably, which allows to the user to have a short list of relevant Skyline Web services that represent a summary of the full skyline, where, each Cluster is represented by one service. For example the cluster N° 1 is represented by the Skyline Web service S_{447}, which is the closest to the center of the cluster.

4.2 QoS Attributes Weighting for the Full DataSet

To select one service among the K-representative Skyline Web service, the user is asked to express his priorities about QoS attributes using linguistic terms. We consider the QoS attributes importance expressed by the user preferences using the linguistic terms (see Table 7). The QoS attributes are grouped in 04 groups, G1 = {Q1, Q2}, G2 = {Q3, Q4, Q7, Q8}, G3 = {Q5}, G4 = {Q6}.

The Fuzzy comparison matrix between groups is generated automatically. The weights of groups are computed and the QoS belongs to the same group have the same weights. After the generation of the comparison matrix and defuzzification, the QoS attributes, Weights are computed (Table 7).

[1] http://www.uoguelph.ca/~qmahmoud/qws/index.html.

Table 6. K-representative Skyline Web services result.

Cluster N°	Services number	Representative Skyline
1	8	S_{447}
2	10	S_{1541}
3	7	S_{20}
4	16	S_{2125}
5	14	S_{1736}
6	12	S_{1295}
7	18	S_{823}
8	6	S_{2104}
6	23	S_{1371}
10	3	S_{236}

Table 7. QoS attributes importance computing.

Attribute	Q1	Q2	Q3	Q4	Q5	Q6	Q7	Q8
Importance	VI	VI	N	N	I	NI	N	N
Groups	G1	G1	G2	G2	G3	G4	G2	G2
Weights	0,28	0,28	0,12	0,12	0,2	0,04	0,12	0,12

As we can see, for eight attributes, the user is asked only to express his preferences about the eight attributes using linguistic terms and the computing of the weights will be done in a simple and intuitive manner. In comparison with Fuzzy AHP which asks many efforts to fulfill a comparison matrix composed on 64 entries which make it difficult to use in practice.

4.3 QoS Matrix Discretization

To compute the Skyline Web services eigenvectors we use the discretization technique to transform continues QoS values into discrete ones, after the normalization and labeling of QoS values, the discretized QoS data of the K-representative Skyline Web Services is given in Table 8.

4.4 Computing Eigenvectors with Respect to Each Attribute

In this step, we leverage the QoS discretized data to generate all fuzzy comparison matrix with respect to each attribute using the linguistic terms dependence (*DEP*) which expresses the relationship between each tow linguistic terms using Triangular Fuzzy Numbers, for the attribute Q1: Response time, the fuzzy comparison matrix is given in Table 9.

Then, we fixed the values, $\alpha = 0{,}5$ in equation N° 6. The α-cut comparison matrix and the eigenvector related to the response time attribute are generated (see Tables 10 and 11).

Table 8. QoS values discretization result.

	Q1	Q2	Q3	Q4	Q5	Q6	Q7	Q8
A_{447}	VI	NI	VI	N	N	N	VI	I
A_{1541}	I	I	NI	VI	NIA	VI	N	VI
A_{20}	VI	N	NIA	VI	NI	NIA	N	VI
A_{2125}	VI	I	NIA	VI	N	VI	I	NIA
A_{1736}	NIA	I	NI	VI	I	N	VI	I
A_{1295}	VI	NIA	NI	NIA	N	N	I	NI
A_{823}	VI	I	N	VI	NI	NIA	N	VI
A_{2104}	N	NIA	N	NIA	VI	NIA	VI	I
A_{1371}	N	VI	I	VI	N	N	I	VI
A_{236}	VI	N	I	VI	NIA	N	NIA	VI

Table 9. Fuzzy comparison matrix with respect to response time.

	S_{447}	S_{1541}	S_{20}	S_{2125}	S_{1736}	S_{1295}	S_{823}	S_{2104}	S_{1371}	S_{236}
S_{447}	1	$\tilde{3}$	$\tilde{1}$	$\tilde{1}$	$\tilde{9}$	$\tilde{1}$	$\tilde{1}$	$\tilde{5}$	$\tilde{5}$	$\tilde{1}$
S_{1541}	$\tilde{3}^{-1}$	1	$\tilde{3}^{-1}$	$\tilde{3}^{-1}$	$\tilde{7}$	$\tilde{3}^{-1}$	$\tilde{3}^{-1}$	$\tilde{3}$	$\tilde{3}$	$\tilde{3}^{-1}$
S_{20}	$\tilde{1}^{-1}$	$\tilde{3}$	1	$\tilde{1}$	$\tilde{9}$	$\tilde{1}$	$\tilde{1}$	$\tilde{5}$	$\tilde{5}$	$\tilde{1}$
S_{2125}	$\tilde{1}^{-1}$	$\tilde{3}$	$\tilde{1}^{-1}$	1	$\tilde{9}$	$\tilde{1}$	$\tilde{1}$	$\tilde{5}$	$\tilde{5}$	$\tilde{1}$
S_{1736}	$\tilde{9}^{-1}$	$\tilde{7}^{-1}$	$\tilde{9}^{-1}$	$\tilde{9}^{-1}$	1	$\tilde{9}^{-1}$	$\tilde{9}^{-1}$	$\tilde{5}^{-1}$	$\tilde{5}^{-1}$	$\tilde{9}^{-1}$
S_{1295}	$\tilde{1}^{-1}$	$\tilde{3}$	$\tilde{1}^{-1}$	$\tilde{1}^{-1}$	$\tilde{9}$	1	$\tilde{1}$	$\tilde{5}$	$\tilde{5}$	$\tilde{1}$
S_{823}	$\tilde{1}^{-1}$	$\tilde{3}$	$\tilde{1}^{-1}$	$\tilde{1}^{-1}$	$\tilde{9}$	$\tilde{1}^{-1}$	1	$\tilde{5}$	$\tilde{5}$	$\tilde{1}$
S_{2104}	$\tilde{5}^{-1}$	$\tilde{3}^{-1}$	$\tilde{5}^{-1}$	$\tilde{5}^{-1}$	$\tilde{5}$	$\tilde{5}^{-1}$	$\tilde{5}^{-1}$	1	$\tilde{5}^{-1}$	$\tilde{5}^{-1}$
S_{1371}	$\tilde{5}^{-1}$	$\tilde{3}^{-1}$	$\tilde{5}^{-1}$	$\tilde{5}^{-1}$	$\tilde{5}$	$\tilde{5}^{-1}$	$\tilde{5}^{-1}$	$\tilde{5}$	1	$\tilde{5}^{-1}$
S_{236}	$\tilde{1}^{-1}$	$\tilde{3}$	$\tilde{1}^{-1}$	$\tilde{1}^{-1}$	$\tilde{9}$	$\tilde{1}^{-1}$	$\tilde{1}^{-1}$	$\tilde{5}$	$\tilde{5}$	1

Table 10. The α-cut Fuzzy comparison matrix with respect to Response time QoS attribute.

	S_{447}	S_{1541}	S_{20}	S_{2125}	S_{1736}	S_{1295}	S_{823}	S_{2104}	S_{1371}	S_{236}
S_{447}	1	[2,4]	[1,2]	[1,2]	[8,10]	[1,2]	[1,2]	[4,6]	[4,6]	[1,2]
S_{1541}	[1/4,1/2]	1	[1/4,1/2]	[1/4,1/2]	[6,8]	[1/4,1/2]	[1/4,1/2]	[2,4]	[2,4]	[1/4,1/2]
S_{20}	[1/2,1]	[2,4]	1	[1,2]	[8,10]	[1,2]	[1,2]	[4,6]	[4,6]	[1,2]
S_{2125}	[1/2,1]	[2,4]	[1/2,1]	1	[8,10]	[1,2]	[1,2]	[4,6]	[4,6]	[1,2]
S_{1736}	[1/10,1/8]	[1/8,1/6]	[1/10,1/8]	[1/10,1/8]	1	[1/10,1/8]	[1/10,1/8]	[1/6,1/4]	[1/6,1/4]	[1/10,1/8]
S_{1295}	[1/2,1]	[2,4]	[1/2,1]	[1/2,1]	[8,10]	1	[1,2]	[4,6]	[4,6]	[1,2]
S_{823}	[1/2,1]	[2,4]	[1/2,1]	[1/2,1]	[8,10]	[1/2,1]	1	[4,6]	[4,6]	[1,2]
S_{2104}	[1/6,1/4]	[1/4,1/2]	[1/6,1/4]	[1/2,1]	[4,6]	[1/6,1/4]	[1/6,1/4]	1	[1/6,1/4]	[1/6,1/4]
S_{1371}	[1/6,1/4]	[1/4,1/2]	[1/6,1/4]	[1/2,1]	[4,6]	[1/6,1/4]	[1/6,1/4]	[4,6]	1	[1/6,1/4]
S_{236}	[1/2,1]	[2,4]	[1/2,1]	[1/2,1]	[8,10]	[1/2,1]	[1/2,1]	[4,6]	[4,6]	1

Table 11. The eigenvector for Fuzzy comparison matrix of the attribute Response time.

	S_{447}	S_{1541}	S_{20}	S_{2125}	S_{1736}	S_{1295}	S_{823}	S_{2104}	S_{1371}	S_{236}	Eigenvector
S_{447}	1	3	1,5	1,5	9	1,5	1,5	5	5	1,5	0,1692
S_{1541}	0,375	1	0,375	0,375	7	0,375	0,375	3	3	0,375	0,0619
S_{20}	0,75	3	1	1,5	9	1,5	1,5	5	5	1,5	0,1570
S_{2125}	0,75	3	0,75	1	9	1,5	1,5	5	5	1,5	0,1461
S_{1736}	0,1125	0,14583	0,1125	0,1125	1	0,1125	0,1125	0,208	0,208	0,1125	0,0125
S_{1295}	0,75	3	0,75	0,75	9	1	1,5	5	5	1,5	0,1363
S_{823}	0,75	3	0,75	0,75	9	0,75	1	5	5	1,5	0,1273
S_{2104}	0,208	0,375	0,208	0,208	5	0,208	0,208	1	0,208	0,208	0,0291
S_{1371}	0,208	0,375	0,208	0,208	5	0,208	0,208	5	1	0,208	0,0416
S_{236}	0,75	3	0,75	0,75	9	0,75	0,75	5	5	1	0,1191
							λ_{max} = 10,91 CI = 0,10 CR = 0,068				

Table 12. The final ranking of K-representative Skyline Web service.

QoS Attributes		Eigenvectors for each QoS attributes with respect to each Skyline Web service										CR <= 0,1
		S_{447}	S_{1541}	S_{20}	S_{2125}	S_{1736}	S_{1295}	S_{823}	S_{2104}	S_{1371}	S_{236}	
RT	0,28	0,1692	0,0619	0,1570	0,1461	0,0125	0,1363	0,1273	0,0291	0,0416	0,1191	0,068
AVAIL	0,28	0,0323	0,1464	0,0642	0,1370	0,1284	0,0182	0,1205	0,0168	0,2758	0,0605	0,067
THR	0,12	0,3110	0,0448	0,0210	0,0193	0,0420	0,0393	0,0912	0,0861	0,1781	0,1672	0,056
SUCC	0,12	0,0342	0,1558	0,1453	0,1359	0,1272	0,0127	0,1321	0,0127	0,1220	0,1220	0,073
RELIA	0,20	0,0986	0,0210	0,0585	0,0928	0,1863	0,0769	0,0398	0,3233	0,0829	0,0199	0,080
COMPLIA	0,04	0,0878	0,2848	0,0216	0,2639	0,0828	0,0781	0,0200	0,0184	0,0735	0,0692	0,052
BEST P	0,12	0,2134	0,0429	0,0402	0,0955	0,1978	0,0899	0,0376	0,1844	0,0845	0,0138	0,054
LATENCY	0,12	0,0693	0,1777	0,1649	0,0129	0,0625	0,0202	0,1536	0,0612	0,1434	0,1342	0,048
IFuzzy AHP Score		0,1550	0,1245	0,1191	0,1400	0,1316	0,0812	0,1279	0,1196	0,1718	0,1095	
Rank		2nd	6th	8th	3rd	4th	10th	5th	7th	1st	9th	

To compute the coherence indicator (CI), the expression (1) is used as follows:

$$IC = (\lambda_{max} - N)/(N - 1) = (10,91 - 10)/(10 - 1) = 0,10.$$

While the Coherence ratio (CR) is calculated as follows according to the expression (2):

$$RC = (IC/ACI) = (0,10/1,49) = 0,068.$$

Then Coherence ratio (CR) is equal to 0,068 which is less than 0.1, this means that weights allocated to QoS properties are consistent.

4.5 Ranking K-Representative Skyline Web Services

After computing the Qos weights and Skyline Web services eigenvectors (see Table 12), the Skyline Web services can be raked using the equation N° 10, the final ranking is given in Table 13.

Table 13. The ranked K-representative Skyline Services.

Rank	1st	2nd	3rd	4th	5th	6th	7th	8th	9th	10th
Service	S_{1371}	S_{447}	S_{2125}	S_{1736}	S_{823}	S_{1541}	S_{2104}	S_{20}	S_{236}	S_{1295}

5 Conclusion

In this paper, the problem of K-representative Skyline Web services ranking is handled, taking into consideration the preferences and the needs of the user. The K-representative Skyline method is used to reduce the number of candidate services to be considered in the decision space giving users a summary about the full Skyline Web services. An improved version of the Fuzzy AHP method is proposed for weighting QoS attributes and ranking of Skyline Web services, introducing triangular fuzzy numbers to deal with vagueness and uncertainly in user's judgments. The experimental evaluation illustrates that our approach can capture the user preferences, involving fewer efforts and retrieves the best ranked K-Representative Skylines Web services. As a future work, we plan to extend this work integrating a comparison study with similar approaches.

References

1. Alrifai, M., Skoutas, D., Risse, T.: Selecting Skyline services for QoS-based web service composition. In: Proceedings of the 19th International Conference one World Wide Web, pp. 11–20 (2010). https://doi.org/10.1145/1772690.1772693
2. Shao, L., Zhang, J., Wei, Y., Zhao, J., Xie, B., Mei, H.: Personalized QoS prediction forweb services via collaborative filtering. In: ICWS, pp. 439–446 (2007). https://doi.org/10.1109/icws.2007.140
3. Benouaret, K.: Advanced techniques for Web service query optimisation. Doctoral dissertation. Claude Bernard Lyon 1 university. Computer science laboratory in Picture and information Systems (2012)
4. Yu, Q., Bouguettaya, A.: Computing service skyline from uncertain qows. IEEE Trans. Serv. Comput. **3**(1), 16–29 (2010). https://doi.org/10.1109/tsc.2010.7
5. Lin, X., Yuan, Y., Zhang, Q., Zhang, Y.: Selecting stars: the k most representative Skyline operator. In: IEEE 23rd International Conference on Data Engineering, ICDE 2007, pp. 86–95. IEEE, April 2007. https://doi.org/10.1109/icde.2007.367854
6. Tao, Y., Ding, L., Lin, X., Pei, J.: Distance-based representative Skyline. In: IEEE 25th International Conference on Data Engineering, ICDE 2009, pp. 892–903. IEEE, March 2009. https://doi.org/10.1109/icde.2009.84

7. Chen, G., Ma, X., Yang, D., Tang, S., Shuai, M., Xie, K.: Efficient approaches for summarizing subspace clusters into k representatives. Soft. Comput. **15**(5), 845–853 (2011). https://doi.org/10.1007/s00500-010-0552-8
8. Inti, S., Tandon, V.: Application of fuzzy preference-analytic hierarchy process logic in evaluating sustainability of transportation infrastructure requiring multicriteria decision making. J. Infrastruct. Syst. **23**(4), 04017014 (2017). https://doi.org/10.1061/(ASCE)IS.1943-555X.0000373
9. Wang, T.C., Chen, Y.H.: Applying fuzzy linguistic preference relations to the improvement of consistency of fuzzy AHP. Inf. Sci. **178**(19), 3755–3765 (2008). https://doi.org/10.1016/j.ins.2008.05.028
10. Kaya, T., Kahraman, C.: A fuzzy approach to e-banking website quality assessment based on an integrated AHP-ELECTRE method. Technol. Econ. Dev. Econ. **17**(2), 313–334 (2011). https://doi.org/10.3846/20294913.2011.583727
11. Büyüközkan, G., Çifçi, G.: A combined fuzzy AHP and fuzzy TOPSIS based strategic analysis of electronic service quality in healthcare industry. Expert Syst. Appl. **39**(3), 2341–2354 (2012). https://doi.org/10.1016/j.eswa.2011.08.061
12. Önüt, S., Kara, S.S., Efendigil, T.: A hybrid fuzzy MCDM approach to machine tool selection. J. Intell. Manuf. **19**(4), 443–453 (2008). https://doi.org/10.1007/s10845-008-0095-3
13. Ouadah, A, Benouaret, K., Hadjali, A., Nader, F.: SkyAP-S3: a hybrid approach for efficient skyline services selection. In: Proceedings of the 8th IEEE International Conference on Service Oriented Computing & Applications (SOCA 2015), Rome, Italy, 19–21 Oct 2015. https://doi.org/10.1109/soca.2015.22
14. Ouadah, A., Benouaret, K., Hadjali, A., Nader, F.: Combining skyline and multi-criteria decision methods to enhance web services selection. In: 12th International Symposium on Programming and Systems (ISPS 2015). IEEE, Algiers, Algeria, 28–30 Apr 2015. https://doi.org/10.1109/isps.2015.7244975
15. Ouadah, A., Hadjali, A., Nader, F., et al.: SEFAP: an efficient approach for ranking skyline web services. J. Ambient Intell. Human. Comput. (2018). https://doi.org/10.1007/s12652-018-0721-7
16. Borzsonyi, S., Kossmann, D., Stock, K.: The Skyline operator. In: ICDE (2001). https://doi.org/10.1109/icde.2001.914855
17. Saaty, R.W.: The analytic hierarchy process-what it simple percentage and how it used simple percentage. Math. Models **9**(3), 161–176 (1987)
18. Chang, D.Y.: Applications of the extent analysis method on fuzzy AHP. Eur. J. Oper. Res. **95**, 649–655 (1996). https://doi.org/10.1016/0377-2217(95)00300-2
19. Zadeh, L.A.: Fuzzy sets as a basis for a theory of possibility. Fuzzy Sets Syst. **1**, 3–28 (1978). https://doi.org/10.1016/s0165-0114(99)80004-9
20. Buckley, J.J.: Fuzzy hierarchical analysis. Fuzzy Sets Syst. **17**(3), 233–247 (1985). https://doi.org/10.1016/0165-0114(85)90090-9
21. Kwong, C.K., Bai, H.: A fuzzy AHP approach to the determination of importance weights of customer requirements in quality function deployment. J. Intell. Manuf. **13**(5), 367–377 (2002). https://doi.org/10.1023/a:1019984626631
22. Tang, Y.C.: An approach to budget allocation for an aerospace company – fuzzy analytic hierarchy process and artificial neural network. Neurocomputing **72**, 3477–3489 (2009). https://doi.org/10.1016/j.neucom.2009.03.020
23. Jain, A.K.: Data clustering: 50 years beyond K-means. Pattern Recognit. Lett. **31**(8) (2010). https://doi.org/10.1016/j.patrec.2009.09.011
24. Garcia, S., Luengo, J., Sáez, J.A., Lopez, V., Herrera, F.: A survey of discretization techniques: taxonomy and empirical analysis in supervised learning. IEEE Trans. Knowl. Data Eng. **25**(4), 734–750 (2013). https://doi.org/10.1109/TKDE.2012.35

25. Dougherty, J., Kohavi, R., Sahami, M.: Supervised and unsupervised discretization of continuous features. In: Machine Learning: Proceedings of the Twelfth International Conference, vol. 12, pp. 194–202, July 1995
26. Al-Masri, E., Mahmoud, Q.H.: QoS-based discovery and ranking of web services. In: IEEE 16th International Conference one Computer Communications and Networks (ICCCN), pp. 529–534 (2007). https://doi.org/10.1109/icccn.2007.4317873

ATL Based Refinement of WS-CDL Choreography into BPEL Processes

Khadidja Salah Mansour(✉) and Youcef Hammal

LSI, Department of Computer Science, FEI, USTHB University, Algiers, Algeria
{ksalahmansour, yhammal}@usthb.dz

Abstract. Web services are often combined together to provide richer features for designing safer and more reliable systems. The composition of services is intended to inter-operate, interact and coordinate multiple services for the achievement of a global goal, or provide new service functions in general. The process for creating such composite services from existing ones is called Web services composition whose description may be achieved through a choreography which globally specifies the interactions between participating services. However, a composition is actually achieved using processes called orchestrators whose actions implement the needed calls to combined services. In this paper, we choose CDL and BPEL as specification languages for choreography and orchestration, respectively. The paper proposes a refinement of CDL choreographies into executable BPEL orchestrations using metamodel-driven translation technique that consists of a set of ATL rules. We propose a solution to transform a given choreography into a set of orchestrations, which exploits CDL and BPEL meta-models. We then propose and implement a set of translation rules using the language ATL, which refines a given CDL specification into a BPEL orchestration processes.

Keywords: Web service composition · Meta-model driven · Translation Choreography · Orchestration · ATL · WS-CDL · WS-BPEL

1 Introduction

Web services have emerged through the efforts of several organizations with a common interest in developing and maintaining an "electronic market". They wanted to be able to communicate more simply and without having to consult each of their transactions in order to be able to interpret their different data. In fact, a Web service exposes its functionalities via an interface described in a computer-readable format (written in the WSDL language). Other computer systems interact, without human intervention, with the web service using a procedure that uses SOAP protocol messages.

Web services have become therefore an essential technique for building loosely coupled distributed systems. Service Oriented Architecture has been widely used in several areas such as e-business systems, e-government, automotive systems, multimedia services, finance and many other areas.

S. Chikhi et al. (Eds.): MISC 2018, LNNS 64, pp. 329–343, 2019.
https://doi.org/10.1007/978-3-030-05481-6_25

As aforementioned, in Web services technology, the basic infrastructure consists of standards: WSDL, SOAP (Simple Object Access Protocol) and Universal Description, Discovery and Integration (UDDI). Web services enable companies to make their information and expertise accessible via the Internet. SOAP is the standard protocol used to interact with Web services. This protocol describes among other things the format of the messages exchanged.

A Web service has the particularity of offering task-specific functionality, and most of the time cannot meet the requirements of users. In order to provide these users with personalized services, it is sometimes necessary to combine them, thus offering new features.

In this context, the main line of research has received a great deal of attention from research organizations and industry, including the combination of services (called component services) into a composite service. The process performing this task is called Web services composition.

As a result, the current trend is to express the logic of a composite Web service using a business process modeling language adapted to Web services. A number of such languages have emerged such as WS-BPEL and WS-CDL.

The composition of services can be described at different levels of abstraction. At a higher level, this is considered as the exchange of messages between the participating services. Such an abstract specification defining public message exchanges, interaction rules and agreements between participating services is called choreography [1, 2]. WS-CDL is one the languages for the specification of choreographies.

A choreography is a decentralized perspective, which describes the public messages exchanged, and defines how abstract participating services should interact with each other. However, at a finer level, it is necessary to define the behavior of each process within the choreography in such a way these could be used in the implementation phase. The most important one among these processes is the orchestrator aiming at the achieving of the collaboration between interacting services.

Hence, an orchestration is a centralized coordination of the participating services, defining the message exchanges with the necessary internal actions, such as translations and invocations of the internal functions. WS-BPEL is the most famous language suited to describe these processes thanks to its standardization and execution engines.

In this paper, we propose and implement a solution to transform a given CDL choreography specification (collaboration between services) into a set of orchestration BPEL processes so that the common objective is preserved. The translation from CDL to BPEL aims to automate the composition process, and to accelerate application development through the reusability of the service functionalities. To achieve our goal, we use a technique driven by meta-models related to languages that support the composition of services starting with a CDL choreography of services and ending with BPEL orchestrations.

We hence use the Atlas Transformation Language (ATL) [3] to write translation rules between concepts of CDL and BPEL meta-models in order to that transform elements of a CDL choreography into equivalent ones in BPEL processes.

The paper is structured as follows. Section 2 describes underlying techniques for the description of Web services compositions. Next, Sect. 3 presents ATL language. Section 4 describes the translation process of CDL Choreography into BPEL orchestrations based on ATL rules. Section 5 then presents related work and gives a comparison between related approaches with our solution proposed in this paper. Finally, Sect. 6 concludes and outlines future directions.

2 Composition of Web Services

2.1 The Process of Web Services Composition

The composition consists in combining the functionalities of several Web services within the same business process (or business process) in order to respond to complex requests that a single service could not satisfy [4]. The business process is a concrete representation of the tasks to be performed in a composition. The course of a composition requires the completion of the following steps:

1. The discovery of Web services that can meet the needs of the composition is usually done manually by sending queries to UDDI directories. Many studies aim to automate this step [5–10].
2. The organization of interactions between composite Web services is distinguished by the terms choreography and orchestration [11].
3. The choreography manages message exchanges with different Web services.
4. Orchestration, on the other hand, organizes the interactions between several Web services. A composition is associated with a specification to manage message exchanges and set up the necessary control structures (iterative loops, conditional operators and exception handling).

2.2 Types of Web Services Composition

Most work on Web Services composition identifies two types of composition: orchestration and service choreography. However, according to the related literature, the definitions of types of composition differ. They are considered as means of conceiving composition [1], while orchestration and choreography are views of the composition of services.

- **Orchestration**: It is a set of processes executed in a predefined order to meet a goal. This type of composition makes it possible to centralize the invocation of component Web services. So, each service is described in terms of internal actions. The contracts between two services are constituted according to the process to be executed. The authors of [2, 12] define the orchestration as an executable process. [12] add that orchestration is a set of actions to be performed through Web services. An execution engine, a Web service acting as a conductor, manages the Web services flow through control logic. To design an orchestration of Web services, one must describe the interactions between the runtime engine and Web services. These interactions correspond to the calls, made by the engine, of action (s) proposed by the component Web services [1]. Web services orchestration is the programming of an engine that invokes a set of Web services according to a predefined process. This engine defines the process as a whole and calls Web services (both internal and external to the organization) according to the order of the execution tasks.

- **Choreography**: A choreography [2] describes composition as a means of achieving a common goal by using a set of abstract Web services. Collaboration between Web services in the collection (as part of the composition) is described by control streams. The choreography specifies a set of Web services to accomplish a common goal. To conceive a choreography, the interactions between the different services must be described. The control logic is supervised by each of the services involved in the composition. The execution of the process is then distributed. In [1] the description of each Web service involved in the choreography includes a description of its participation in the process. Thus, these services can collaborate using exchanged messages to find out whether a particular service can help in the execution of the request. Each Web service can communicate with another service.

3 ATL (Atlas Transformation Language)

As we are interested in translating CDL choreographies into BPEL processes, we make use of a method based on ATL language.

ATL is a model driven translation language developed by OBEO [2] and INRIA [3], which can be specified as both a meta-model and a concrete textual syntax. It also provides both declarative and imperative constructs and represents a hybrid model translation language [3]. The architecture of ATL is a layered architecture: ATLAS Model Weaving (AMW), ATL and ATL Virtual Machine (ATL VM)

An ATL translation program is composed of rules and an action block representing the built imperative and associated with the rules representing the declarative part.

Transformation rule is the basic construct in ATL used to express the transformation logic and it describes the transformation of a source model to a target model. It begins with the word rule followed by the name of the *rule*. It has the following form (see Fig. 1):

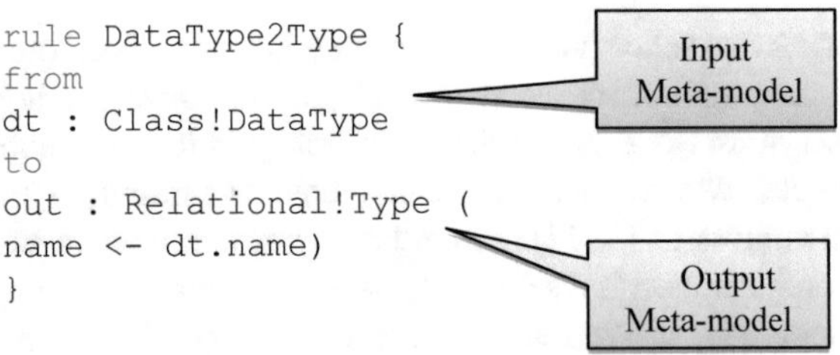

Fig. 1. Sample ATL rule

4 Translation CDL into BPEL Based on ATL

In this section, we expose the translation approach based on ATL illustrated by Fig. 2.

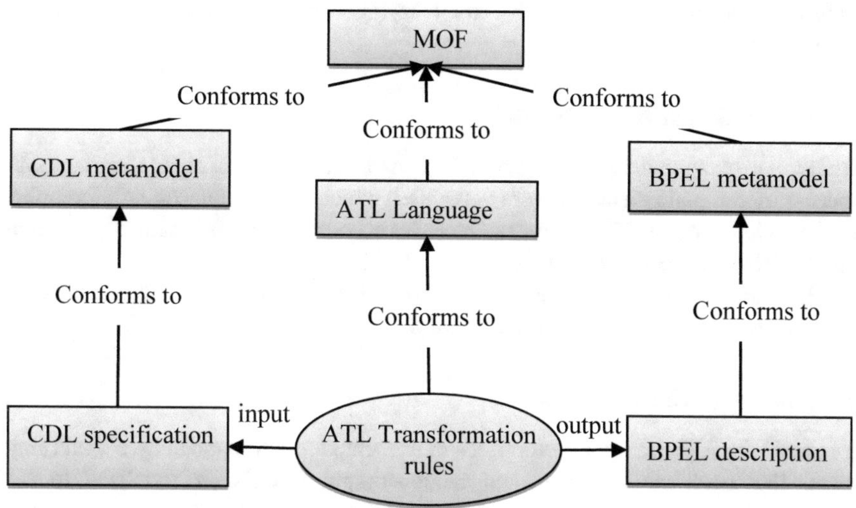

Fig. 2. The ATL based translation approach

4.1 Mapping CDL-BPEL

The CDL to BPEL translation mappings are implemented as a translation rule in the ATL language.

The translation rule defines how a language concept in CDL is mapped into a relevant language concept in BPEL. In both CDL and BPEL meta-models, there are a number of concepts with the same name.

Description language WS-BPEL. WS-BPEL [1] (Business Process Execution Language for Web Services) provides a language for the formal specification of business processes and their interactions (WS-BPEL). IBM, BEA Systems, Microsoft, SAP AG and Siebel Systems all contributed to the development of BPEL4WS (referred to as WS-BPEL). WS-BPEL supports modeling of two types of processes: executable and abstract processes. BPEL4WS is the first Web services composition language adopted by the Web Services community. This is mainly because this language has great expressiveness in the definition of the executable process [13].

Description language WS-CDL. CDL is a language resulting from the standardization efforts of the W3C working group on Web services choreography. The purpose of this language is to describe the relationships between Web services during a choreographic type composition like Web service standards [14].

We classify the choreography into two types:

Structural choreography: A structural choreography represents the choreography of concepts that are responsible for establishing collaboration and message exchanges between collaborating parties.

Behavioral choreography: A behavioral choreography represents the choreography of the concepts that are responsible for carrying out various activities between the collaborating parties.

4.2 Generation of Meta-Models

Initially, we generated the CDL and BPEL meta-models from their schemas. Many unnecessary template elements were generated, which would unnecessarily complicate the translation process. Therefore, we have based our method on manually developed meta-models (see Figs. 3 and 4).

ATL transformation rules are unidirectional, from which they work as read-only on source models and write-only on target model.

4.3 Chain of Transformation

To execute the ATL transformation rules in this transformation (see Fig. 5), this engine expects that each model (i.e., input models/output models) is serialized to XML Metadata Interchange (XMI) format. Source and target models and meta-models may be expressed in XMI [3].

CDL model (XML) to CDL model (XMI). This translation is used to convert the CDL model (XML format) to a CDL model (XMI format) conforming to the CDL meta-model. We represent this translation by *T1*.

CDL model (XMI) to BPEL model (XMI). The core of the translation of this approach executes the translation rules ATL.

This ATL engine reads the CDL (XMI) model as input and generates the BPEL (XMI) model according to the ATL translation rules we have proposed. We represent this translation in *T2*.

BPEL model (XMI) to the BPEL process (XML). After executing the ATL rules on the input CDL (XMI) model, the ATL engine generates a BPEL model in XMI format that conforms to the BPEL meta-model.

The BPEL model (XMI) will not be able to be executed by the orchestration engines for that we carry out another *T3* translation, which will transform in turn the BPEL (XMI) model in BPEL (XML) process.

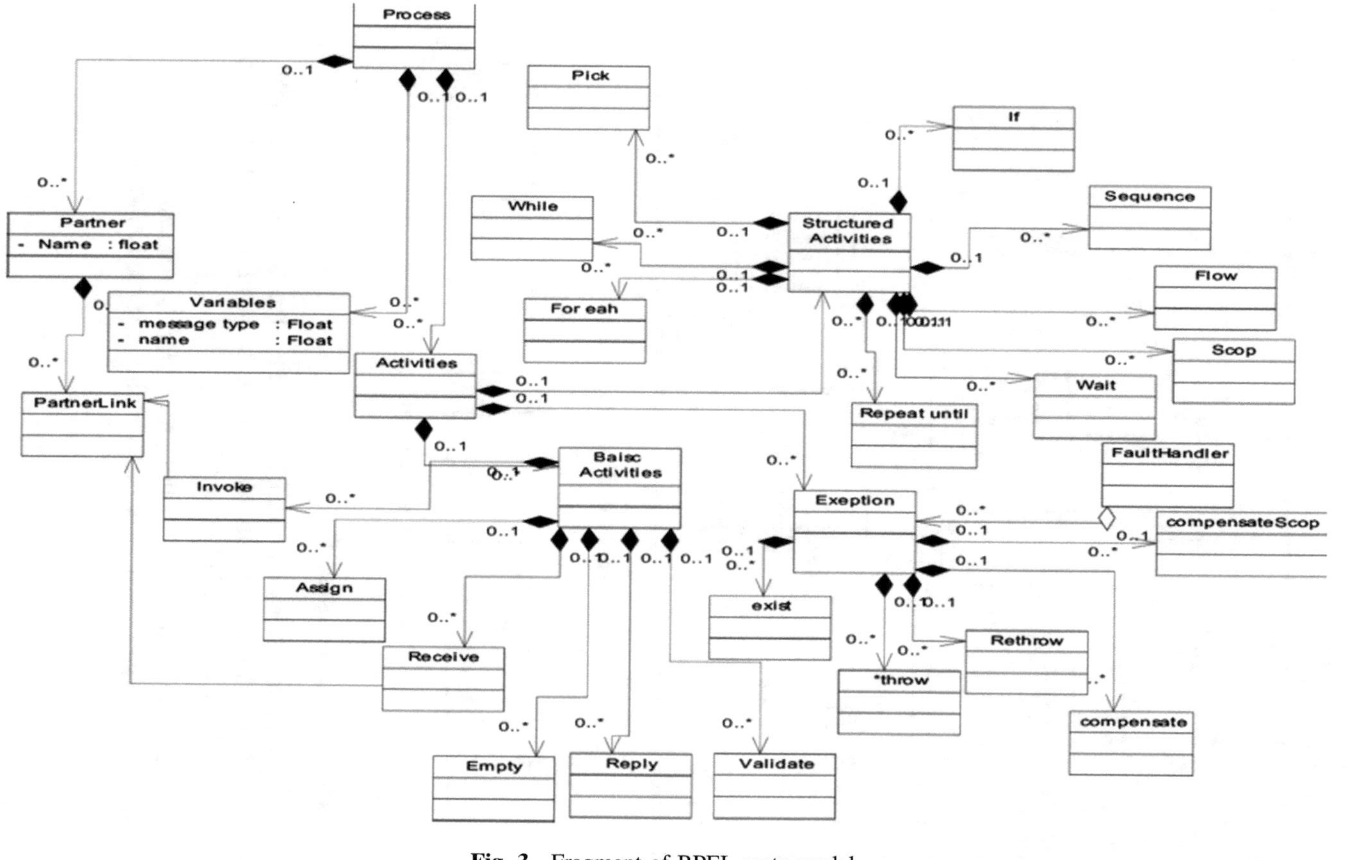

Fig. 3. Fragment of BPEL meta-model

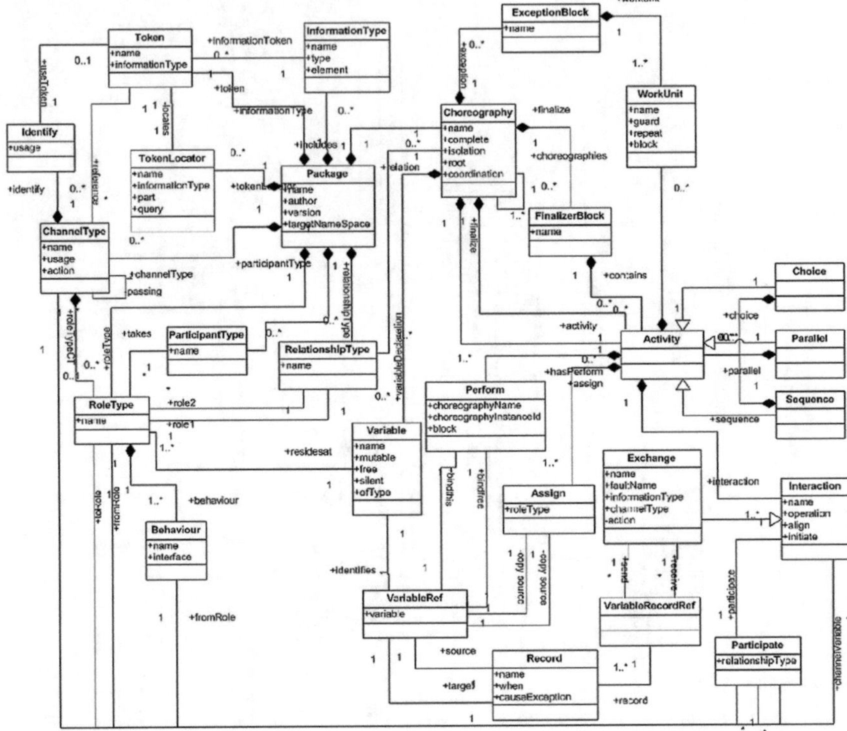

Fig. 4. Fragment of CDL meta-model

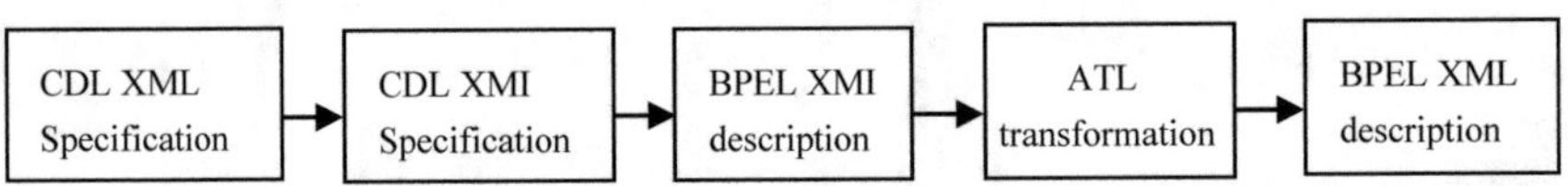

Fig. 5. Steps of the transformation

4.4 ATL Rules of the Translation

In this section we present some rules of CDL to BPEL transformation mappings in ATL language. To avoid naming confusion, we use cdl: and bpel: prefixes to indicate which meta-models the concepts belong to:

- ***cdl: roleType* to *bpel: process***: In BPEL, each process represents a role in the collaboration, so we generate a "*bpel: process*" for each "cdl: *roleType*"

```
rule roleType2MM1process{
  from s: MM!RoleType -- (not s.isConnectedToChan nelType()
         --and s.name=thisModule.orchestrator --)
  to t: MM1!Process(
      name <- s.name + 'Process',
      targetnameSpace <- s.gettargetNameSpace(),
      Variables <- s.residesat,
      ----> collect(v | thisModule.varTovar(v)),
      partnerlinks <- s.takes,
      messageExchanges <- s.messageExchanges,
      correlationSets <- s.ChannelType
      --activity <- s.activity
                 )
  }
```

- ***Cdl:channelType*** **to** ***bpel:correlationSet***: A *"cdl: channelType"* is used as a communication point in the collaboration. But, it can also be used to derive the *"bpel: correlationSet"* for the exchange of stateful messages. Therefore, we map each *"cdl: channelType"* with *"cdl: identity"* into a *"bpel: correlationSet"*.

```
rule channeltype2correlationset
{
  from v : MM!ChannelType--,
           v2:MM!Token
  to  bv : MM1!CorrelationSet (
        name<- v.name,
        properties <- v.getproperties()
                                    )
}
```

- ***Cdl: sequence*** **to** ***bpel: sequence***: *"Cdl: sequence"* is used to specify the order of closed activities and such behavior is *"bpel: sequence"*.

```
      rule sequence2sequence --activity seque =>activity
    {
      from v : MM!sequence
      to  bv : MM1!sequence (
               messageExchanges <- v.messageExchanges,
               activity <- v.sequence,
               activity <- v.activity,
               assign2 <- v.assign
                              )
    }
```

- ***Cdl: parallel*** **to** ***bpel: flow***: A *"cdl: parallel"* activity is used to specify the concurrent execution of joined activities. Similarly, the parallel behaviour is specified by the "*bpel: flow*" activity, so we map each "cdl: parallel" to the "*bpel: flow*" activity.

```
rule Parallel2Flow
{
from  v : MM!Parallel--,
          -- v2:MM!Activity
to bv : MM1!Flow(
        activity <- v.messageExchanges,
        activity <-v.parallel,
        activity <- v.activity,
    assign2 <- v.assign
    --messageExchanges <- v.interaction
    )
}
```

- ***Cdl: choice* to *bpel: ifelse:*** *"Cdl: choice"* does not include any conditions to decide which activity to perform. There is no direct mapping from *"cdl: choice"* to a construct in BPEL that can specify an unconditional branch. The conditional branch concept specified by *"bpel: ifelse"* implements *"cdl: choice"*. Since both are decision-making constructs, we map each choice of *"cdl: choice"* into a construct of *"bpel: ifelse"*.

```
 rule Choice2if
{
  from v  : MM!Choice
  to   bv : MM1!If (
             condition <- 'manually',
             If <- v.choice,
             assign2 <- v.assign
                     )
}
```

The rules for mapping from CDL to BPEL are summarized in the following Table 1.

The approach has been implemented using java. Figure 6 illustrates a screenshot of this application showing the most important functions of the translation process.

5 Related Work

A number of research projects targeting the translation of choreography into orchestration have been reported in the literature. We present herein related work throughout what follows and we will briefly outline some of them.

Kim and Huemer [15] proposed an approach based on ebXML standard for choreography specification which is translated into BPEL process. These authors proposed guidelines which represent architectural relationships between choreography and orchestration that have not been addressed for the translation, but did not provide any concrete implementation of the latter.

In [16], Mendling and Hafner describe the translation CDL into BPEL, which it based on XSLT. The choreography is seen through two aspects, namely global choreography and local choreography. Global choreography refers to the exchange of

Table 1. Mapping CDL to BPEL

	CDL	BPEL	Details
Structural mapping	RoleType	Process by role	The attribute bpel: targerNamespace is derivated by cdl: package
	participantType	PartnerLink	
	relationshipType	PartnerLinkType	
	Variable	Variable	The attribute bpel:messageType is derivated by cdl:type
	ChannelType	correlationSet	bpel:properties is derivate by cdl:name by cdl:token
Behavioral mapping	Sequence	Sequence	
	Parallel	Flow	
	Choice	if-else	
	repeat = false, block = false	if-else	
	repeat = true	while	
	action = request	Invoke	
	action = request	Receive	
	action = respond	Reply	
	Assign	Assign	
	Finalize	compensationHandler	
	NoAction	Empty	
	SilentAction	sequence with nested empty	
	Timeout	pick	

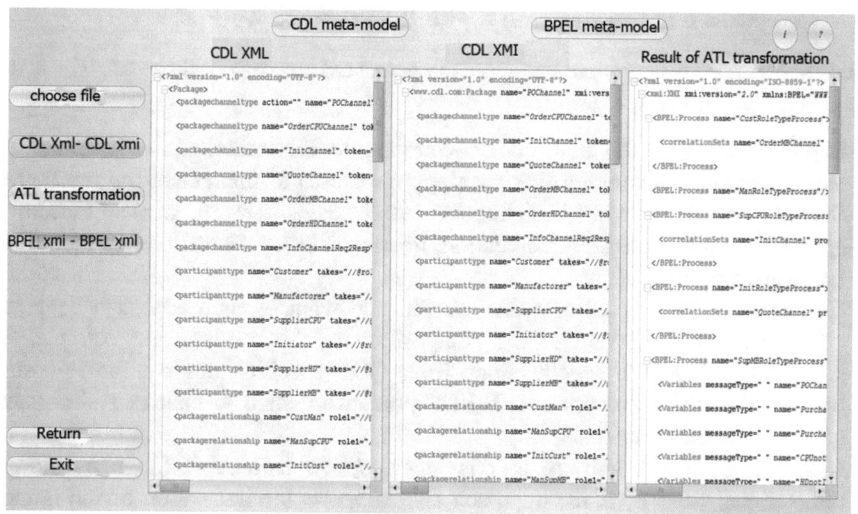

Fig. 6. Example of transformation

messages from a global perspective and is mentioned as a coordination protocol, while the local choreography corresponds to the exchange of messages from the point of view of a single party. The principal advantage in this approach is the bidirectional translation, that is, the translation is defined CDL into BPEL and the authors have explain the mapping CDL to BPEL. However, there is no information of translation mapping for the BPEL *invoke activity*, which is one of the import interactivities.

Another approach which generates a BPEL process from a given CDL specification is given in Rosenberg, Enzi et al. [17]. The translation methodology also takes into account Service Level Agreements (SLAs), which are defined as annotations to the CDL specification and are transformed into strategies that can be applied by a BPEL engine at runtime. In addition, BPEL specifications are also generated for all generated processes. The choreography of translation from CDL to BPEL is inspired by [18]. The main advantage for this approach is the use of the concept of endpoint projection, which avoids the generation of unnecessary constructs in the BPEL process and therefore optimizes the translation process but they do not mention the architectural relationship between choreography and orchestration.

Unlike other approaches Weber, Haller et al. study first the relation between choreography and the executable business process then present their CDL translation in BPEL in the context of a virtual organization, possibly also applicable to other domains [19]. Research also introduces the concept of information gap; a term used to describe the different levels of detail between choreography and orchestration, and indicates that the sum of orchestrations contains more knowledge than the choreography involved. Given a choreography, the approach generates executable processes for each role, with the respective WSDL specification. The translation is implemented as Java Web Services. The con-birth database stores the private information of the individual services and is implemented as a relational database.

5.1 Comparison of Existing Work with the Proposed Approach

In this subsection, we compare the approach presented above with those of [16, 17, 19] according the following criteria:

- Implementation of the translation: according to this criterion, we consider the implementation technology that each approach uses to implement the translation. Sendall and Kozaczynski distinguish three categories of model translation approaches: direct model manipulation, intermediate representation and translation language support [20].
- Generation of additional artifacts: we indicate whether each approach generates other documents relevant to the composition of the service. For example, the generation of WSDL specifications from CDL specifications can be important because the BPEL process uses WSDL port information to interact and communicate with other services.
- Bidirectional translation: under this criterion, we indicate if the translation is bidirectional (that is, if the approach also supports the translation of the orchestration into a choreography). We consider bidirectional translation as an important criterion because in a collaboration, collaborating participants may also be

interested in including new participants in the collaboration. In this case, collaborating participants may publish their interface information so that new participants can use the interface, to establish collaboration. In order to generate the interface from the participants, their orchestration must be transformed into a choreography and published in the register. Hence a bidirectional translation to generate a choreography from an orchestration.

- Additional translation details: because of the difference in information content in choreography and orchestration, service-specific details must be added in the generated BPEL process. Under this criterion, we indicate whether such service-specific details are injected into the translation process or manually added to the target model beforehand.

The following table summarizes the comparison of the above approaches with the proposed approach. (See Table 2).

Table 2. Comparison of existing work with proposed work

Criteria	Mendling and Hafner (2005)	Rosenberg, Enzi et al. (2007)	Weber, Haller et al. (2008)	Our approach
Implementation of the translation	Implementation based on XSLT	DOM4J API and XSLT implementation	Implementation based on java Web service	Meta-model translation, implementation based on ATL
Generation of additional artifacts	Definition files, partnerLinkTypes and propertyAlias	WSCDL	WSCDL	WSCDL
Bidirectional translation	Yes	No	No	No
Additional Translation Details	Manually specified in the generated BPEL process	Manually specified in generated BPEL process	added in translation process	Specified manually in the generated BPEL process

6 Conclusion

In this work, we proposed an approach for transforming a choreography into a set of linked orchestrations using technology based on meta-models. To achieve this translation, we chose CDL and BPEL respectively as specification languages for choreography and orchestration. The presented translation approach takes the CDL model and executes the translation rules in the ATL engine to generate the BPEL model. However, to execute the ATL translation rules, this engine expects that each model (i.e. input models/output models) is serialized to the XML Metadata Interchange (XMI) format but the input/output models (CDL and BPEL) models are in XML. Thus, we implemented a translation chain that transforms a CDL specification into a BPEL process, aiming finally at translating the resulting process into a formal language for validation and verification.

For future work, we plan to use formal verification methods to implement and execute an application to validate the choreography translation into an orchestration process. We plan also to propose a reverse approach to get the bidirectional translation.

References

1. Peltz, C.: Web services orchestration and choreography. IEEE Comput. **36**(10), 46–52 (2003)
2. Barros, A., Dumas, M., Oaks, P.: A critical overview of the web services choreography description language. Bus. Process Trends White Paper [en ligne] (2005)
3. ATL: a QVT-like transformation language. In: OOPSLA 2006: Companion to the 21st ACM SIGPLAN Conference on Object-Oriented Programming, Systems, Languages, and Application (OOPSLA) Portland, Oregon, USA, pp. 719–720. ACM (2006)
4. Berardi, D., Calvanese, D., Giacomo, G.D., Lenzerini, M., Mecella, M.: A foundational vision of e-services. In: Bussler, C., Fensel, D., Orlowska, M.E., Yang, J. (eds.) WES. Lecture Notes in Computer Science, vol. 3095, pp. 28–40. Springer (2003)
5. Benatallah, B., Hacid, M.-S., Rey, C., Toumani, F.: Semantic reasoning for webservices discovery. In: WWW2003 Workshop on E-Services and the Semantic Web, Budapest, Hungary (2003)
6. Paolucci, M., Kawamura, T., Payne, T., Sycara, K.: Importing the semantic web in UDDI. In: Proceedings of E-Services and the Semantic Web Workshop (2002)
7. Paolucci, M., Kawamura, T., Payne, T.R., Sycara, K.P.: Semantic matching of web services capabilities. In: Horrocks, I., Hendler, J.A. (eds.) Semantic Web Conference. Lecture Notes in Computer Science, vol. 2342, pp. 333–347. Springer (2002)
8. Aversano, L., Canfora, G., Ciampi, A.: An algorithm for Web service discovery through their composition. In: ICWS 2004 Proceedings of the IEEE International Conference on Web Services (ICWS 2004), p. 332. IEEE Computer Society, Washington DC, USA (2004)
9. Benatallah, B., Hacid, M.-S., L´eger, A., Rey, C., Toumani, F.: On automating web services discovery. VLDB J. **14**(1), 84–96 (2005)
10. Bernstein, A., Klein, M.: Discovering services: towards high-precision service retrieval. In: Bussler, C., Hull, R., McIlraith, S.A., Orlowska, M.E., Pernici, B., Yang, J. (eds.) WES. Lecture Notes in Computer Science, vol. 2512, pp. 260–276. Springer (2002)
11. Cabral, L., Domingue, J.: Mediation of semantic Web services in IRS-III. In: First International Workshop on Mediation in Semantic Web Services (MEDIATE 2005) held in conjunction with the 3rd International Conference on Service Oriented Computing (ICSOC 2005). Amsterdam, The Netherlands (2005)
12. Benatallah B., Dijkman R., Dumas M., Maamar Z.: Service composition: concepts, techniques, tools and trends. In: Stojanovic Z., Dahanayake A. (eds.) Service Oriented Software Engineering: Challenges and Practices, pp. 48–66. Idea Group Inc (IGI) (2005)
13. Wohed, P., van der Aalst, W., Dumas, M., ter Hofstede, A.: Analysis of web services composition languages: the case of BPEL4WS. In: Proceedings of the 22nd International Conference on Conceptual Modeling, Oct. 2003. LNCS, pp. 200–215. Springer, Chicago, IL, USA (2003)
14. Kavantzas, N., Burdett, D., Ritzinger, G., Fletcher, T., Lafon, Y., Barreto, C.: Web services choreography description language version 1.0. W3C Candidate Recommendation (2005). http://www.w3.org/TR/ws-cdl-10/

15. Workflow Management Coalition (WFMC).: Workflow standard: Workflow process definition interface—xml process definition language (xpdl). Technical report (WFMC-TC 1025), Workflow Management Coalition, Lighthouse Point, Florida, USA (2002)
16. Mendling, J., Hafner, M.: From WS-CDL choreography to BPEL process orchestration. J. Enterp. Inf. Manag. (JEIM) **21**(5), 525–542 (2008)
17. Rosenberg, F., Enzi, C., et al.: Integrating quality of service aspects in top-down business process development using WS-CDL and WS-BPEL. In: Eleventh IEEE International EDOC Enterprise Computing Conference, EDOC 2007, October 2007, pp. 15–26. IEEE Computer Society, Annapolis, Maryland, USA (2007)
18. Mendling, J., Hafner, M.: From inter-organizational workflows to process execution: generating BPEL from WS-CDL. In: On the Move to Meaningful Internet Systems 2005: OTM Workshops, pp. 506–515. Springer (2005)
19. Weber, I., Haller, J., et al.: Automated derivation of executable business processes from choreographies in virtual organisations. Int. J. Bus. Process Integr. Manag. **3**(2), 85–95 (2008)
20. Sendall, S., Kozaczynski, W.: Model translation: the heart and soul of model-driven software development. IEEE Softw. **20**, 42–45 (2003)

Correction to: Spectral Band Selection Using Binary Gray Wolf Optimizerand Signal to Noise Ration Measure

Seyyid Ahmed Medjahed[1](✉) and Mohammed Ouali[2]

[1] Centre Universitaire Ahmed Zabana, Relizane, Algérie
sa.medjahed@gmail.com, seyyid.ahmed@univ-usto.dz
[2] Thales Canada Inc., 105 Moatfield Drive, North York, ON M3B 0A4, Canada
mohammed.ouali@usherbrooke.ca

Correction to:
Chapter "Spectral Band Selection Using Binary Gray Wolf Optimizerand Signal to Noise Ration Measure" in: S. Chikhi et al. (Eds.): *MISC 2018*, LNNS 64, 2019.
https://doi.org/10.1007/978-3-030-05481-6_6

In the original version of the book, the following belated corrections has been updated:

In chapter "Spectral Band Selection Using Binary Gray Wolf Optimizer and Signal to Noise Ration Measure", the affiliation "Thales Canada Inc., 105 Moatfield Drive, North York, ON, M3B 0A4, Canada and Computer Science Department, University of Sherbrooke, Sherbrooke, QC, J1K2R1, Canada" of author "Mohammed Ouali" has been changed as "Thales Canada Inc., 105 Moatfield Drive, North York, ON, M3B 0A4, Canada".

The correction chapter and the book have been updated with the change.

The updated original version of this chapter can be found at
https://doi.org/10.1007/978-3-030-05481-6_6

S. Chikhi et al. (Eds.): MISC 2018, LNNS 64, p. C1, 2019.
https://doi.org/10.1007/978-3-030-05481-6_26

Author Index

S. Chikhi et al. (Eds.): MISC 2018, LNNS 64, pp. 345–346, 2019.
https://doi.org/10.1007/978-3-030-05481-6

Printed in the United States
by Baker & Taylor Publisher Services